Design and Synthesis of Organic Molecules Based on Molecular Recognition

Edited by
Georges van Binst

Chairmen of the Conference
Ephraim Katchalski-Katzir and Vladimir Prelog

With 91 Figures

Springer-Verlag
Berlin Heidelberg New York Tokyo

Proceedings of the XVIIIth Solvay Conference on Chemistry
Brussels, November 28–December 01, 1983

Editor

Professor Georges Van Binst
Vrije Universiteit Brussels,
Fakulteit der Wetenschappen, Organische Chemie
Pleinlaan 2, B-1050 Brussels

Chairmen of the Conference

Ephraim Katchalski-Katzir
The Weizmann Institute of Science
Rehovot, Israel

Vladimir Prelog
Swiss Federal Institute of Technology
Zurich, Switzerland

ISBN-13: 978-3-642-70928-9 e-ISBN-13: 978-3-642-70926-5
DOI: 10.1007/978-3-642-70926-5

Library of Congress Cataloging-in-Publication Data. Solvay Conference on Chemistry (18th : 1983 : Brussels, Belgium). Design and synthesis of organic molecules based on molecular recognition. 1. Bioorganic chemistry--Congresses. 2. Binding sites (Biochemistry)--Congresses. 3. Molecular association--Congresses. 4. Chemistry, Organic--Synthesis--Congresses. I. Binst, G. van (George), 1934- .II. Title. QP514.2.S58 1983 574.19'29 85-30319

Typesetting: Schwetzinger Verlagsdruckerei GmbH; printing: Beltz Offsetdruck, Hemsbach
Bookbinding: J. Schäffer, Grünstadt.
2152/3140-543210

Preface

A topic for a Solvay Conference should be general enough to conform with the great tradition of previous conferences. On the other hand it should not be so broad that it can not be covered during the limited time of the conference. After discussion of several such topics, "Design and Synthesis of Organic Molecules Based on Molecular Recognition" was chosen.

According to Albert Einstein we live in an era of perfect methods and confused aims. For example in organic chemistry the known synthetic methods allow us to prepare an astronomical number of compounds; the gap between the possible and the relevant becomes larger every day. After discovery and classification of the main types of reactions, the study of the selectivity of reactions becomes of paramount importance. One can learn quite a lot from Nature, which uses molecular recognition to achieve selectivity in a degree so far unattainable by mere mortals. To analyze the structural features applied by Nature, to accomplish high molecular recognition, and to simulate these features by synthesis have recently become therefore favorite occupations of chemists. The purpose of this conference was to bring together two groups, the analysts and the syntheticists, to discuss the present status of the knowledge.

This monograph contains the chairman's introduction in which he has summarized the main points at issue and the contributions of the renowned scientists who participated. The organizers hope that it will stimulate further research in this fascinating field.

Vladimir Prelog

Invited Speakers

Breslow, R.
Prof. of Chemistry – Department of Chemistry, Columbia University, New York, NY (USA)

Cram, D. J.
Prof. of Chemistry – Department of Chemistry and Biochemistry, University of California, Los Angeles, CA (USA)

Franklin, T. J.
Dr. in Chemistry – Pharmaceutical Division, Imperial Chemical Industries, Macclesfield, Cheshire (United Kingdom)

Hruby, V. J.
Prof. of Organic Chemistry – College of Liberal Arts, University of Arizona, Tucson, AZ (USA)

Huber, R.
Prof. of Chemistry – Max Planck Institute of Biochemistry, Martinsried (West Germany)

Jencks, W. P.
Prof. of Biochemistry – Department of Biochemistry, Brandeis University, Waltham, MA (USA)

Karplus, M.
Prof. of Chemistry – Harvard University, Cambridge, MA (USA)

Lahav, M.
Prof. of Chemistry – Weizmann Institute of Science, Rehovot (Israel)

Lehn, J.-M.
Prof. of Chemistry – Institut Le Bel, Louis-Pasteur-University, Strasbourg (France)

Luisi, P. L.
Prof. of Organic Chemistry – Technical-Chemical Laboratory, Swiss Federal Institute of Technology, Zürich (Switzerland)

Scheraga, H. A.
Prof. of Chemistry – Baker Laboratory of Chemistry, Cornell University, Ithaca, NY (USA)

Schwyzer, R.
Prof. of Molecular Biology – Swiss Federal Institute of Technology, Zurich (Switzerland)

Invited Participants

Arigoni, D.
Prof. of Organic Chemistry – Swiss Federal Institute of Technology, Zurich (Switzerland)

Belew, J.
Prof. of Chemistry – Baylor University, Waco, TX (USA)

Blaschke, G.
Prof. of Organic Chemistry – Pharmaceutical Institute, University of Münster (West Germany)

Fonken, G. J.
Vice-President for Academic Affairs and Research – University of Texas, Austin, TX (USA)

Frauenfelder, H.
Prof. of Physics – Department of Physics, University of Illinois, Urbana, IL (USA)

Gil-Av, E.
Prof. of Organic Chemistry – Department of Organic Chemistry, Weizmann Institute of Science, Rehovot (Israel)

Kellogg, R. M.
Prof. of Organic Chemistry – Department of Organic Chemistry, University of Groningen (Netherlands)

Kessler, H.
Prof. of Organic Chemistry, Johann-Wolfgang-Goethe-University, Frankfurt/Main (West Germany)

Lerner, R. A.
Dr. in Chemistry – Scripps Clinic and Research Foundation, La Jolla, CA (USA)

Patchornik, A.
Prof. of Organic Chemistry – Weizmann Institute of Science, Rehovot (Israel)

Schulten, K.
Prof. of Biophysics – Technical University of Munich (West Germany)

Williams, R. J. P.
Prof. of Inorganic Chemistry – Inorganic Chemistry Lab., University of Oxford (United Kingdom)

Wulff, G.
Prof. of Organic Chemistry – Inst. Org. Chemistry, University of Dusseldorf (West Germany)

Wüthrich, K.
Prof. of Biophysics – Swiss Federal Institute of Technology, Zurich (Switzerland)

Auditors

Anteunis, M. J.
Prof. of Chemistry – State University Gent (Belgium)

David-Geuskens, C.
Dr. in Chemistry – Free University Brussels (Belgium)

De Coen, J.-L.
Dr. in Chemistry – Free University Brussels (Belgium)

Desreux, J. F.
Dr. in Chemistry – State University Liège (Belgium)

Durant, F.
Prof. of Chemistry – University Faculty Namur (Belgium)

Engelhard, M.
Dr. in Chemistry – Max-Planck Institute for Nutrition Physiology – Dortmund (West Germany)

Englert, A.
Dr. in Chemistry – Free University Brussels (Belgium)

Figeys, H.
Prof. of Chemistry – Free University Brussels (Belgium)

Geuskens, G.
Prof. of Chemistry – Free University Brussels (Belgium)

Ghosez, L.
Prof. of Chemistry – Catholic University Louvain (Belgium)

Grosjean, H.
Prof. of Biochemistry – Free University Brussels (Belgium)

Hamers, R.
Prof. of Biology – Free University Brussels (Belgium)

Janssen, M.
Dr. in Chemistry – Janssen Pharmaceutica, Beerse (Belgium)

Laszlo, P.
Prof. of Chemistry – State University Liège (Belgium)

Loffet, A.
Dr. in Biology – Bioproducts, UCB (Belgium)

Nasielski, J.
Prof. of Chemistry – Free University Brussels (Belgium)

Pipeleers, D. G.
Prof. of Physiology – Free University Brussels (Belgium)

Reisse, J.
Prof. of Chemistry – Free University Brussels (Belgium)

Schram, E.
Prof. of Biochemistry – Free University Brussels (Belgium)

Stien, M.-L.
Dr. in Chemistry – Free University Brussels (Belgium)

Swillens, S.
Prof. of Medicine – Free University Brussels (Belgium)

Vauquelin, G.
Dr. in Chemistry – Free University Brussels (Belgium)

Wodak, S. J.
Dr. in Chemistry – Free University Brussels (Belgium)

Zanen, J.
Dr. in Chemistry – State University Mons (Belgium)

Scientific Secretary

Van Binst, G.
Prof. of Organic Chemistry – Free University Brussels (Belgium)

Tourwé, D.
Dr. in Chemistry, Workleader in Organic Chemistry – Free University Brussels (Belgium)

De Cock, E.
Dr. in Chemistry, Scientific Collaborator – Free University Brussels (Belgium)

Jans, A.
Dr. in Chemistry, Scientific Collaborator – Free University Brussels (Belgium)

Wynants, C.
PhD Student in Organic Chemistry – Free University Brussels (Belgium)

Zeeuws, R.
Dr. in Biochemistry, Scientific Collaborator – Free University Brussels (Belgium)

Debuysscher-Dewaersegger, L.
Dr. in Chemistry, Scientific Collaborator – Free University Brussels (Belgium)

Administrative Secretary

Galland, N.
Golard, C.
Peeters, I.
Wellens, S.

Table of Contents

Topic 4. Design of New "Guests"

Reports

Communications

Concluding Remarks

Introduction

Design and Synthesis of Organic Molecules Based on Molecular Recognition

Ephraim Katchalski-Katzir

The Weizmann Institute of Science, Rehovot, and The Center for Biotechnology, Tel-Aviv University, Ramat Aviv, Israel

A great deal of work has been carried out by chemists, physicists and physical chemists in order to gain a better understanding of the principles determining specific molecular interaction [1–3]. The biologist, of course, is well aware of specific biological reactions in practically every field of study in which he is involved. Specific biological reactions occur in many of the vital life processes, such as the interactions between enzymes and their corresponding substrates and inhibitors, between hormones and their receptors, between antibodies and antigens, and between cell and cell. Powerful physical techniques such as X-ray and NMR have made it possible to determine on an atomic level the conformation of some important biopolymers – enzymes, antibodies and nucleic acids – and to shed new light on the atomic structure of the molecular domains responsible for the specific interactions exhibited by these macromolecules. Furthermore, analysis of the conformation of specific biopolymer-ligand complexes, using the above techniques, has opened up new possibilities for the evaluation of the forces and energetics involved, and led to the elucidation of some of the molecular mechanisms that characterize biocatalysis.

In view of the achievements of the molecular biologists with regard to the factors determining biological recognition it is not surprising that organic chemists have been quick to utilize the information gained in an attempt to design and synthesize receptor "host-molecules" which can accommodate corresponding "guest-molecules" in a rather specific manner. An impressive variety of such macrocyclic "host-molecules" having remarkably specific binding characteristics has already been synthesized. The next challenge awaiting the organic chemist now lies in the construction of new "host-molecules" of chemical and physical interest as well as in elucidating the nature of the factors involved in the "lock and key" substrate versus receptor complementarity.

The well-known fact that enzymes can bind substrates selectively while markedly enhancing their chemical modification has induced chemists to design and prepare compounds capable of mimicking these and other aspects of enzyme catalysis. This interesting approach led to the development of a new branch of chemistry, biomimetic chemistry, which covers all of the attempts by chemists to imitate, by classical techniques of organic chemistry, natural reactions and enzymic processes as a means of improving the potentialities of organic chemistry.

Among the participants at this conference we have molecular biologists, organic chemists, and physical chemists, theoreticians and experimentalists, all of whom are interested in the basic and practical aspects of molecular recognition. Once the technical lectures start, one can anticipate that the molecular biologists will be fascinated by the achievements of the synthetic organic chemists who have succeeded in constructing new and interesting macrocyclic host molecules; that the organic chemists will find themselves overwhelmed by the wisdom of nature in designing the most complex biopolymer molecules which are capable of carrying out highly specific reactions; that the theoreticians will shed new light on the nature of the factors determining molecular recognition, and that all of us will raise new ideas concerning the design and synthesis of useful organic molecules based on molecular recognition.

Involvement of Macromolecules in Specific Biological Interactions

Interestingly, in all specific biological interactions at least one and often both of the interacting components are represented by high molecular weight materials, i.e., by biopolymers, such as proteins and nucleic acid. Biopolymers thus seem to represent essential components in biological recognition. High molecular weight compounds store within themselves a wealth of information, some of which does not exist in the corresponding low molecular weight compounds. They contain a definite, characteristic sequence of residues; they can form well-defined three-dimensional structures; they can bring together in space the appropriate atoms or groups of atoms responsible for catalysis, recognition or specific binding, and they can under proper conditions undergo major or minor conformational changes of paramount biological importance. No wonder, therefore, that nature has chosen to employ proteins or nucleic acids in many of the life processes in which specific biological recognition is required.

Although one can perceive in general terms the advantages of biopolymers in systems where biological recognition occurs, it is less easy to elucidate quantitatively, in classical chemical and physical terms, the nature and character of the forces determining specific biological interactions. The detailed understanding of such forces is of the greatest interest, since it promises to lead to a better insight into many life processes; it will undoubtedly provide the organic chemist with sound guidelines for the construction of appropriate hosts to given guests as well as for the production of new synthetic catalysts.

On the Forces Involved in Molecular Recognition

The understanding of the physics and chemistry of the binding between biopolymers and their corresponding ligands has regrettably not kept abreast of the experimental developments. The lack of progress can be explained primarily by the fact that this binding, a highly complex process, represents the outcome of a large number of simultaneously occurring molecular processes.

Several types of attractive forces between individual molecules can contribute to the interactions occurring in biopolymer systems; these include Coulomb forces, van der Waals forces, hydrogen bridges and hydrophobic bonding.

Electrostatic interactions act between charged groups over relatively long distances and Coulombic energies decay in inverse proportion to the distance between the charged groups. Van der Waals forces, on the other hand, act at short range, and their influence is not felt beyond the distance of a few molecular diameters; Van der Waals energies decay in inverse proportion to the sixth power of the distance between the interacting groups. Hydrogen bonds occur between a proton donor AH, where A is an electronegative atom (O, N, S, Cl, Br, I), and an acceptor group B, which is a lone electron pair or the π-electron orbital of an unsaturated bond. Dissolution of hydrocarbons in water is therefore an entropically unfavorable process. Thus, if two new hydrocarbon molecules associate with each other, part of the "frozen water" surrounding the hydrocarbons is liberated and the ensuing increase in entropy is the driving force behind the process of hydrophobic binding.

Complementarity, Specificity and Dynamics in Protein Recognition

Following the suggestion of Levitt [4–7], I would describe protein recognition in terms of complementarity, specificity and dynamics.

1. Complementarity. This is revealed by: a) The observation that the interface between an enzyme and its substrate, between an antibody and its corresponding antigen, or between two proteins bound together, is stabilized by the same types of interactions as those stabilizing globular protein molecules, i.e., there is close packing of atoms, a maximum of hydrogen bonds, complementary pairing of charges and buried hydrophobic groups; b) The shapes of enzyme and substrate, or antigen and antibody, which give a close-packed interface without internal holes; c) Pairing of oppositely charged groups – this is important for reasons of specificity.

2. Specificity. a) Specificity is achieved through favorable interactions between a ligand and the specific high molecular weight compound to which it binds. The nature of the interatomic forces places certain constraints on specificity. b) Charged substrates can be recognized with high specificity, since pairing between the wrong charges is extremely unfavorable for interaction. Because the force between charges decreases slowly with separation, unfavorable interactions cannot be avoided as small atomic shifts occur. c) Large substrates will not fit into pockets designed for small substrates, giving good discrimination. The reverse, however, is not true, since small substrates can fit into a large pocket.

3. Dynamics. a) All atoms move as an inescapable consequence of thermal energy at room temperature. b) Precise recognition occurs in spite of atomic motion, not because of it. It should be noted, however, that charge interactions which are not sensitive to the precise atomic positions are only slightly affected by thermal vibrations. c) In certain cases dynamics can play a constructive role by allowing access of substrates to an active site that is blocked in the static structure. d) Under appropriate conditions, marked conformational alterations can occur in proteins and probably also in other biopolymers; these might play an important biological role in some cases, such as those discussed below.

On the Structural Aspects of Protein Recognition

Clefts of the Active Site of Enzymes

The use of X-ray techniques to elucidate the three-dimensional structure of enzymes shows that many of them possess a characteristic concave cleft at the active site. Concavities of this type have been observed, for example, in the case of lysozyme [8, 9] trypsin [10], yeast hexokinase [11], liver alcohol dehydrogenase [12] and citrate synthase [13]. It is thus reasonable to assume that the interaction between an enzyme and its substrate, inhibitor or cofactor usually occurs not in bulk water but rather in a shielded proteic cleft whose specific microenvironment is induced by the amino acid residues forming the cleft. Hydrophobicity, electrostatics, solvation and a relatively low dielectric constant prevailing within the cleft no doubt play a decisive role in determining the nature and rate of the reaction catalyzed by the enzyme.

Various aspects of the possible influence of the microenvironment in enzymic reactions and of the particular apolarity of the enzymic active site have been discussed in the literature [14]. Worth noting in this connection are the data showing that a given residue in a protein can be made superreactive by its microenvironment.

The presence of specific concavities within proteins, particularly enzymes, has facilitated the development of affinity chromatography. This technique is based on the use of immobilized ligands which may have either a very narrow or a rather wide range of specificity. The ligands are capable of complementing their corresponding protein concavities, and thus of forming complexes which retard to a greater or

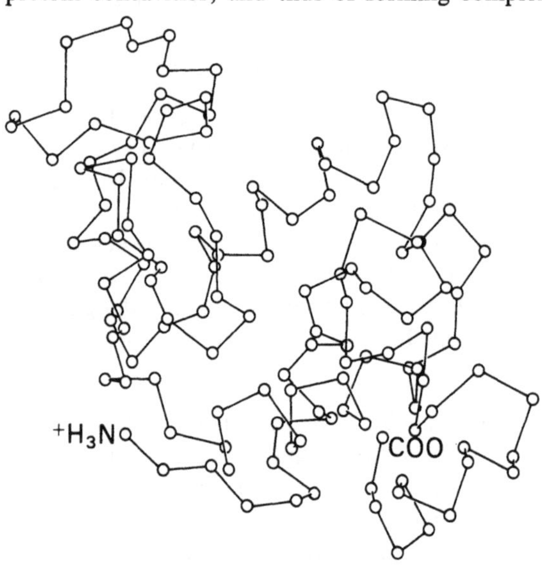

Fig. 1. Three-dimensional structure of lysozyme. Only the α-carbon atoms are shown. The active site cleft appears at the upper right of the figure [8]

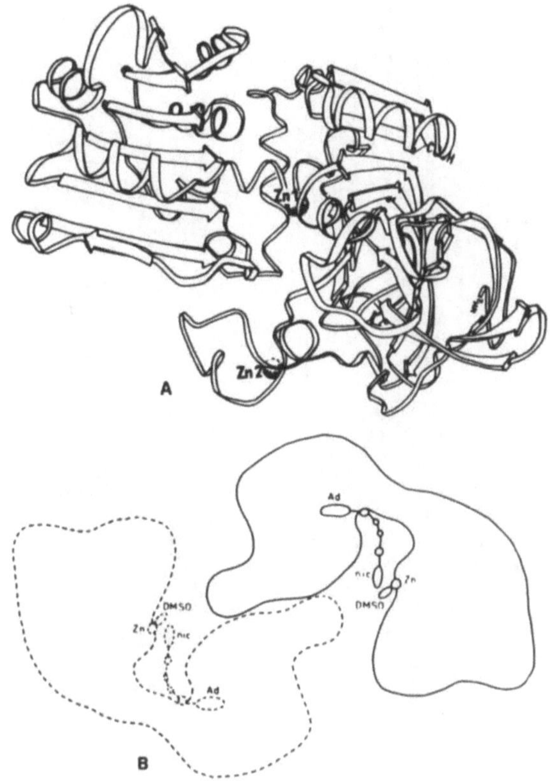

Fig. 2. A Schematic diagram of one subunit of horse liver alcohol dehydrogenase. Zn1 is the active-site zinc. **B** Schematic diagram of a section through the horse liver alcohol dihydrogenase dimer. The catalytic zinc atoms are shown, with the inhibitory substrate analogue DMSO and coenzyme molecules indicated. The dimer has two active sites, each composed of parts of both subunits [12]

lesser extent the movement of a protein along a given affinity chromatography column. Among the general ligands often discussed one should mention enzyme cofactors, dyes, and calmodulin. The preparation and use of immobilized cofactors such as NAD, NADP, AMP, ATP, cyclic AMP, pyridoxal phosphate and the flavins have been described previously [15].

The use of dyes such as Cibacron blue and related dyes as general ligands in affinity chromatography is relatively new [16]. These triazine dyes bind with considerable specificity and significant affinity to nucleotide-dependent enzymes and to a series of other proteins, their binding to enzymes often resembling the binding of native cofactors. Kinetic, spectrophotometric and chromatographic studies have shown, for example, that most of the NAD^+ and $NADP^+$-dependent dehydrogenases and many ATP-dependent enzymes interact with Cibacron blue F3G-A. The interaction is accompanied by inhibition of the enzyme activity, identifiable as competitive with respect to the corresponding nucleotide substrate. The dye-enzyme complex can be readily cleaved, either nonspecifically by a high concentration of salt or specifically by a low concentration of the appropriate nucleotide ligand.

It is worth mentioning that dye-ligand affinity chromatography has found application within a relatively short time in the isolation and purification of a series of dehydrogenases, phosphotransferases, t-RNA synthetases, restriction nucleases, initiation factors of protein biosynthesis, α-fetoprotein, clotting factors II and X and complement factors.

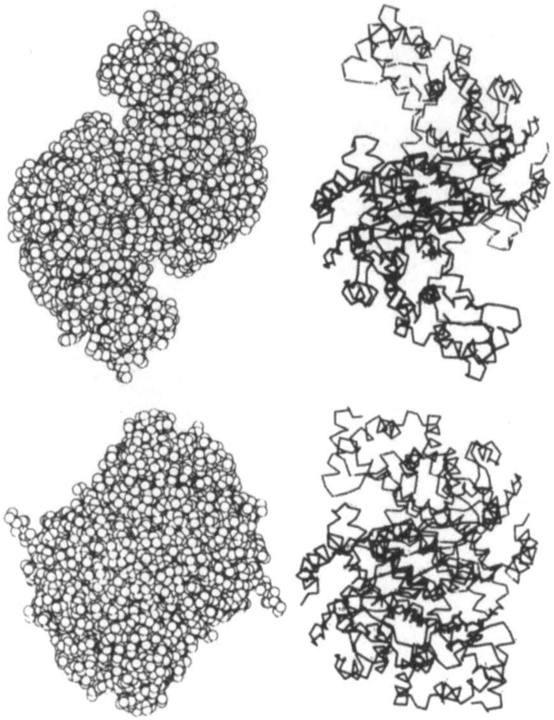

Fig. 3. Open and closed forms of citrate synthase [13]

Complementarity in Protein Recognition

In the previous section I have mentioned the importance of complementarity, as defined above, in determining the specific interactions between an enzyme and its substrate, an antibody and its corresponding antigen, or between two proteins bound together. A considerable amount of data is now available on the complementarity existing on the atomic level in certain well-defined biological systems. In the following, however, I have chosen only one representative example to illustrate the structural aspects of recognition, and to demonstrate the complex problems facing the theoretician in his endeavor to evaluate quantitatively the interaction between a protein and its specific ligand. I refer to the recent attempt of James et al. [17, 18] to elucidate the covalent and noncovalent interactions in the active site of protease A derived from *Streptomyces griseus* as seen at 1.8 Å resolution.

The enzyme's three-dimensional structure was determined by standard techniques. The structure of complexes between the enzyme and two tetrapeptide derivatives, Ac-Pro-Ala-Pro-PheOH and Ac-Pro-Ala-Pro-TyrOH, was also elucidated. Details of the interaction between these peptides and *Streptomyces griseus* protease A (SGPA) were carefully investigated, and the authors discussed the conformational changes occurring in the enzyme as a result of interactions with the substrate as well as possible catalytic mechanisms of this serine protease.

The amino acid sequences of the enzyme showed that homology with the trypsin family of serine proteases occurs only in the immediate vicinity of the acitve site

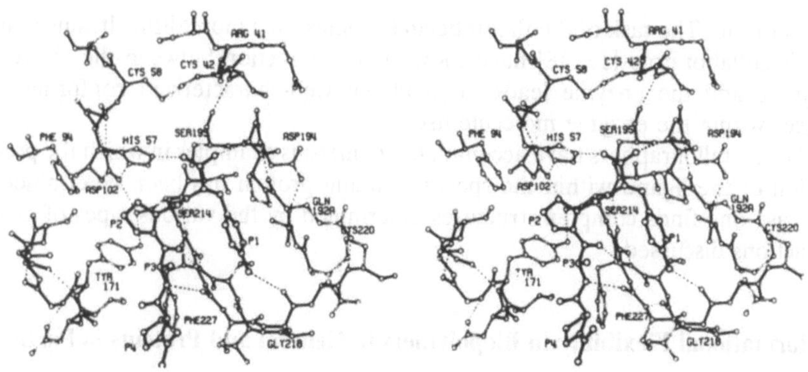

Fig. 4. Detailed molecular interactions of the tetrapeptide Ac-Pro-Ala-Pro-Tyr-OH bound to the active site of *Streptomyces griseus Protein A*. H-bonds are indicated by dashed lines [18]

residues. The three-dimensional crystal structure of SGPA at 2.8 Å resolution further revealed that almost 60% of the amino acid residues were topologically equivalent to the residues of α-chymotrypsin, in spite of the limited sequence homology. Thus, considerable information on the location in space of the atoms within the protein molecules of SGPA was now available, and since the enzyme binds specifically with the above two tetrapeptides it was possible to elucidate the atomic interactions occurring during the binding process and to predict the magnitude of the forces between enzyme and substrate. It is here, however, that one runs into complex theoretical problems. SGPA has a specific primary site for binding aromatic residues in the P_1 position, and it is the C-terminal amide bond between Pro and Phe-OH or between Pro and Tyr-OH which is split by the enzyme. The position of the tetrapeptides within the enzyme molecule has been determined. However, the complexity of the atomic interactions is already revealed by the fact that the P_1-Tyr residue makes several van der Waals contacts with segments of the polypeptide chain – more than forty contacts, in fact, at distances of less than 4 Å. In addition, the Tyr residue has five hydrogen bonds to groups on the enzyme.

The carboxylate moiety is involved in two strong hydrogen bonds, one from $N^{\varepsilon 2}$ of His^{57} to one of the oxygen atoms (2.82 Å), and the other from the peptide NH of Gly 193 to the other carboxylate oxygen atom (2.81 Å). The NH of P_1-Tyr forms a hydrogen bond with the C=O of Ser 214 (3.03 Å). The phenolic OH of the tyrosine ring is weakly hydrogen-bonded to the main chain carboxyl oxygen atom of Gly 218 (3.14 Å).

In addition to the specific interactions described above, there is an important hydrophobic S_2 binding site which accommodates the P_2-proline residue. The number of non-bonded contacts made by atoms of the proline ring with atoms of this S_2 binding site is relatively small – twelve compared to more than forty for the P_2-tyrosine. The S_3 binding site is formed by the two main chain hydrogen bonds from P_3Ala (the NH and CO) to Gly 216 (CO-NH) in an approximately antiparallel β sheet.

The large number of atomic interactions between the atoms of the substrate and the corresponding atoms of the enzyme gives some indication of the problems encountered in calculating the forces involved in the specific interactions under consideration. The task of the theoretician becomes even more difficult, since James and his collaborators [17, 18] have shown that interaction between the above tetrapeptide and the enzyme leads to small but well-characterized conformational changes within the enzyme molecule itself.

The crystallographers have accumulated numerous examples in which the precise position of the ligand within the specific binding protein has been determined; in each case one finds complex structures determined by the various types of atomic interactions discussed.

Conformational Flexibility in Biopolymers in General and Proteins in Particular

Early X-ray diffraction analyses of DNA, t-RNA and many of the globular proteins led crystallographers to assume that the molecular structure of these biopolymers should be represented by rigid three-dimensional structures. Gradually, however, evidence accumulated to show that internal conformational motions prevail not only in the double helix of DNA and the different RNAs [19] but also in many of the globular proteins [20–22]. Moreover, macromolecular conformational fluctuations as well as reversible conformational changes seem to play an important role in many of the biological reactions in which recognition by and of biopolymers occurs.

In the following I shall discuss in some detail the conformational flexibility of a few representative proteins. Right at the start I should point out that the static appearance of protein molecules derived from X-ray diffraction analysis gives rather a deceptive picture. Protein molecules are flexible, often differently so in different parts. Amino side chains may be flexible, peptide segments may be mobile and structural domains may move as rigid bodies relative to other parts of the molecule.

Evidence for mobility within proteins comes from a variety of physical methods; single crystal X-ray or neutron diffraction, electron microscopy, and spectroscopic techniques such as NMR, fluorescence depolarization, Mössbauer spectroscopy and H-exchange studies. Theoretical approaches such as potential-energy minimization and molecular-dynamics calculations may also be used to study flexibility. An illustration of the frequency range of the various thermal motions detected in proteins is given in Table 1.

Even when we restrict our attention to examples of protein flexibility characterized by motion of large parts of the molecule, we find that the functional roles of flexibility are quite diverse, as suggested by Huber and Bennett [23]: a) Flexibility might serve a regulatory function by controlling substrate binding. Interaction with a rigid, stereochemically complementary surface would be stronger than binding to a flexible segment that must be stabilized before it can provide optimal noncovalent interactions. Substrate binding could in some cases be regulated by an allosteric linkage of the transition between flexible and rigid states to other environmental factors. b) Some proteins are composed of domains that are structurally separated but covalently linked by segments that allow them to move and function independently. c) In some enzymes domains may move in relation to one another during the catalytic cycle. These enzymes occur both in an open form which presumably allows substrate binding and product release, and in a closed form in which the substrate is shielded from solution and is properly aligned in relation to catalytic groups.

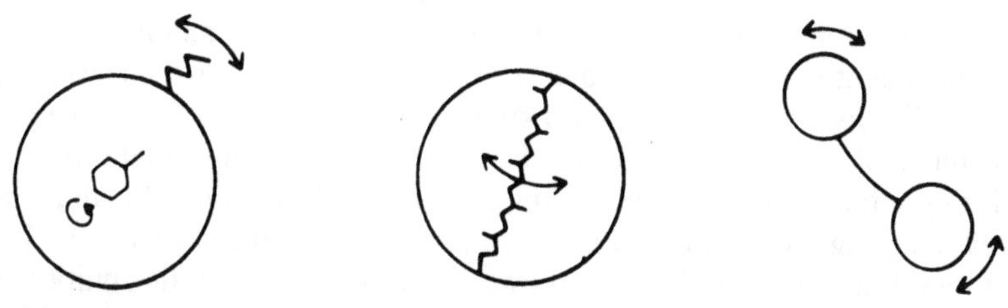

Fig. 5. Scheme illustrating three levels of flexibility of proteins: motion of amino acids, peptide segments and domains [23]

Huber and his collaborators [23] have convincingly demonstrated how segment flexibility is being used to regulate enzymatic activity of the trypsin-trypsinogen system (for review see [10]). Activation of trypsinogen occurs upon cleavage and release of the N-terminal hexapeptide of the zymogen. Trypsinogen cannot bind peptide, but it binds the basic pancreatic trypsin inhibitor (PTI) with considerable strength. This binary complex binds with high affinity to exogenous peptides related to Ile-Val, showing that trypsinogen in the trypsinogen-PTI complex assumes a trypsine-like conformation in solution, as was found to be the case for the crystalline trypsinogen-PTI complex. Further crystallographic studies have shown that a part of the trypsinogen molecule, known as the activation domain, is disordered and is not visible in the electron density map. In trypsin, however, the activation domain is an ordered one, and participates in the formation of the substrate binding site.

Table 1. Thermal motions in proteins

Frequency range (s^{-1})	Type of motion	Proteins studied in detail
$10^{10} - 10^{12}$	(1) Surface group rotations (–CH, –CHNH$^+$ etc.) (2) Wobble of N^- and C^- terminal residues (3) α-CH oscillations	BPTI, MCBP, lysozyme, RNase, myoglobin
$5 \times 10^8 - 5 \times 10^{10}$	Interior side-chain oscillations (aliphatic and aromatic)	BPTI, MCBP, lysozyme
$10^7 - 10^9$	Rotational diffusion (may be anisotropic)	all
$10^7 - 10^8$	Backbone warp	BPTI, MCBP, lysozyme
$10^4 - 10^8$	180° Aromatic ring flip	BPTI, lysozyme, MCBP
? – 10^4	Opening of secondary structure (^1H-^2H exchange)	BPTI, RNase
? – 10^6	Conformational changes and translocations	

BPTI, bovine pancreatic trypsin inhibitor; MCBP, muscle calcium binding protein

A different type of flexibility has been noted in immunoglobulins, in which independent rigid domains move relative to one another with considerable freedom [24, 25]. Antibody molecules are known to be bivalent, i.e., capable of binding two ligand molecules per molecule of antibody. Little conformational change occurs during interaction with the binding regions of the immunoglobulins; however, the relatively rigid domains can move towards one another. This is the kind of elbow motion that occurs in the well-known Fab segments. Such flexibility seems to allow the immunoglobulin molecule to adapt to the variable disposition of antigenic sites on cell surfaces.

The relatively unhindered domain motion in immunoglobulins may be contrasted with that observed in some enzymes, in which the domain motion occurs upon substrate binding during the catalytic cycle. The phenomenon has been established by crystal structure analysis of the different forms of yeast hexokinase [11], liver alcohol dehydrogenase [12] and citrate synthase [13], and appears to occur also in glyceraldehyde-3-phosphate-dehydrogenase.

The observation that kinases generally have a bilobal structure roughly similar to that found in hexokinase has led to the suggestion that analogous domain motions are a common feature of this class of enzymes.

Conformational changes of the last type have been beautifully demonstrated [13] in the case of citrate synthase, a condensing enzyme that catalyzes the reaction between acetyl coenzyme A and oxaloacetate to form citrate. The molecule is a dimer of two identical subunits each containing 437 amino acid residues. It is a large globular molecule formed almost entirely of α-helices. The two subunits pack tightly via eight helices in an antiparallel arrangement. Each subunit consists of two well-characterized domains: a large one mediating the dimer aggregation, and a small one of about 110 residues comprising the rest of the molecule. The small domain has a much less rigid structure than the large one, and appears to respond to the domain arrangement and to functional states of the enzyme by changes in tertiary structure. Citrate is bound in a cleft between the large and small domains, and upon domain closure is completely enveloped in a highly polar pocket. CoA-SH is bound to the small domain, and the cysteamine part comes very close to the bound citrate. Only in the closed conformational form of the enzyme is the CoA binding site completely formed. Domain closure thus provides a better binding site for the cofactor. The importance of electrostatic forces in the binding of citrate and oxaloacetate by citrate synthase is revealed by the fact that both of these negatively charged substrate molecules are surrounded by four histidine residues and three arginine residues. Kinetic studies show that citrate synthase has an ordered mechanism in which oxaloacetate is bound first, whereupon a conformational change occurs leading to very strong binding of acetyl CoA. Catalytic action, condensation of oxaloacetate and acetyl CoA to form citryl CoA, and hydrolysis of citryl CoA to citrate and CoA all proceed in the closed form.

An understanding of the atomic and molecular forces prevailing in the conformational changes illustrated above might pave the way for prediction of the conditions favoring tight binding between a biologically active protein and its corresponding ligand on the one hand, and conditions under which the ligand is weakly bound and can be readily released from the protein on the other hand.

Monoclonal Antibodies as Highly Specific Protein Tools for Antigen Site Recognition

The original technique developed by Köhler and Milstein [26] for the biosynthesis of monoclonal antibodies has been greatly extended during the last few years, and a very large number of murine monoclonal antibodies have been prepared. It is now possible to prepare monoclonal antibodies which are directed to different antigenic sites, and which bind to them with different binding constants. No wonder then that the use of monoclonal antibodies is expanding rapidly in diagnosis as well as in affinity chromatography. The use of antibodies in diagnosis and therapeutics has been impressively illustrated [27, 28]. In addition, Lerner [29] has described his remarkable finding that the short chains of amino acids assembled in the laboratory might mimic a site on the surface of a viral protein and could thus give rise to antibodies of predetermined specificity that confer immunity against the virus. Since synthetic or native low molecular weight peptides might adopt different conformational structures in solution, it is somewhat surprising to learn that such peptides can induce the formation of antibodies to protein segments, and that the structure of such antibodies is relatively frozen within the three-dimensional structure of the protein. This immunological effect might result, in my opinion, from well-characterized receptors which are present in some of the cells participating in the complex immunological system, and which induce in the synthetic interacting peptide a definite conformational structure similar to that found in the corresponding antigenic site of the protein antigen.

Monoclonal antibodies are of great interest, not only as diagnostic and pharmaceutical tools but also as high-molecular-weight specific reagents which can modify, by noncovalent binding, the conformation and biological activity of protein, nucleic acid, and other complex antigens. In this connection it is interesting to recall that monoclonal antibodies to a native active β-galactosidase can invoke enzymatic activity in a defective, non-active enzyme produced by an *E. coli* mutant [30]. Enzyme activation in this case undoubtedly results from the induction of an active enzyme conformation in the defective enzyme by a monoclonal antibody showing high affinity for the native active enzyme.

In my own laboratory at Tel-Aviv University we have been testing the effects of different monoclonal antibodies on the activity of carboxypeptidase A [31]. We have found some monoclonal antibodies that inhibit the esterase and peptidase activities of the enzyme, some that inhibit the esterase but not the peptidase activity, others that block the peptidase but not the esterase activity, and yet others that bind to the enzyme but do not markedly affect either of its catalytic activities. Some of the monoclonal antibodies seem to modify the conformation of carboxypeptidase A, thus greatly altering the kinetic parameters of the enzyme.

Binding and Information Transfer by Peptides

So far I have dealt with the manner in which proteins recognize complementary small or large molecules. In all of these cases the protein – enzyme, receptor, or antibody – acts as host to the corresponding guest molecules. Let me now turn to some comments concerning some specific guest molecules, namely, peptides having biological activity. Some speculations on the mode of action of peptides with biological activity are presented in this Symposium by Schwyzer and Hruby. Schwyzer has shown in his paper on "Antenna Function of Membranes and their Regioselective, Conformation Selective and Stereoselective Interactions with Regulatory Peptides" that receptors for flexible peptides can distinguish between functionally different segments of the peptide chain (domains). He terms these segments "message" (responsible for triggering the receptor response) and "address" (responsible for selecting receptor types). The lipid phase of the target cell membrane is postulated to capture peptide agonists directly from the intercellular fluid, and thus to enhance the probability of receptor encounter (antenna function). The membrane-bound peptides may thus creep over the surface until they reach the surface of the receptor and transmit the biological message. The proposed mechanism seems to explain the facilitation of the peptide's interaction with the receptor as a result of its transfer from a region (solution in a biological fluid) in which it is subjected to three-dimensional diffusion to a region (biological membrane) where it undergoes a two-dimensional diffusion. It follows that under the proper conditions the reduction in dimensionality of the biological diffusion might markedly enhance the interaction of the peptide hormone with its receptor.

Particularly interesting is the evidence indicating that the peptide-lipid interaction resulting from hydrophobic and electrostatic forces induces regioselectivity and conformation-selectivity within the bound ACTH molecules. The C-terminal of ACTH, the pharmacological "address" segment (residues 11–24), remains on the membrane surface in an extended conformation, whereas the N-terminal, the pharmacological "message" segment (residues 1–9), assumes a helical conformation that penetrates into the hydrophobic membrane layer.

In a series of ingenious experiments on ACTH and dynorphin Schwyzer has also shown that their pharmacological potency and hydrophobic membrane interactions are critically dependent on the covalent linkage of "message" and "address" to form amphiphilic molecules. Amphiphilicity is caused in these cases by the amphiphilic primary structures, in contrast to the amphiphilicity resulting from the secondary and tertiary structure of some other peptide hormones. Such an amphiphilicity was postulated recently by Kaiser and Kézdy [32] for hormones such as insulin and β-endorphin [33].

Hruby discusses in his lecture the flexibility of peptide hormones and neurotransmitters possessing different conformations depending on environment. He further points out that conformational flexibility could be functionally important. It might facilitate recognition of and interaction with several different cellular receptors, enable adjustment to evolutionary changes in receptors, and lead to recognition by different enzymes and thus result in rapid degradation and turnover. A legitimate question thus arises: have the conformations found in solution any relevance to the biological activities of the peptides under investigation? A careful analysis of the mode of hormonal action, the structure of known receptors, and the mode of action of different peptide hormone analogs and homologs led Hruby to suggest that peptide hormones possessing specific binding and transduction properties probably attain unique well-defined conformations once bound to the recognition site of the receptor. Indeed, utilization of conformational restriction by side-chain covalent cyclization, transannular stabilization of conformation by geminal dialkyl substitution in cyclic systems, etc., provided an excellent opportunity for investigating the topographical requirements for peptide hormone and neurotransmitter interactions. The remarkable success of Hruby and his collaborators in synthesizing α-melanotropin superpotent agonist analogs, oxytocin and glucagon antagonists, and δ-receptor selective enkephalins, fully justifies his claim that the peptide-hormone, once it has become bound to its corresponding receptor, attains a fixed, well-characterized conformation.

Binding Energies

The free energy of binding – of hormone to receptor, of antigen to antibody, of substrate or inhibitor to enzyme – determines the thermodynamic characteristics of molecular recognition. Free energies of binding reflect, directly or indirectly, the strength of binding of the guest ligand to its host molecule. The energetics of the primary interactions between agonists, antagonists and their receptors are discussed in this Symposium by Franklin. It is worth noting that in his calculations of the change in free energy occurring as a result of binding, Franklin has correctly taken into consideration the intrinsic free energy of binding, the free energy of substrate destabilization, the free energy of conformational change occurring in the receptor or enzyme, as well as the intrinsic entropy changes exclusive of solvation effects. Conformational modifications cannot and should not be ignored when estimating the various thermodynamic parameters involved in the molecular interactions that occur as a result of molecular recognition.

Jencks discusses at some length the economics of binding energies. Particularly noteworthy are his thermodynamic analysis of the various chemical steps involved in muscle contraction and relaxation, as well as the new insights into catalysis by enzymes and the mode of action of receptors. One should also take note of his comment on coupling and directionality, i.e., concerning the processes in which the free energy of hydrolysis of an energy-rich compound is utilized to provide directionality in structures and flows within cells and tissues. In his lecture, Jencks offers answers to the questions that he himself once posed, namely, what everyone wanted to known about tight binding and enzyme catalysis but never thought of asking [34].

Enzymes in Reverse Micelles

So far I have been dealing with molecular recognition in biological systems. Before turning to the design and synthesis of organic molecules based on the information acquired from molecular recognition in biological systems, let me comment briefly on the work of Luisi on the behavior of biopolymers in reverse micelles [35].

The reverse micelles chosen for study are those formed in a mixture containing n-octane representing the continuous phase, water at low percentage (0.5 to 2%), and bis(2-ethyl-hexyl)-sodium sulfosuccinate acting as surfactant. The reverse micelles prepared can accommodate different guest biopolymers such as proteins, nucleic acids and others, and thus act as hosts providing a special microenvironment for their guests. Hydrophilic enzymes, such as lysozyme, chymotrypsin and liver alcohol dehydrogenase, as well as hydrophobic enzymes such as hypoxygenase, could be solubilized in Luisi's system, which is optically transparent and thus suitable for spectroscopic study.

The enzymes embedded within the reverse micelles retain full enzymic activity; it is noteworthy, however, that many of them are found to undergo an increase in "conformational rigidity", i.e., in the percentage of secondary structure (e.g. helix content), and that the effect is more marked the lower the water content of the micellar system. The special microenvironment in which biopolymer molecules are enclosed might therefore exert a marked effect on their conformation, stability and activity.

Molecular Recognition on Crystal Surfaces

The work of Lahav could be described as an extension at the molecular level of the work of Louis Pasteur on the relationship between the morphology of a crystal and its internal symmetry. In his studies on the growth and dissolution of organic crystals in the presence of tailor-made inhibitors, Lahav [36, 37] has yielded a most impressive model for molecular recognition in non-living systems, of interest to the organic chemist and the crystallographer. He describes here his work on the interaction between a crystallizing substrate and its environment in general, and stereospecific (tailor-made) growth inhibitors in particular. Lahav notes that the stereochemical similarity between the additive and the crystal structure of one of the enantiomorphic substrates present in solution has a marked effect on its rate of crystallization. Kinetic and mechanistic studies have shown further that the chiral inhibitor added is enantioselectively adsorbed on the surface of the growing crystal possessing the same absolute configuration. Once adsorbed it causes a drastic decrease in the growth rate of the crystal, thus shifting the crystallization equilibrium in a racemic mixture towards the unaffected enantiomorphous phase.

In the best cases it was possible to achieve [(R, S) glutamic acid and (S) lysine; (R, S)-threonine + (S) glutamic acid; (R, S) asparagine + (S) aspartic acid] resolution with quantitative enantiomeric excess. The growth of the affected crystals in these instances was delayed up to several days with respect to that of the unaffected ones. In all of the systems investigated it could be demonstrated that the additive is occluded throughout the bulk of the affected crystals in amounts ranging typically from 0.05 to 1.5% w/w of substrate; it was found in much smaller amounts, if any, in the crystals of the enantiomorph.

Lahav's thoroughgoing analysis of the kinetics of crystal growth inhibition has enabled the morphology of a growing crystal to be modified in the presence of tailor-made inhibitors. Furthermore, the method developed has made possible the direct assignment of absolute configuration of chiral molecules. These findings open up promising new avenues of theoretical and practical interest concerning molecular recognition in the solid phases.

Microcyclic Receptors

Enzyme molecules have been found to contain well-defined clefts possessing predetermined atomic structures which facilitate the specific binding of corresponding substrates and cofactors; in addition, cyclic transport antibiotics, i.e., native ionophores such as valinomycin, alamethicin and nonactin, capable of binding ions specifically, have been isolated and chemically identified. As a result, organic chemists have been encouraged to modify native cyclic compounds, e.g. cyclodextrins, and then to investigate their ability to bind ions and other guest molecules in a specific manner. It is worth mentioning that a considerable amount of work has already been carried out on the conformation of native ionophores before and after their binding with their specific ions. In the case of valinomycin, for example, it was also possible to determine the types of bonds formed during complexation and to describe the conformational changes occurring as a result of binding with the host ion [38, 39].

The primary conformation of valinomycin, its observed structure when complexed and when uncomplexed with K^+, which it binds a thousand times more strongly than Na^+, is shown in Fig. 6.

A great number of synthetic microcyclic receptors have been prepared during the last few years by Lehn and his collaborators in Strasbourg and by Cram and his colleagues in Los Angeles. Synthetic microcyclic receptors containing an intramolecular cavity (crypt) were found to display remarkable binding properties toward a variety of charged substrates, spherical cations and anions, small molecules, functional chains, complexes, etc. [40]. They thus provide new elements for an analysis of the factors involved in the "lock and key" substrate versus receptor complementarity within the cryptate inclusion complex formed.

Some representative synthetic microcyclic receptors prepared by Lehn and his collaborators [41] are presented in Fig. 7. These, particularly the macrocyclic polyethers of the "18-crown-6" type, form complexes with cations such as K^+ and Na^+, as well as with primary ammonium salts, RNH_3^+, by inclusion of the anchor group NH_3^+ into the central circular cavity of the ring.

Lehn postulates that the receptor cryptate surrounds the substrate with binding sites of suitable hardness and softness. The balance between structural rigidity and flexibility is another important factor for selectivity. Rigidity confers more "lock and key" complementarity to the receptors, whereas flexibility introduces conformational factors allowing for "induced fit" allostery and cooperativity.

The Strasbourg school has shown that macropolycyclic receptor molecules may perform two other functions in addition to recognition, namely, molecular catalysis and membrane transport, if appropriate reactive sites or structural units are introduced into the receptor. For example, a chiral microcyclic receptor which binds primarily ammonium salts has been modified so as to perform reactions on the bound substrate. Figure 8 describes the cryptate thus synthesized, which is capable of carrying out accelerated hydrogen transfer inside the receptor-substrate complex between a derivative bearing dihydropyridine groups and bound pyridinium salts [42]. The extension of Lehn's work to chiral microcyclic compounds is of particular interest since these chiral microcyclic polyethers of the "18-crown-6" type are capable of structural and chiral recognition of host substrates.

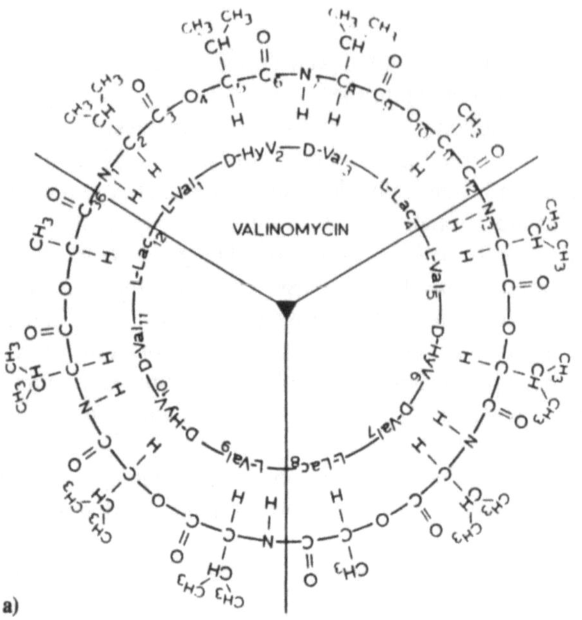

a)

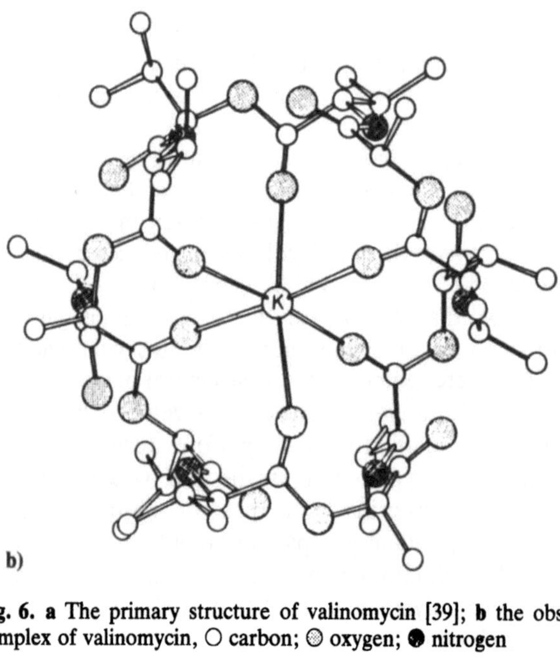

b)

Fig. 6. a The primary structure of valinomycin [39]; **b** the observed structure of the potassium complex of valinomycin, ○ carbon; ◎ oxygen; ● nitrogen

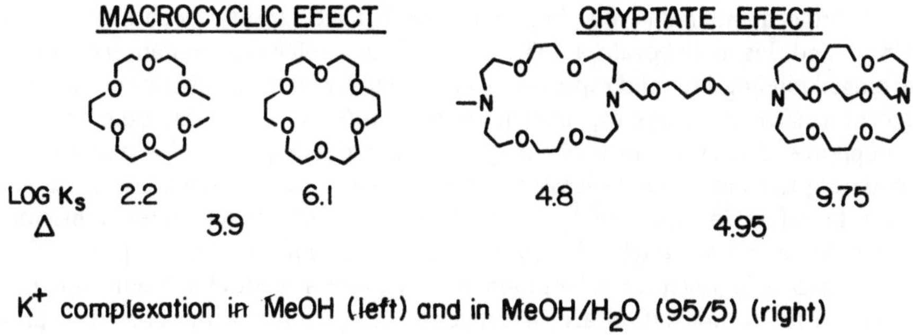

a)

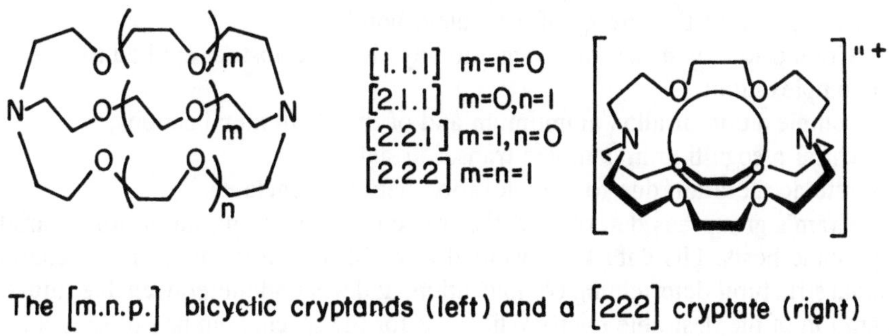

b)

Fig. 7a, b. Representative cryptands prepared by Lehn and his collaborators [41, 42]

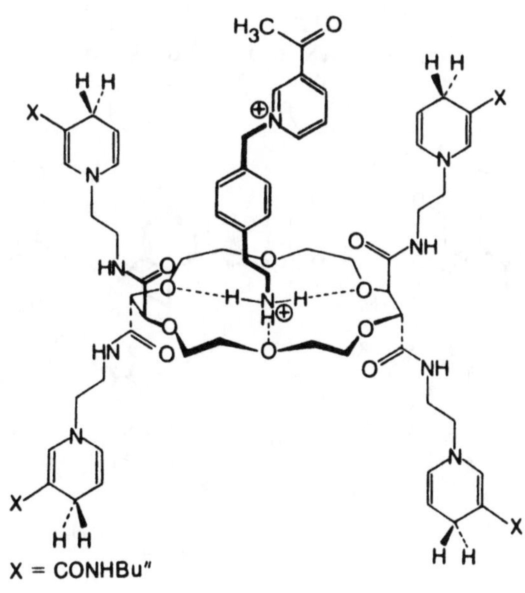

Fig. 8. Lehn's crown ether model of NADH [42]

A large number of cyclic host compounds have recently been synthesized by Cram and his collaborators [43]. These host molecules contain convergently arranged binding sites and represent according to Cram the synthetic counterparts of receptor sites of enzymes, nucleic acids, proteins of the immune system or ionophores. The cyclic hosts of the type presented in Fig. 9 are designed to accommodate guest molecules possessing divergently-arranged binding sites; the guests seem therefore to represent the synthetic counterparts of substrates, inhibitors or drugs. Metal cations might also serve as guests. Complexes are composed of hosts and guests held together in solution in a definite structural relationship. Forces available for complexation are: hydrogen-bonding; ion pairing; pi-acid to pi-base attraction; metal ion to ligand attraction; Van der Waals attraction; and entropic component of desolvation. Multiple binding sites are needed for the structuring of complexes, since the binding energy and orienting power at a single contact site is low compared to the energy of a covalent bond.

Cram discusses structural recognition by his cyclic hosts as applied to three kinds of complexation:

1) complexation of alkyl ammonium and of the alkali metal cations;
2) chiral recognition in complex transition states;
3) rate acceleration due to complexation between reactants.

Cram's group has determined the free energy of complexation for a variety of lipophilic hosts. His data have yielded a wealth of information on the conformational structural determining complexation, and will undoubtedly enable future preparation of the desirable microcyclic hosts for the specific binding of Li^+, Na^+, K^+, Cs^+ or Mg^{2+}, as well as various organic ammonium derivatives.

The potential usefulness of the microcyclic compounds synthesized so far has been demonstrated by their ability to extract guest picrate salts from water into

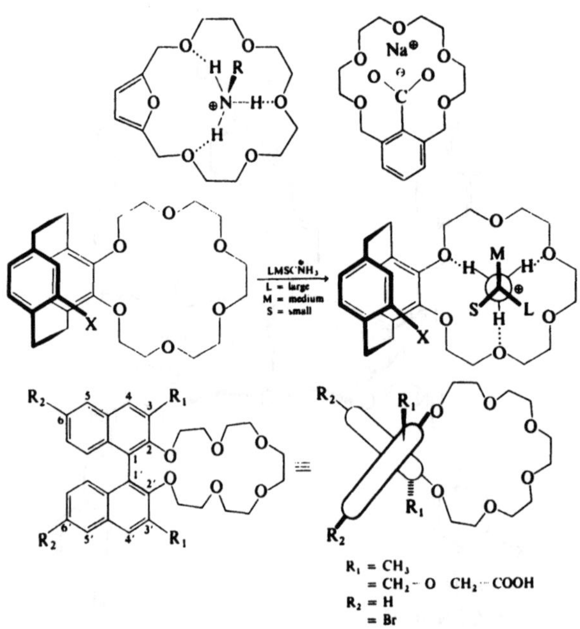

Fig. 9. Examples of host molecules [43]

chloroform containing dissolved host microcyclic compounds. Of particular interest, however, are the recent findings of Cram and his collaborators that some of the chiral microcyclic hosts might be used for ligating organometallics and thus provide asymmetric environments for carbon-carbon bond-making reactions.

Cram's group, like that of Lehn, is attempting to imitate enzyme catalysis via kinetic acceleration through host-guest complexation. A series of transacylase mimics have been synthesized [44] which contain the complexing site, the proton transfer catalyst and the nucleophile found in the chymotrypsin active site. Relatively rapid rates of acylation were observed. It is obvious, however, that additional information will have to be acquired and more model compounds synthesized before the mode of action of an enzyme such as chymotrypsin can be clarified.

Biomimetic Chemistry

Biomimetic chemistry is a branch of organic chemistry, developed mainly by Breslow and his colleagues [45] at Columbia University, which attempts to imitate natural reactions and enzymatic processes as a way of extending the potentialities of organic chemistry. Breslow points out that biomimetic chemistry is not so much concerned with enzyme models as aids in understanding enzymes as it is with the possibility that the imitation of natural chemistry may lead to new synthetic processes.

A whimsical representation of the factors involved in selective enzymatic oxidation of different paraffinic chains was put forward some time ago by Breslow [45]. In his illustration a monster holds the paraffinic chain tightly, and the oxidation (dehydrogenation) of methylenes 9 and 10 is induced by a carefully selected rigid reagent as a result of the close approach of the reagent to these two methylene groups. Breslow ingeniously designated various benzophenone carboxylic acids as his reagents of choice. The benzophenone carboxylic esters have a great advantage in that benzophenone can be photoexcited to its triplet state in which it is able to attack nonactivated methylene and methine groups, but cannot attack itself. The excited benzophenone leads to dehydrogenation at specific C–C bonds within the paraffinic chain. When a relatively short chain of twelve carbons was used in Breslow's biomimetic system, there was a rather selective functionalization of the 11th carbon with some attack on carbon 10. Extension of the experimental data to chains containing 14 to 15 carbons proved that the main oxidation took place at carbon 10 to 12, and that in no case was there oxidation at carbons up to carbon [8]. Impressive results were achieved by Breslow and his colleagues in the oxidation of steroids by covalently bound benzophenone acetic acid. In some of their biomimetic systems highly specific steroid oxidations were obtained, with remarkably high yields of final products.

Extension of the above ideas enabled Breslow and his collaborators to develop ingenious methods for geometrically directed functionalizations involving template-directed chlorination of inactivated C–H bonds [46]. Various examples of this approach are presented in Breslow's lecture. Of particular interest are the biomimetic reactions in which cyclodextrins are used to host the compounds to be chemically modified. In one of his early papers Breslow reported [47, 48] that anisole can be

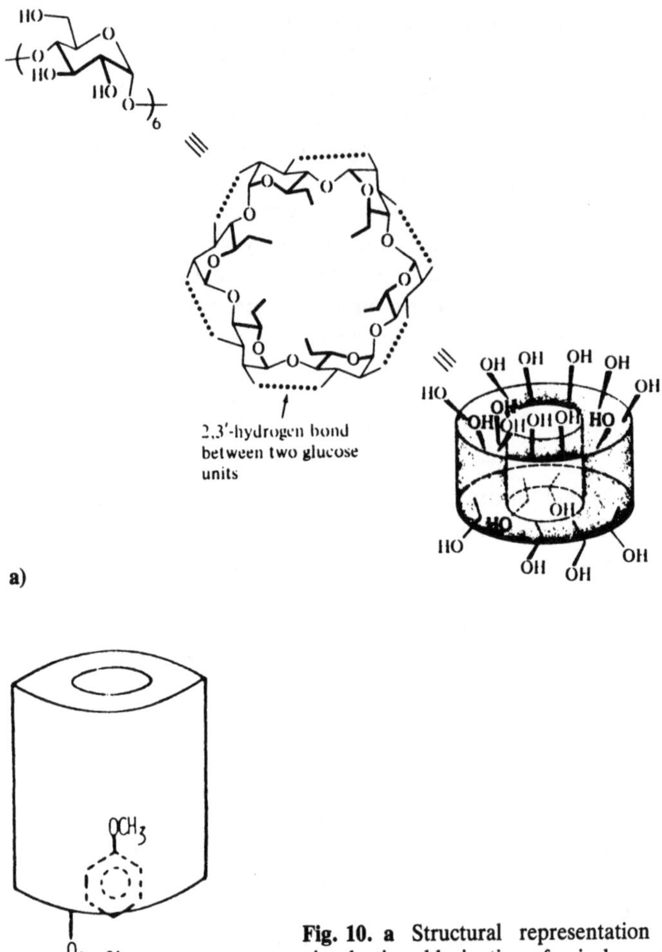

Fig. 10. a Structural representation of α-cyclodextrin, **b** regioselective chlorination of anisole encapsulated in a cyclodextrin cavity [48]

embedded in the cavity of α-cyclodextrin (the compound with six glucoses in its ring) and that within this complex aromatic chlorination is completely selective for the para position. By contrast, anisole in free solution is chlorinated rather randomly in both the ortho and the para positions. It was further shown recently that the chirality of the glucose units of the cyclodextrins can be utilized to carry out stereoselective chemical reactions in a host-guest complex in which cyclodextrins serve as hosts. Such stereoselectivity was found for example in a system in which the reversible transamination of amino acids in the presence of pyridoxal phosphate was imitated.

Finally, I would like to endorse Breslow's concluding comment that much remains to be done with catalysts and reactions of the kind that he has described. The selectivities achieved, however, demonstrate that even the known geometric factors can be used to control enzyme-like processes and achieve useful selectivities.

Asymmetric Hydrogenation

The asymmetric hydrogenation techniques developed by organic chemists during the last ten to fifteen years are of great practical and theoretical interest. Molecular biologists are particularly interested in such techniques because of the stereospecificity of many of the biologically catalyzed reactions. The work on asymmetric hydrogenation started with Osborn et al. [49], who showed that chloro-*tris*(triphenylphosphine) rhodium [$RhCl(PPh_3)_3$] is an excellent soluble hydrogenation catalyst for unhindered olefines. This discovery was followed by the preparation of optically active phosphines [50] by replacement of the triphenylphosphine in Wilkinson's catalyst by a known asymmetric phosphine. Hydrogenation of prochiral olefines of the type given in Eq. (1) with the aid of the chiral phosphine rhodium catalysts yielded optically active compounds. When a bis(phosphine) ligand called DiPAMP complexed with rhodium was used, the asymmetric catalysis in Eq. (2) was accomplished in quantitative yield with a 95% excess of the desired isomer over the racemic mixture (for review see [51]).

$$R_1CH{=}CR_2R_3 \xrightarrow{H_2} R_1CH_2CHR_2R_3 \qquad (1)$$

$$\underset{\displaystyle NHCOCH_3}{RCH{=}\underset{|}{C}\text{–}COOH} \cdot \xrightarrow{H_2} \underset{\displaystyle NHCOCH_3}{RCH_2\text{–}\underset{|}{C}^*H\text{–}COOH} \longrightarrow \underset{\displaystyle NH_2}{RCH_2\underset{|}{C}^*H\text{–}COOH} \qquad (2)$$

This technology has become the basis of a commercial process for synthesizing the amino acid DOPA (3,4-dihydroxyphenylalanine), a drug used for treating Parkinson's disease.

Asymmetric hydrogenation provides a viable alternative to biochemical methods for the preparation of amino acids, substituted lactic acids, and other materials derived from the variation of carboxyl function. The newly developed asymmetric hydrogenation procedures are of interest not only from the practical but also from the theoretical point of view.

Finally, it is worth mentioning that Halperin [52], who carefully analyzed the hydrogenation of olefinic substrates such as α-amino acrylic acid and derivatives with the aid of rhodium complexes containing chiral phosphine ligands, reached the conclusion that the stereoselection is dictated not by the preferred initial binding of the substrate to the chiral catalysts but rather by the much higher reactivity of the minor diastereomer of the catalyst-substrate adduct corresponding to the less favored binding mode. Halperin's findings are of particular significance in the context of enzymic reactions where the opposite interpretation is often assumed, namely, that the stable adducts formed by optimal fitting of the substrate to the enzyme are intermediates in the enzymic reactions.

Concluding Remarks

Nature has provided us with innumerable examples of specific interactions, many of which play a vital role in determining life processes. As the molecular biologist continues to unravel the basic principles of molecular recognition – complementarity, specificity and dynamics – one can expect that the emerging information will be utilized by the synthetic organic chemist with growing efficiency. Remarkable progress has already been achieved in the synthesis of microcyclic host molecules, cryptates, which combine specifically with ions as well as with various guest molecules. Biomimetics is making great strides, encouraging one to hope that it will one day be possible to design and synthesize systems which can efficiently imitate the action of biocatalysts in a stereospecific manner.

I am quite sure that we will learn a good deal from the discussions at this symposium; in particular, I believe that our molecular biologists will learn from the achievements as well as the failures of our organic chemists, that our organic chemists will continue to benefit from the findings of the molecular biologists, and that our theoreticians will be able to shed new light on observed phenomena and suggest new approaches and techniques for the design and synthesis of organic molecules based on molecular recognition.

References

1. Chapeville, F., Haenni, A.-L., editors (1980) Chemical recognition in biology. Molecular Biology, Biochemistry and Biophysics *32,* 1–430, Springer Verlag, Berlin, New York
2. Katchalski-Katzir, E. (1983) Some general considerations on the recognition by and of proteins. In: Affinity Chromatography and Biological Recognition (Chaiken, I. M., Wilchek, M., Parikh, I., editors), pp. 7–26, New York: Academic Press
3. Sussman, J., Traub, W., Yonath, A., editors (1981) Structural Aspects of Recognition and Assembly in Biological Macromolecules. Vol. *1,* 1–485. Vol. *2,* 486–961. Rehovot, Philadelphia: Balaban International Science Services
4. Levitt, M. (1982) The Weizmann Institute of Science, private communication
5. Levitt, M. (1982) Ann. Rev. Biophys. Bioeng. *11,* 251–271
6. Levitt, M. (1983) J. Mol. Biol. *168,* 595–620
7. Levitt, M. (1983) J. Mol. Biol. *168,* 621–657
8. Phillips, D. C. (1966) Sci. Amer. *215,* 78–90
9. Ford, L. O., Johnson, L. N., Mackin, P. A., Phillips, D. C., Tjian, R. (1974) J. Mol. Biol. *88,* 349–371
10. Bode, W. (1979) J. Mol. Biol. *127,* 357–374
11. Bennett, W. S., Jr., Steitz, T. A. (1980) J. Mol. Biol. *140,* 211–230
12. Eklund, H., Samama, J.-P., Wallen, L., Braenden, C.-I., Akeson, A., Jones, T. A. (1981) J. Mol. Biol. *146,* 561–587
13. Remington, S., Wiegand, G., Huber, R. (1982) J. Mol. Biol. *158,* 111–152
14. Luisi, P. L. (1979) Naturwissenschaften *66,* 498–504
15. Jackoby, W. B., Wilchek, M. (editors) (1974) Affinity Techniques, Enzyme Purification: Part B, Methods in Enzymology *34,* 1–755, New York: Academic Press
16. Kopperschläger, G., Böhme, H.-I., Hofman, E. (1982) Cibacron blue F3G-A and related dyes as ligands in affinity chromatography. In: Advances in Biochemical Engineering (Fichter, A., editor), *25,* 101–138, Berlin: Springer-Verlag
17. James, M. N. G., Brayer, G. D., Delbaere, L. T. J., Sielecki, A. R., Gertler, A. (1980) J. Mol. Biol. *139,* 423–438
18. James, M. N. G., Sielecki, A. R., Brayer, G. D., Delbaere, L. T. J., Bauer, C. A. (1981) Covalent and noncovalent interaction in the active site of SGPA seen at 1.8 Å resolution. In: Structural Aspects of Recognition and Assembly in Biological Macromolecules. Vol. 1 (Sussman, J., Traub, W., Yonath, A., editors), pp. 1–18, Rehovot, Philadelphia: Balaban International Science Services
19. Sarma, R. H., editor (1981) Biomolecular Stereodynamics, Vol. *1,* 1–470, New York: Adenine Press

20. Sarma, R. H., editor (1981) Biomolecular Stereodynamics, Vol. 2, 1–498, New York: Adenine Press
21. Katchalski-Katzir, E. (1983) Biopolymers *22,* 579–586
22. Goodman, M., editor (1983) Peptides, Polypeptides and Proteins: Interactions and their Biological Implications (Symposium), Biopolymers *22,* 1–586
23. Huber, R., Bennett, W. S. (1983) Biopolymers *22,* 261–279
24. Huber, R., Deisenhofer, J., Colman, P. M., Matushima, M., Palm, W. (1976) Nature (London) *264,* 415–420
25. Marquart, M., Deisenhofer, J., Huber, R., Palm, W. (1980) J. Mol. Biol. *141,* 369–391
26. Köhler, G., Milstein, C. (1975) Nature *256,* 495–497
27. Haber, E. (1983) Antibodies as *in-vivo* diagnostic and therapeutic agents. In: Affinity Chromatography and Biological Recognition (Chaiken, I. M., Wilchek, M., Parikh, I. (editors)), pp. 317–330. New York: Academic Press
28. Smith, T. W., Butler, V. P., Jr., Haber, E., Fozzard, H., Marcus, F. L., Bremner, W. F., Schulman, I. C., Phillips, A. (1982) N. England J. Med. *307,* 1357
29. Lerner, R. A. (1983) Sci. Amer. *248,* 48–56
30. Celada, F., Strom, R. (1983) Biopolymers *22,* 465–473
31. Solomon, B., Moav, N., Pines, G., Katchalski-Katzir, E. (1984) Mol. Immunol. *21,* 1–11
32. Kaiser, E. T., Kézdy, F. J. (1984) Science *223,* 249–255
33. Blanc, J. P., Taylor, J. W., Miller, R. J., Kaiser, E. T. (1983) J. Biol. Chem. *258,* 8277–8284
34. Jencks, W. P. (1980) What every one wanted to know about tight binding and enzyme catalysis, but never thought of asking. In: Molecular Biology, Biochemistry and Biophysics (Chapeville, F., Haenni, A. L., editors), Vol. *32,* 3–25
35. Luisi, P. L., Meier, P., Wolf, R. (1980) Properties of enzymes solubilized in hydrocarbons via reverse micelles. In: Enzyme Engineering, Vol. *5* (Weetall, H. H., Royer, G. P., editors), New York: Plenum Publishing Co
36. Addadi, L., Berkovitch-Yellin, Z., Domb, N., Gati, E., Lahav, M., Leiserowitz, L. (1982) Nature *296,* 21–26
37. Berkovitch-Yellin, Z., Addadi, L., Idelson, M., Leiserowitz, L., Lahav, M. (1982) Nature *296,* 27–34
38. Ovchinikov, Y. A. (1979) Europ. J. Biochem. *94,* 321–336
39. Smith, G. D., Duax, V. L., Langs, D. A., De Titta, G. T., Edmonds, J. W., Rohrer, D. C., Weeks, C. M. (1975) J. Amer. Chem. Soc. *97,* 7242–7247
40. Weber, E., Vögte, F. (1981) Crown-type compounds – An introductory overview. In: Host-Guest Complex Chemistry I. Topics in Current Chemistry *98,* 1–41 (Vögte, F., editor). Berlin: Springer-Verlag
41. Wipff, G., Kollman, P. A., Lehn, J. M. (1983) J. Mol. Struc. *93,* 153–165
42. Lehn, J.-M. (1979) Pure and Applied Chemistry, Vol. *51,* 979–997
43. Cram, D. J., Trueblood, K. N. (1981) Concept, structure, and binding in complexation. In: Host-Guest Complex Chemistry I (Vögte, F., editor). Topics in Current Chemistry *98,* pp. 43–106, Berlin: Springer-Verlag
44. Cram, D. J., Katz, H. E. (1983) J. Amer. Chem. Soc. *105,* 135–137
45. Breslow, R. (1972) Chem. Soc. Rev. *1,* 553–580
46. Breslow, R. (1980) Acc. Chem. Res. *13,* 170–177
47. Breslow, R., Campbell, P. (1969) J. Amer. Chem. Soc. *91,* 3085
48. Breslow, R., Campbell, P. (1971) Bioorg. Chem. *1,* 140–156
49. Osborn, J. A., Jardine, F. H., Young, J. F., Wilkinson, G. (1966) J. Chem. Soc. A, 1711–1732
50. Gallagher, N. J., Jenkins, I. D. (1968) In: Topics in Stereochemistry, Vol. 3, Chapter 1 (Allinger, N. L., Eliel, E. L., editors), New York: Wiley & Sons
51. Bosnich, B., Fryzuk, M. D. (1981) Asymmetric synthesis mediated by transition metal complexes. Topics in Stereochemistry *12,* 119–154 (Geoffry, G. L., editor), New York: Wiley & Sons
52. Halperin, J. (1982) Science *217,* 401–407

Discussion

List of interventions

1 Breslow
  1.2 Katzir
2 Scheraga
  2.1 Prigogine
  2.2 Scheraga
  2.3 Prigogine
3 Williams
4 Hamers
5 Luisi
  5.1 Breslow
  5.2 Luisi
  5.3 Breslow
  5.4 Katzir
  5.5 Prigogine
6 Franklin
7 Cram
8 Kessler
9 Lehn

1 Intervention of Breslow

What makes us think that we can imitate enzymes with small molecules? I came to the conclusion some time ago that if this is going to be possible – and I frankly think it will be – it will be because nature has optimized the entire system, and what is optimized is a system and not a molecule. It is quite possible that the macromolecules which nature makes are not the only and perhaps even not the best catalysts. But of course when you impose the extra requirement that the genetic information be one-dimensional, this is a tremendous limitation. It is a limitation which makes it certain that you must assemble a polymer which must then spontaneously fold. And if you want to pass on 2-dimensional or 3-dimensional information, the genetic requirements will be so complex that it may be better to make these molecules in order to simplify genetics. So, if it does turn out that we make better catalysts, or at least comparable catalyst enzymes, I think it is just because the whole system has been optimized by nature, whereas we have the option of designing a molecule, and not worrying about how it would reproduce in a self-reproducing way.

1.2 Katzir

Let me rephrase your remark. In designing specific catalysts the organic chemist proceeds in the same way as the engineer, i.e. he considers all of the materials and synthetic techniques available to him and selects those which are most suitable to achieve his goals. Nature, on the other hand, works by a tinker mechanism. And since genetic engineering has provided living organisms with a foolproof procedure for synthesizing linear polypeptides, this procedure is utilized by living organisms in the synthesis of the biologically desired specific catalysts. Thus the organic chemist seems to function in a multidimensional system, whereas living organisms are restricted to the linear genetic code dimension.

2 Intervention of Scheraga

Would Prof. Prigogine please comment on his statement that correlation of Van der Waals forces leads to forces of a long-range nature?

2.1 Prigogine

I have referred to the relation between the concentration of the molecule at a given point compared to its concentration somewhere else. When you have a reaction going on, you ask: is there a correlation between the excited state at some point and at a point at some distance? And you can show that there are correlations having a long-range character whenever the reaction is non-linear. Because essentially you produce two molecules simultaneously. Therefore there is a correlation between the two, and these correlations will act like attraction or like repulsion. And in fact these types of correlations have now been seen by very detailed computer experiments. I don't know if there have been any experiments in the laboratory. You find this already in much simpler systems. For example: you simply take a block of matter, you heat it on one side, you cool it from the other. The very fact that you impose a gradient of temperature leads to long-range correlation of temperature. Non-equilibrium makes the system coherent.

2.2 Scheraga

Would the same idea apply to the Van der Waals interactions between the various atoms within a protein molecule as it folds?

2.3 Prigogine

If there is some reason to believe that the system is in some flow, in some gradient, yes.

3 Intervention of Williams

The point raised by Prof. Breslow concerning the philosophical approach to limitations on biological reactions needs to be extended. Biology is limited by the abundance and also by the availability of the elements. This has curious consequences in the use of elements in biology which organic chemists can ignore (here organic chemists includes inorganic chemists). That the inorganic elements (with their peculiar abundances) are essential to the biological system is seen by the way in which virtually all small molecules are catalyzed by metallo-enzymes, e.g. CO_2, O_2, H_2, CH_4, C_2H_5OH, NO_3^-, SO_4^{2-}, N_2, even H_2O.

The concept of a macrocyclic system for selectivity also has an interesting echo in inorganic chemistry. The first examples are from Pauling's discussion in the period 1930–1940 of radius ratio fitting of ions in holes.

The discussion was later extended in the period 1950–1960 to complex ion selectivity with an echo in biology from porphyrin, corrin and chlorin. However, there kinetic factors are used even in the insertion step: nature has some tricks left which we have not exploited.

4 Intervention of Hamers

I would like to comment on F. Jacob's remark that biological systems are assembled by tinkering. Tinker mechanisms have an intrinsic flexibility which extensively engineered systems do not have. Biological systems evolve, and what is optimal today is not perfection tomorrow. Perfection does not exist. Tinker mechanisms are probably the cheapest way to solve problems which will change with time. To start from scratch is an expenditure in time and energy which leads to solutions that will be made obsolete by the evolution of the problem to be solved.

5 Intervention of Luisi

1. I would like to comment on the term "optimization" used this morning in connection with the evolution of biological systems. This term may convey the idea that molecular evolution has as a goal the optimization of molecules and biopolymers towards some kind of ultimate target of "perfection" (a term also used in connection with enzyme evolution). However, if you consistently accept molecular darwinism (and you do not have to), molecular evolution is based rather upon conservation of the biological autonomy as a response to environmental or other changes.

Conservationism, rather than optimization is the basic principle of the evolutionary machinery. The organisms which survive are those which are able to modify themselves so as to conserve their basic biological entity under the next set of conditions. These accumulated modifications are the basis of change and evolution. Evolution does not move upwards (optimization, perfectionism), but towards all possible (casual) directions.

2. The second remark concerns the use of the word model, which is often used in quite different connotations. The most proper use of the word, I believe, is one according to which it is the biological system which is the model, and the synthetic chemist looks at this model in order to obtain a hint for a working project (just as a photographer looks at his subject as a "model"). The term model is also used the other way around, i.e. to define the synthetic product which mimics the biological system. In addition, the term model has acquired yet another, broader meaning to signify a scheme or a pattern.

Thus statistics uses the word to signify an equation or a set of equations which describe a given physical model. These three major uses of the word are now well imprinted in the literature, and it would be of no use to try to straighten out the situation. We should however be aware of this confusion of terminology. Organic chemists often use the expression "We build a model in order to understand better the function of the biological system". I think that up to now there has been no simple instance where this has happened, where the biochemists have learned something about functions on the basis of synthetic models. I agree with the other definition of "model", whereby you look at the biological system, and looking at it you build something in your own specific field. I suspect that the information never goes in the opposite direction.

5.1 Breslow

The trouble with the statement you just made is that it is obviously open to disproof. There are lots of models that have taught something to the biochemists. All of the coenzyme mechanisms were discovered by work with models. That has taught the biochemists something, I hope.

5.2 Luisi

Yes, but that has really nothing to do with the biological function in essence. You don't learn why a dehydrogenase works on the basis of the organic chemistry involved. Organic chemistry remains in its own domain.

5.3 Breslow

I don't know what you mean by *why*. I don't know *why* it works, but *how* it works one can certainly learn. In fact, when you say you have explained something, what you really mean is that you have related it to other things that you believe you know better. That's really what an explanation is. And what this model work is doing is generating some of the chemistry that one hopes one knows a little better, and which one can use in order to understand more difficult systems. The assumption that everything you need to know was already around is just not right. Much of the model work is aimed at that sort of thing.

5.4 Katzir

I would even stick my neck out on behalf of the organic chemists and say that many of them, deep in their hearts, probably think: "Maybe in the end we will beat an enzyme and build something better." Why not?

5.5 Prigogine

I believe that our experience has shown quite strongly the limit of the idea of optimization. If you have a hydrodynamical system in non-equilibrium, you see beautiful structures appearing. And for a long time people would say that they had optimized something, heat flow or conductivity or any other property. But in fact this has not been shown to be true. In other words, in general you cannot speak about some global optimization of structures. The point is rather that you are in a non-linear situation; biochemistry, like hydrodynamics, is a highly non-linear field, and therefore many structures are possible. But the structure which appears is essentially the one which can develop. There must be some possibility; it cannot be against the environment. To speak about optimization is perhaps too optimistic a point of view.

6 Intervention of Franklin

In considering the nature of enzymes, their catalytic power must be balanced against essential features such as cellular or extracellular localization and control by other ligands over their catalytic activity. Their structure may represent in some instances a compromise between their various characteristics.

7 Intervention of Cram

The biological model provides the organic chemist with inspiration, challenge and identification of interesting problems. Biologists and biochemists are explorers of the products of molecular evolution. Organic chemists are designers and inventors, who look to the biological model for interesting research objectives. The trial and error (tinkering) processes of molecular evolution took over three billion years to find their current expression – the organic chemist has been involved in his biological modeling for only about fifty years. The organic chemist is not limited to mimicking the biological model; he only uses it as a point of departure. The organic chemist has certain advantages over the evolutionary system.

The organic chemist is not limited to 0 to 60 °C, to water or fats as solvents, to functional groups compatible with water. The organic chemist can ignore self replication, self delivery, toxicity, biodegradability, self regulatory properties which evolution has had to deal with. The organic chemist can pick his investigative "shots" – he can design his compounds, he can control his solvent and temperature, what goes into solution and how long they stay there. I predict that within ten years, organic chemists will have synthesized nonpeptide catalysts that mimic the desirable properties of several enzyme systems. Within twenty five years, organic chemists will have designed and synthesized catalytic systems that do things never possible to accomplish with enzyme systems. But the organic chemist will derive from the biological model the inspiration to start to do these things.

8 Intervention of Kessler

I want to mention another important aspect of difference between organic chemistry and biochemistry. After the conference we will probably have a more detailed understanding of molecular recognition. But this is just the first step. In accepting the lock and key picture we have to consider that this is just a one dimensional view of a multidimensional world. Working with hormones and its receptors we are faced with the fact that the key is soft and reacts differently with different receptors. This kind of variability is important in order for living systems to create flexibility in regulations (feedback effects and so on). Our present approach is to compare with a conventional mathematical computer; actually we have to deal with a human brain.

9 Intervention of Lehn

With respect to the relation between chemistry and biology, one should point out that chemistry is not restricted to the chemical processes of biology. One could cite here a very early statement by Marcelin Berthelot, paraphrasing what he wrote in 1860, that nature does not exhaust all the possibilities of chemistry.

Topic 1

Molecular Recognition in Biochemical Processes

Specific Reversible Interactions in Life Processes: Antenna Function of Membranes and Their Regioselective, Conformation Selective and Stereoselective Interactions with Regulatory Peptides

Robert Schwyzer

Institute for Molecular Biology and Biophysics, Swiss Federal Institute of Technology, CH-8093 Zürich, Switzerland

Introduction

Many drugs and regulatory peptides, including hormones and neurotransmitters, exert their biological activity by interacting reversibly with specific receptors on the target cell surface [1]. The few receptors that have been studied in some chemical detail, e.g. the acetylcholine receptor and the insulin receptor, are membrane-associated proteins with strong affinities for their agonists (range of K_d about 10^{-8} to 10^{-10} M). The receptor-agonist interaction has been likened to the association of enzymes with their substrates or allosteric effectors [2] and similar molecular recognition mechanisms implied, especially an induced complementary fit that optimizes Van der Waals, hydrogen bond, and Coulomb interactions between receptor and peptide agonist [3]. In particular, receptors for flexible peptides can distinguish between functionally different segments of the peptide chains (domains) [4]. Such segments are called *message* (responsible for triggering the receptor response) and *address* (responsible for selecting receptor types).

There are only about 1,000 to 10,000 receptors per target cell, occupying about 1% of the membrane surface [5]. The lipid phase of the cell membrane is important for the function and the topological deployment of the receptor proteins [6]. Particular lipids may even be involved in functional receptor lipid aggregates [7].

The question therefore arises whether the lipid phase of the target cell membrane also influences the functionality of peptide agonists directly, perhaps by capturing them from the intercellular fluid and enhancing the probability of receptor encounter (antenna function), or by inducing, regioselectively and stereoselectively, conformations and topological arrangements of the flexible peptides that would facilitate their complementary interaction with the receptor recognition site.

Although many experiments involving peptides and dissolved lipids or lipid micelles have been reported, the question of the molecular details of an interaction between regulatory peptides and the anisotropic ultrastructure of lipid bilayers and its relation to the biological activity of the peptides has remained essentially unanswered.

Table 1. Primary Structures of Adrenocorticotropin, Dynorphin, and Enkephalin Peptides

Peptide	Amino Acid Sequence[a] Message	Address	Net Charge[b]
1 $ACTH_{1-24}$	+ SYSMEHFRWG–	KPVGKKRRPVKVYP-	6+
2 $ACTH_{5-24}$	+ EHFRWG–	KPVGKKRRPVKVYP-	6+
3 $ACTH_{7-24}$	+ FRWG–	KPVGKKRRPVKVYP-	7+
4 $ACTH_{11-24}$		+ KPVGKKRRPVKVYP-	6+
5 $ACTH_{1-13}$ amide	+ SYSMEHFRWG–	KPVnh$_2$	2+
6 $ACTH_{1-10}$	+ SYSMEHFRWG-		0
7 $Dynorphin_{1-13}$	+ YGGF–	LRRIRPKLK-	5+
8 $Dynorphin_{1-4}$	+ YGGF-		0
9 Leuenkephalin	+ YGGF–	L-	0

[a] Abbreviations according to the IUPAC-IUB recommendations, viz. E = glutamic acid, F = phenylalanine, G = glycine, H = histidine, I = isoleucine, K = lysine, L = leucine, M = methionine, P = proline, R = arginine, S = serine, V = valine, W = tryptophan, and Y = tyrosine residues; nh$_2$ = acid amide group

[b] Approximate value at neutral pH

We have recently studied the interactions of adrenocorticotropin, dynorphin, and enkephalin peptides (Table 1) with membranes prepared from phosphatidyl choline, phosphatidylserine, phosphatidic acid, and cerebroside sulfate lipids either singly or in combination, using essentially three methods:

Capacitance minimization (CM). This method [8] detects the adsorption of charged peptides to planar lipid bilayers composed of neutral lipids (e.g. lecithin). Amounts of adsorbed charges and thermodynamic parameters can be estimated [9].

Infrared attenuated total reflection spectroscopy (IR-ATR). This method [10] can be used to study lipid bilayer membranes and peptides associated with them in the dry and in the hydrated states [11, 12]. Position and shape of the IR bands allow the detection of peptide secondary structures; their orientation with respect to the lipid molecules can be determined with polarized radiation. Adsorption from and escape into the aqueous surroundings, penetration through the membranes, and ^{1}H-^{2}H exchange can easily be studied.

Vesicle-mediated hydrophobic photolabeling (VMHL). The extremely hydrophobic photolabel, [^{125}I]3-trifluoromethyl-3-(*m*-iodophenyl)diazirine, [^{125}I]TID, was introduced by Brunner and Semenza for the study of proteins in biological membranes [13]. By carefully observing certain limitations of the method it is possible to characterize hydrophobic contacts between peptides and lipid vesicles (liposomes), and to distinguish them from the labeling caused by hydrophobic peptide-[^{125}I]TID aggregates in solution and on the vesicle surfaces [14–16]. Hydrophilic peptides or peptide segments that are only adsorbed to the vesicle surface (e.g. by electrostatic interactions) are very weakly labeled.

Results

Membrane antenna function. CM and IR-ATR studies revealed that $ACTH_{1-24}$ is readily adsorbed to lecithin (PC) and to dioleoylphosphatidylcholine (DOPC) bilayers from 0.1 mM solutions [17, 18]. With liposomes carrying a negative surface charge corresponding roughly to that of adrenal cortex target cells [19], an apparent dissociation constant $K' = 20\ \mu M$ was estimated by VMHL and equilibrium dialysis [14, 15]. Adam and Delbrück [20] and Berg and Purcell [5] calculated that at dissociation constants below about $K' = 7$ mM, two-dimensional diffusion, of the peptide in the plane of the membrane (reduction of dimensionality [20]) is favored over three-dimensional diffusion as a mechanism of receptor search. The surface area occupied by one $ACTH_{1-24}$ molecule at saturation of the anionic liposomes is estimated as about 10 nm^2, somewhat less than the area occupied by one receptor. As only a few percent of the total receptor population must be occupied by the agonist to evoke a full response, and assuming that a large fraction of the membrane-associated agonist molecules rapidly encounter a receptor [5, 20], a bulk concentration of only about 10^{-10} M $ACTH_{1-24}$ would suffice to produce a very considerable biological effect. This is well within the range observed in pharmacological assays and in receptor binding studies (10^{-8}–10^{-10} M).

Dynorphin and the enkephalins have not yet been studied in this respect, but dynorphin is expected to behave much the same as $ACTH_{1-24}$ (see below).

Regioselectivity and conformation selectivity. IR-ATR and VMHL gave consistent results with $ACTH_{1-24}$ that led to the following model [14, 15, 18]: $ACTH_{1-24}$ is adsorbed to membranes by hydrophobic and electrostatic forces. Its C-terminal, pharmacological "address" segment (residues 11–24) remains on the membrane surface in an extended conformation in which the planes of the peptide bonds are oriented perpendicularly to the plane of the membrane. The N-terminal, pharmacological "message" segment (residues 1–9) assumes a helical conformation (similar to that observed in trifluoroethanol [21–23] as a membrane-mimicking solvent [24]) that penetrates into the hydrophobic membrane layers (Gly^{10} or Gly^{10}-Lys^{11}-Pro^{12} might serve as a hinge in the peptide chain). This specific type of interaction is favored by low $ACTH_{1-24}$/lipid molar ratios and occurs by adsorption of the peptide from aqueous solutions in which it has a random secondary structure. $ACTH_{1-10}$, despite its relative hydrophobicity, does not interact with membranes and is not labeled. It forced into dry membranes, it assumes a β-pleated sheet structure oriented perpendicularly to the membrane planes, but immediately escapes on contact with water. $ACTH_{11-24}$ interacts electrostatically with anionic membranes, but not hydrophobically. If sandwiched in between hydrated layers of neutral membranes, it shows similar conformational characteristics as the corresponding segment in $ACTH_{1-24}$ [25].

The shorter ACTH peptides, Nos. 2–6 (see Table 2), also interact with anionic vesicles, as demonstrated with VMHL (and equilibrium dialysis of No. 4). The labeling rate reflecting the penetration of the whole or the residual message (as in Nos. 2 and 3) correlated excellently with their agonistic and antagonistic potencies (EC_{50} and IC_{50} values) for *in vitro* steroidogenesis and inhibition of a synaptosomal protein kinase [26].

The extremely potent opioid peptide, dynorphin$_{1-13}$, contains message and address elements [27] in a segmental arrangement similar to that of ACTH$_{1-24}$. VMHL revealed that the message enters the hydrophobic membrane layers, but that the address remains in the aqueous surroundings [14, 16]. IR-ATR studies are in progress.

Enkephalin peptides (e.g. No. 9, Table 1) and their C-terminal amides have been shown by IR-ATR (unpublished) and VMHL to interact weakly, but apparently conformation-selectively with membranes. The interaction is specific for lipid head groups, but a peptide regioselectivity has not yet been established [14, 16].

Stereoselectivity. The chiral center of lipids is located in the head group region. Thus, it is possible that peptide-membrane interactions are stereoselective. The stereoselectivity may be the same as that observed pharmacologically, preferring the natural peptide enantiomers, or it may be the opposite.

Dr. Panagiota Minakakis prepared and characterized very pure samples of all-D enkephalin (D-Tyr-Gly-Gly-D-Phe-D-Leu) and its amide. Dr. Shunsaku Kimura elaborated a new method for isolating the [^{125}I]TID-labeled peptides in a pure state, free of radioactive by-products. It was found that liposomes containing cerebroside sulfates interact about 2.5–3 times more strongly with the active, natural, all-L leuenkephalin than with the inactive all-D enantiomer. The stereoselectivity for the enkephalin amides was much less pronounced. This may be explained by the charge difference: the weak forces resulting from a molecular complementary of membrane surface and leuenkephalin (net charge 0) may be obscured by a stronger ionic interaction between the anionic membrane surface and the leuenkephalin amide (net charge 1+). Such an explanation is consistent with the molecular details of enkephalin-phosphatidylserine binding (importance of the peptide cationic center [28]). Thus, the cerebroside sulfate membrane selects the same enantiomer (all-L) as does the receptor. Membrane stereoselectivity has already been reported for cholesterol/epicholesterol [29, 30] and for morphine [31] and other opiates [32], but not for peptides.

Conclusions

Lipid bilayer membranes serve not only as matrices for receptors and other membrane-bound proteins, but interact with many neuropeptides and peptide hormones in a manner that allows multiple correlations with the biological activity of the peptides.

The interaction can be strong enough to enable the lipid bilayer to act as an antenna for the capture of certain peptides from the intercellular space and to facilitate their encounter with receptors by reduction of dimensionality in the biological diffusion process. The interaction can be selective enough to endow lipid bilayers with the capability of serving as devices for a favorable structural, conformational, and topological arrangement of the peptides prior to their encounter with the receptors. The interaction may be stereoselective, preferring (in the case studied here) the natural, much more active enantiomer, also aiding receptor interaction.

In the cases of adrenocorticotropin-(1–24)-tetracosapeptide and dynorphin-(1–13)-tridecapeptide, the receptor-triggering message segments enter the membrane hydrophobic layers, while the strongly potentiating, charged address segments remain outside. Both the pharmacological potency and the hydrophobic membrane interaction are critically dependent on the covalent linkage of message and address to form amphiphilic molecules. The amphiphilicity discovered for ACTH and dynorphin is caused by *amphiphilic primary structures,* in contrast to the amphiphilicity postulated for e.g. β-endorphin [33] and insulin [34], which is *caused by amphiphilic secondary and tertiary structures.* Amphiphilicity is a new parameter which must be taken into account when discussing quantitative structure-activity relationships.

The conformation transitions and the orientations of the peptides interacting with membranes may facilitate correct receptor interactions. In the case of ACTH, a helical structure of the message segment has already been postulated to explain action-specific receptor binding [35]. Thus, the membrane interaction may produce a secondary structure and an orientation of the helix that fit the requirements of a receptor exposing its recognition site inside the hydrophobic layer of the outer leaflet of the lipid bilayer. Enkephalin interactions may be favorable for other expositions and secondary structure requirements of their receptors.

In any case, the discovery of regiospecific, conformation specific and stereoselective membrane interactions of peptides is opening up a whole new area of receptor research, currently one of the most active fields in molecular biology.

Acknowledgment

I thank my coworkers and colleagues, Drs. U.-P. Fringeli, H.-U. Gremlich, S. Kimura, P. Minakakis, and D. F. Sargent for their careful work and enthusiastic collaboration. Thanks are due to Prof. Dr. K. Mühlethaler, who let us use his infrared spectrometer, and to the Swiss National Foundation, the Bonizzi-Theler Stiftung, and the ETH for financial aid.

References

1. Ariëns, E. J., Beld, A. J., Rodrigues de Miranda, J. F., and Simonis, A. M. (1979) in: The Receptors (O'Brien, R. D., ed.) Plenum Press, N.Y., Vol. I, pp. 33–91
2. Monod, J. (1966) Endocrinology *78,* 412–425
3. Schwyzer, R. (1963) Pure Appl. Chem. *6,* 265–295
4. Schwyzer, R. (1980) Proc. R. Soc. Lond. B *210,* 5–20
5. Berg, H. C., and Purcell, E. M. (1977) Biophys. J. *20,* 193–219
6. Weinstein, J. N., Blumenthal, R., van Renswoude, J., Kempf, I. C., and Klausner, R. D. (1982) J. Membrane Biol. *66,* 203–212
7. Smith, A. P., Lee, N. M., and Loh, H. H. (1983) Trends Pharmacol. Sci. *4,* 163–164
8. Schoch, P., Sargent, D. F., and Schwyzer, R. (1979) J. Membr. Biol. *46,* 71–89
9. Schoch, P., and Sargent, D. F. (1980) Biochim. Biophys. Acta *602,* 234–247
10. Harrick, N. J. (1967) Internal Reflection Spectroscopy, Interscience publ., N. Y.
11. Fringeli, U.-P. (1977) Z. Naturforsch. *32c,* 20–45
12. Fringeli, U.-P., and Günthard, H. H. (1981) in: Membrane Spectroscopy (Grell, E., ed.) Springer Verlag, Heidelberg, pp. 270–332
13. Brunner, J., and Semenza, G. (1981) Biochemistry *20,* 7174–7182
14. Schwyzer, R., Gremlich, H.-U., Gysin, B., and Fringeli, U.-P. (1983) in: Peptides 1982 (Blaha, K., and Malon, P., eds.) Walter de Gruyter & Co., Berlin, pp. 55–71

15. Gysin, B., and Schwyzer, R. (1984) Biochemistry *23*, 1811–1818
16. Gysin, B., and Schwyzer, R. (1983) Arch. Biochem. Biophys. *225*, 467–474
17. Gremlich, H.-U., Sargent, D. F., and Schwyzer R. (1981) Biophys. Struct. Mech. *8*, 61–65
18. Gremlich, H.-U., Fringeli, U.-P., and Schwyzer R. (1983) Biochemistry *22*, 4257–4263
19. Seltzmann, T. P., Finn, F. M., Widnell, C. C., and Hofmann, K. (1974) J. Biol. Chem. *250*, 1193–1196
20. Adam, G., and Delbrück, M. (1968) in: Structural Chemistry and Molecular Biology (Rich, A., and Davidson, N., eds.) Freeman & Co., San Francisco, pp. 198–215
21. Greff, D., Thoma, F., Fermandjian, S., Löw, M., and Kisfaludy, L. (1976) Biochim. Biophys. Acta *439*, 219–231
22. Nabedryk-Viala, E., Thiery, C., Calvet, P., Fermandjian, S., Kisfaludy, L., and Thiery, J. M. (1978) Biochim. Biophys. Acta *536*, 252–262
23. Thoma, F., Dive, V., Lam-Than, H., Piriou, F., Lintner, K., Fermandjian, S., Löw, M., and Kisfaludy, L. (1981) Biochimie *63*, 907–910
24. Pitner, T. P., and Urry, D. W. (1972) J. Amer. Chem. Soc. *94*, 1399–1400
25. Gremlich, H.-U., Fringeli, U.-P., and Schwyzer, R. (1984) Biochemistry *23*, 1808–1810
26. Gysin, B., and Schwyzer, R. (1983) FEBS Lett. *158*, 12–16
27. Chavkin, C., and Goldstein, A. (1981) Proc. Natl. Acad. Sci. USA *78*, 6543–6547
28. Jarell, H. C., Deslauriers, R., McGregor, W. H., and Smith, I. C. P. (1980) Biochemistry *19*, 385–390
29. Huang, C.-H. (1976) Nature *259*, 242–244
30. Chatterjie, N., and Brockerhoff, H. (1978) Biochim. Biophys. Acta *511*, 116–119
31. Abood, L. G., and Hoss, W. (1975) Europ. J. Pharmacol. *32*, 66–75
32. Loh, H. H., Law, P. Y., Osterwald, T., Cho, T. M., and Way, E. L. (1978) Fed. Proc. FASEB *37*, 147–152
33. Kaiser, E. T., and Kézdy, F. J. (1983) Proc. Natl. Acad. Sci. USA *80*, 1137–1143
34. Birdi, K. S. (1976) J. Colloid Interface Sci. *57*, 228–232
35. Schwyzer, R. (1977) Ann N.Y. Acad. Sci. *297*, 3–26

Discussion

List of interventions

1 Hruby
  1.1 Schwyzer
2 Franklin
  2.1 Schwyzer
3 Katzir
  3.1 Schwyzer
4 Schulten
5 Karplus
  5.1 Schwyzer
6 Gil-Av
7 Wulff

1 Intervention of Hruby

Presumably stereospecific recognition of peptide hormone neurotransmitters would be primarily of importance in your conceptual framework for recognition. However, could it be possible to extend its significance for transduction in the context of efficacy, potency enhancement or loss, or even perhaps for antagonist activity? I ask this because there has been much speculation that peptide receptor interaction might involve lipid phase changes which may in some cases be related to transduction or other related biological processes.

1.1 Schwyzer

If I may take your last comment, what we actually see in IR-ATR of the enkephalins is something like a phase change which is induced by the all-L enkephalins but not by the all-D enkephalins. You see changes in the lipid signal especially in the hydrophobic chain parts and this has been interpreted by others as indicating a higher degree of crystallinity of the membrane. We have now prepared hybrids using the address of ACTH and the enkephalin part of dynorphin. The subtype specificity of the opioid response has changed considerably. The binding to the delta receptors is lowered but the stimulation of the delta receptors is increased. So here we have a divergence of binding and stimulation. It is possible however, that by putting in another address, the message will be positioned at another angle in the membrane.

2 Intervention of Franklin

1. Have you been able to construct peptide antagonists by concentrating on the address region of the hormones i.e. by making compounds which could block the initial interaction of the peptide with the address recognition region of the cell membrane?

2. Do you think that such antagonists could even compete successfully with the natural agonists *in vivo* because of the much greater binding energy of the agonist?

2.1 Schwyzer

1. There was a slide in which we correlated the antagonist potency with the vesicle mediated hydrophobic photolabeling. The peptide that had a weak antagonistic potency was one in which we had only the address portion which interacts weakly with neutral membranes. It will of course interact with membranes of normal body cells, which usually have a negative surface charge, and this is a reason for its low antagonistic activity. But the others have a hydrophobic portion which gives additional anchorage strength, thus improving the antagonistic properties by enhancing amphiphilicity.

2. This is a difficult question. In a cell assay, 11–24 ACTH is an antagonist at a concentration 700 times higher than that of the agonist. This will probably not apply to *in vivo* experiments.

3 Intervention of Katzir

Dr. Schwyzer, you mentioned that peptide hormones, such as ACTH, are removed from the blood via nonspecific binding to cell membranes, where they are induced to creep at random until they reach their corresponding receptors, with which they bind specifically. Do they move on the surface like a helicopter jumping from spot to spot, or like a centipede, by a wave type of motion, moving a few of its legs from one spot to another while retaining the others on the surface? Because of energetic considerations I am enclined to think that the latter is the case.

Reduction in dimensionality is involved in the above-described mechanism of search for a receptor by its agonist. In this connection I would like to mention our recent theoretical and experimental studies on the mode of action of enzymes on insoluble linear and two-dimensional substrates. It should be noted that whereas there is a wealth of information about the mode of action of soluble enzymes on soluble substrates, and even about the action of insoluble enzymes on soluble substrates, little is known about the mode of action of soluble enzymes on insoluble substrates, e.g. cellulase acting on cellulose, amylases acting on starch, and collagenases acting on collagen. We solved this problem theoretically using appropriate computer techniques, and we then employed the fluorescence bleaching technique in order to test our results experimentally. The enzymes under consideration were labeled by a fluorescent dye and then adsorbed onto their corresponding insoluble substrates. A narrow laser beam at a suitable wavelength was used to bleach the fluorescent dye, yielding a black spot under the UV-illuminated microscope. The gradual return of fluorescence at this spot allowed us to evaluate the rate of creep of the enzyme on the surface of the insoluble substrate.

Finally, I should like to point out that it might be possible to use enzymes embedded within a biological membrane as probes for detecting the nature of the membrane-environment surrounding the enzyme.

Electrostatic fields, local hydrogen ion concentration, ionic strength, dielectric constants, and strong macromolecular interactions have a marked effect on the characteristic kinetic parameters of enzymes.

Comparison of the kinetic parameters of an enzyme when exposed to a well-defined medium (e.g. well-characterized solution) with those derived for the enzyme embedded in a particular membrane should thus shed light on the nature of the membrane-environment surrounding the enzyme.

3.1 Schwyzer

Positively charged peptides accumulate on negative membranes according to their Boltzmann distribution. Lateral diffusion may be envisaged to take place in an aqueous layer, the thickness of which is related to the Debye length. However, amphiphilic peptides are also anchored more or less strongly in the outer hydrophobic membrane layers and may perform a "creeping" motion in which their more hydrophobic end would move in and out of the hydrophobic membrane layer, but the peptide would still remain in the "Debye layer".

4 Intervention of Schulten

I would like to address the question raised in Dr. Schwyzer's lecture about how the extent of the diffusion space controls the time required by biomolecules to search for a target by diffusive motion. This problem can be addressed by employing the theory of first passage times, which allows one to evaluate the time spent before the first collision with the target. For purposes of calculation one can assume diffusion processes of different spatial dimensions. The table below shows the first passage times for a particle undergoing Brownian motion in a finite space of diameter 2 R when the target has a diameter 2 XR. The times are given in units of R^2/D, where D is the diffusion coefficient.

Table 2. Reaction Times

Dimension	X = 0.1	X = 0.01	X = 0 (ratio of diameters)
1-dim.	0.3300	0.333	1/3
2-dim.	0.79	1.93	– ln x
3-dim.	2.74	32.73	1/3 x

The table shows, for example, that for a target diameter which is one hundredth of the diameter of the whole diffusion space, the reaction time in dimension three is about 100 times longer than the reaction time in dimension one, i.e. a biochemical reaction can be accelerated by a factor of 100 when the diffusion is confined to a one-dimensional space. The table also shows that this factor increases to infinity for decreasing target sizes. Evaluation of the entries in the above table can be generalized to situations where force fields, e.g. originating from the surface potential of a biological membrane, guide diffusion processes along pathways of lower effective dimension [K. Schulten, Z. Schulten and A. Szabo, J. Chem. Phys. *74*, 4426 (1981)].

The theory of first passage times has also been employed by us recently in order to introduce a new variant of the method of photobleaching to measure diffusion in membranes. This variant [R. Peters et al., Proc. Natl. Acad. Sci. USA *78*, 62 (1981)] employs continuous radiation at an intermediate light level and allows one to monitor in living cells the lateral transport of membrane constituents of very small concentrations, e.g. cell surface receptors.

5 Intervention of Karplus

From both experimental and simulation studies it is now clear that the membrane is not a fixed system. The head groups are moving 2 to 3 Å out of the membrane and the lipid chains can curve all the way around and see the surface.

5.1 Schwyzer

If the chains can fold back and the tails come out at the surface then of course the label could also make contacts with the surface. What we do see in one of the address portions is that if we have a negative charge on the lipid bilayer and the address has parts which can be labeled, e.g. tyrosines, no labeling is found. At very high concentrations and lower surface areas we get some labeling. One can also sometimes get clusters of hydrophobic peptides: we find this for ACTH 1–10. The labeling of ACTH 1–10 is concentration dependent.

From about $1 \cdot 10^{-6}$ M we get a strong increase of the solubilization of the label in the peptide and a stronger labeling. If one adds lipids at concentrations higher than $1 \cdot 10^{-6}$ M, the percent labeling goes down almost to zero. With the full dynorphin or ACTH you see the opposite, i.e., no solubilization of the label at very high concentration (10^{-3} M) and also no labeling. But the moment you add the label dissolved in the lipid phase you get strong labeling.

6 Intervention of Gil-Av

The stereoselective results reported by Prof. Schwyzer prove a number of points.

1. They demonstrate again the fact, known from studies on chromatographic resolution, that partitioning of enantiomers between two phases, one of which is chiral, can lead to a considerable degree of chiral differentiation. These relatively recent results are apparently not yet generally known, particularly, as mentioned by the lecturer, in the membrane field.

2. The magnitude of the effects reported are of the same order as those found, e.g., in the partitioning of N-acyl-α-amino acid esters between the gas phase and diamides such as N-lauroyl-L-valine t-butylamide.

3. The intermediacy of diastereomeric host-guest complexes, held together by polar forces, hydrogen bonds, etc., as postulated by Prof. Schwyzer in his mechanism for the effects observed in bilayers with chiral head groups, has been widely accepted to in explaining chromatographic separations of enantiomers on chiral phases.

6 Intervention of Wulff

In the construction of synthetic polymeric receptor sites an antenna function of the polymeric matrix is of interest. The antenna binding is independent of the final binding and its binding constant should influence the two-dimensional diffusion on the surface very strongly.

Functional Significance of Flexibility in Proteins

Robert Huber and William S. Bennett, Jr.

Max-Planck-Institut für Biochemie, D-8033 Martinsried bei München, Federal Republic of Germany

The structural basis and the functional implications of large-scale flexibility are discussed for three systems: trypsin-trypsinogen, immunoglobulins, and citrate synthase. The trypsin-trypsinogen system provides an example in which an order-disorder transition is used as a means to regulate enzymatic activity. Immunoglobulins demonstrate how flexibly linked domains may be used to allow the binding of ligands with diverse arrangements. In citrate synthase, domain motion forms an active site that is shielded from solvent. Analogous large-scale flexibility has been observed in a number of other systems.

R. Huber and W. S. Bennett (1983) Biopolymers *22,* 261–279

Discussion

List of interventions

1 Luisi
2 Katzir
  2.1 Huber
3 Williams
  3.1 Huber

1 Intervention of Luisi

I should like to comment on the term flexibility, as I think that there is often some confusion in the literature about its meaning. The confusion arises from the fact that there are two connotations of the term flexibility, which may get mixed up: the kinetic (or dynamic) and the thermodynamic (or equilibrium state) connotation. Take a molecule which exists in two states, A and B, differing in free energy by 5 Kcal: then the molecule exists overwhelmingly (99.9% or so) in the more stable state. Assume further that there is a barrier of 3 Kcal in the pathway leading from A to B (the latter being more stable). Then the rate of transformation of A to B is exceedingly rapid ($k_{A\rightarrow B} - 10^8\ sec^{-1}$) and the rate of the backward reaction B to A which is determined by an energy barrier of 5 + 3 Kcal, is of the order of 10^5–$10^6\ sec^{-1}$. This trivial example shows that we can have systems which are "rigid" in a thermodynamic sense and extremely dynamic at the same time, i.e. flexible in a kinetic sense. This difference should be taken into account whenever we talk about flexibility.

2 Intervention of Katzir

How can you change the conformation of a part of a biopolymer without making large intramolecular movements?

2.1 Huber

In the large domain of citrate synthase there are small movements. In the small domain a large change occurs, which can be described as a rigid body movement in a first approximation. However, on detailed inspection one finds that there are small structural changes all over the molecule, particularly in the small domain, but also some in the large domain. So it is only to a first approximation a two state model. One must postulate this because the environment of the small domain in the open state faces the solvent much more than in the closed state, where it is in contact with the large domain. This must be reflected in the internal conformation, and this is actually what we see. Nevertheless, I think it is useful to talk of the two state model.

3 Intervention of Williams

All proteins are flexible. We can see flexibility even in cytochrome C on changing the iron atom from Fe^{2+} to Fe^{3+}. The flexibility extends to the surface. In lysozyme the protein closes on the substrate and there are small motions throughout the two domains. The question is therefore about the value of such movements, since all proteins move in all reactions. In other words we must try to define the functional value of mobility, i.e. its "purpose".

3.1 Huber

That is exactly what we have: parts which are more rigid and parts which are more flexible.

Communications

Antibodies of Predetermined Specificity in Biology and Medicine

Richard Alan Lerner

Scripps Clinic and Research Foundation, La Jolla, CA (USA)

Introduction

The diversity and exquisite specificity of antibodies has captured the imagination of scientists since the time of Ehrlich. And since it is possible to make use of the vast immunological repertoire to make antibodies specific for virtually any protein it might have been reasonable to ask, what more does one want? As is usually true in science, we are learning that there is indeed much, much more. The first new major advance was the development by Kohler and Milstein [1] of the hybridoma technology. This development allowed the purification of individual specificities from the set of responses that an animal makes upon being presented with a collection of epitopes in one or even a mixture of molecules. Thus in one elegant stroke we were capable of having unlimited quantities of antibodies reactive with a single epitope. But, what epitope? It turns out that even with monoclonal antibodies this is a difficult question to answer and one which can only be approached by an often complex protocol of biochemical or genetic experiments. The basic problem, of course, is that the inducing immunogen is complex and the sorting out of specificities is a retrospective exercise.

First, since the region in the protein with which anti-peptide antibodies react is known in advance to the experimenter, they can be said to be of predetermined specificity. This has become useful in the search for protein products predicted from the nucleic acid sequence of newly described nucleic acid clones. Also, the predetermined specificity of these antibodies allows protein chemists and cell biologists to carry out precise structure-function experiments, and to orientate proteins in cellular compartments and subcellular organelles.

Recently, it has become clear that antibodies to most regions of a protein can be induced by immunizing with short synthetic peptides. This advance portends a change in the way we generate and think about antibodies from a retrospective to a prospective science. Indeed, the synthetic immunogenic technology has already had an impact on diverse areas of biology and medicine.

Second, protein reactive anti-peptide antibodies are being used to study the chemistry and structure of antigen-antibody union. A subset of the general problem of the chemistry of antigen-antibody union is the chemistry of virus neutralization. The synthetic immunogen technology has already led us a long way toward answers in this area.

Third, and perhaps in the long run the most important, the frequency with which protein-reactive anti-peptide antibodies can be generated has led to a conceptual and experimental merger between the antibody and protein dynamics problems.

The details surrounding these three broad issues form the subject of this paper.

The Nature of Immunogen Determinants of Intact Proteins

For as long as the antibody problem has existed, there have been two obvious routes to a solution. One could study either the antigen or the antibody.

To understand the way that studies proceeded, one needs to reflect on the scientific climate of immunology in the late 1940's and early 1950's. There were a number of things we knew (reviewed by Edelman [2]). Working with haptens, Landsteiner had provided evidence for molecular complementary in antigen-antibody union [3], and there was evidence that most antibodies were multivalent [4]. Tiselius [5] had shown that antibodies were heterogenous in charge and others had demonstrated heterogeneity in their binding constants. But little was known about the structure of antibodies and it was not until the late 1950's that the structure of immunoglobulins began to be revealed [2, 7].

Considering this lack of knowledge about the chemistry of antibodies it is no surprise that prior to 1950 studies in immunology centered around the antigens. In general, the approach which most workers took was to make antibodies to an intact protein and then determine the effect of proteolysis or denaturation on its antigenicity. As we will detail in this section, these studies were to lead to two general conclusions about the antigenicity and immunogenicity of proteins. These were that only a limited portion of an intact protein is immunogenic and that antigenic determinants of proteins were most often constructed from amino acids distant from each other in the sequence but brought into proximity by the tertiary folding of the protein chain. These determinants were often called "conformational" or "discontinuous" and more recently have been referred to as assembled topographical determinants [8].

In 1980, signals began to appear which suggested the possibility for a new technology [9, 10]. Green and her colleagues carried out an experiment suggesting that one could use synthetic immunogens to generate antibodies of predetermined specificity which were reactive with virtually any region of a protein [11]. They used the influenza virus hemagglutinin as a test object because the complete nucleotide sequence of its gene was available [12] and its crystallographic structure was known at high resolution [13].

A series of peptides covering 75% of the HA1 chain were chemically synthesized and antibodies were made to each. Antibodies to almost all (eighteen of twenty) peptides reacted with the intact molecule [11]. Since in its folded state the HA1 molecule displays a number of secondary structures including α-helix, extended chains, and β-sheets, it was clear from this study that reactivity of antipeptide antibodies is independent of secondary structure or location in the molecule. The studies on the HA of the influenza virus made two main points. First, the use of chemically synthesized immunogens to generate antibodies of predetermined specificity was general and could be applied to most regions of proteins. Second, the immunogenicity of an intact protein is less than the sum of the immunogenicity of its pieces [14].

As more and more data concerning anti-peptide antibodies have been gathered, a theoretical construct has begun to emerge. At first, there were two rather simplistic notions to explain the results. The first idea could be called the side chain constellation theory. Basically, the notion was that anti-peptide antibodies "saw" a collection of side chains as if they were a set of haptens on a protein backbone. It was as if antibodies read the sequence of a protein. Given that any type of protein-protein binding, whether it involves antibodies or enzymes, is an interaction between specific shapes, this was never a strong idea. However, to test this notion, we synthesized the C-terminal 20 amino acid of the influenza virus HA1 chain in four different ways: from L amino acids in an N → C and C → N direction and from D amino acids in the N → C and C → N direction (Houghten and Lerner, unpublished). The D isomer polymer in the C → N direction (retro-D) is particularly interesting in that, except for the ends, the relative positions of the side chains approximate the L isomer in the N → C direction but the positions of the carbonyl group and the amide bonds are reversed. In other words, we have a polymer with the same sequence but the wrong shape. Antibodies were made to each of the four peptides and tested for reactivity with the peptides as well as with the folded HA1 protein. All four peptides raised anti-peptide antibodies, but each reacted only against the peptide against which they were generated: only the antibody against the L, N → C peptide reacted with the HA1. The absolute lack of cross-reactivity between any of these four anti-peptide antibodies argues strongly against notions that depend on anti-peptide antibodies reacting with a constellation of side chains.

The second theory evoked a stochastic model. Here, the idea was that after peptide immunization, antibodies were made against the multiple peptide conformations but only the small fraction of antibodies against the conformation shared between the peptide and protein were reactive with the folded protein. Thus, the success of the technology was postulated to be more a testimony to the sensitivity of immunological assays which could detect a small percentage of proper antibodies than to something more fundamentally interesting. However, peptides in solution have thousands to hundreds of thousands of conformations and as more and more success for different peptides was achieved, the stochastic idea became less tenable. In other words, the scientific community was doing the statistics and the collective answer was against the stochastic model. To test the stochastic model in a formal way, Niman et al. [15] used monoclonal anti-peptide antibodies as a way of estimating the frequency with which small peptides induce antibodies that react with folded proteins. They made monoclonal antibodies to six chemically synthesized peptides from three proteins. The frequency with which the peptides induced protein-reactive antibodies was at least four orders of magnitude greater than expected from previous experimental work and vastly different from what would be predicted by calculating the possible number of peptide conformers in solution. These results suggested that sufficient structural information is contained in peptides as small as 13 amino acid residues to induce protein reactive antibodies at a high frequency. The key point was that the frequencies observed were inconsistent with any stochastic model.

If the stochastic model is also not correct, then, with some caveats which we will discuss below, we must begin to think in terms of antibodies reacting with conformations in proteins which are different from that of the native structure. These notions open up the exciting possibility of a merger between the protein dynamics and the antibody problems. In the first instance one can imagine that a given site in a protein is mobile and that as it passes through a conformation which the antibody recognizes, union takes place. A second possibility is that the antibody-antigen union itself induces a shape change. The end result of both models is the same in that antibody is bound to a protein with the site of union in a conformation different from the native one. But the two models are vastly different insofar as their implications for protein structure are concerned. The first model views proteins as highly dynamic structures, whereas the latter makes no such assumptions and assumes that antibody-antigen union has sufficient energy to distort the target structure. The difficulty with the latter model is that it is somewhat circular in that antibodies can only distort that which they can recognize, and distant conformations would not be efficiently recognized. Thus, all facts considered, it seems likely that for some percentage of the time, local disorder occurs on short segments of the protein, allowing reaction with anti-peptide antibodies. This local disorder model suggests that peptides which elicit antibodies recognizing intact proteins are located in areas of relatively great conformational mobility in the intact protein.

Whichever model one favors, it is important to note that peptide immunogens do not have the same possibility for induced fit that occurs in systems with a limited number of receptors (i.e., peptide hormones) because the immune system is a system of diversity, and unless other factors pertain (see below), "fixing" the correct conformation would seem to have no better probability than fixing the incorrect one.

The Repertoire Should be Tapped Further: Concept of Immunological Catalysis

The advent of antibodies of predetermined specificity will condition us to think in terms of binding to specific sites in proteins rather than simply to proteins. One wonders if the next step can be taken and antibodies can be produced which bind to the same structures in proteins as do enzymes. Since the effectiveness of enzymes depends upon the stabilization of minor equilibrium states we might expect antibodies recognizing these same states to carry out catalytic functions. This is a wonderful possibility since one could fish in the immunological repertoire for any kind of enzyme so long as the substrate were sufficiently large to be immunogenic. The basic question is whether the diversity of the immune system is any match for the millions of years of molecular design which go into the evolution of an enzyme. I, for one, would bet on the diversity of the immune system. There is a real impetus to test those notions because, since we can now make antibodies to almost any position on a protein, success would be tantamount to having site-specific enzymes for proteins. Such enzymes would be analogous to restriction endonucleases except that the specificity would be induced by the experimenter. But, how to accomplish this? Two notions come to mind. The first assumes that, as previously discussed, anti-peptide antibodies work via an induced fit mechanism and thus distort the site at which they bind. If one could supply additional energy the protein might hydrolyze at the site of binding. Although possible, this approach seems unlikely to succeed easily. An alternative approach is to make antibodies to intermediates in catalysis to tip the equilibrium in favor of hydrolysis of the peptide bond. The principle, then, is one of immunological catalysis.

Antibodies and enzymes are both proteins whose function depends on their ability to bind specific target molecules. Enzymatic reactions differ from immunological reactions in that the binding of substrate to enzyme leads to chemical catalysis. Enzymes catalyze the hydrolysis of proteins by combining with the protein in a *transition state* of the reaction. It is generally assumed that an enzymatic reaction is accelerated with respect to the non-enzymatic reaction because of the enzyme's ability to reduce the free energy of the transition state, and thus, the free energy of activation, of the reaction [16, 17]. The enzyme might accomplish this by binding a transition state geometry more strongly than the corresponding substrate(s) or product(s). This means that an enzyme's instrinsic binding energy is much greater than can be measured from the binding of substrates or products. Essentially, the enzyme's binding energy is utilized to perform the chemical reaction [18].

The basic idea behind immunological catalysis contemplates the design of antibodies of predetermined specificity that will stabilize transition states of peptide bond hydrolysis upon binding to the specified antigen. This should result in a reduction in the activation energy for the hydrolysis reaction, thus meeting a criterion for catalysis. Antibodies which display this property might be obtained by immunization with synthetic peptide analogs that are chemically modified to resemble the bonding characteristics of a substrate undergoing peptide bond hydrolysis – that is, *transition state analogs* of this reaction. The mechanism by which an antibody might catalyze the hydrolysis of a bound substance may be thought of in terms of an "induced fit" model. As loosely-bound substrate distorts to conform to the binding geometry of the antibody, stress can be relieved by chemical reorganization of a single amide bond such that this reorganization leads toward hydrolysis of the bond.

Toward the goal of designing immunological catalysts, we have begun to synthesize chemical analogs of polypeptides and depsipeptides which incorporate phosphonamidate or phosphonate moieties at specific sites to mimic the transition state for hydrolysis of the amide or ester bond (Tramontano, A., Lin, Bor-Sheng, and Lerner, R. A., experiment in progress). These are reasonable candidates for this experiment because it is known that such phosphonamidates are, in fact, transition state analogs in the inhibition of proteolytic enzymes [19].

Initially, we plan to investigate a system designed for the hydrolysis of a simple p-nitrophenyl ester.

Hydrolysis of the amide bond of polypeptides or proteins will require analogs which bear the phosphonamidate moiety (Fig. 1). Methods for the synthesis of these compounds are being explored. Phosphonamidates described for the inhibition of certain proteases [19, 20] can also be modified for induction of immunological catalysts. Since short polypeptide chains can induce antibodies which recognize the protein at specific sites, we may expect that if an antibody specified for a transition state analog catalyzes the hydrolysis of a short polypeptide chain, it will also catalyze the cleavage of a protein with that particular sequence of amino acids somewhere along its length. The implication of these expectations is that we could confer the

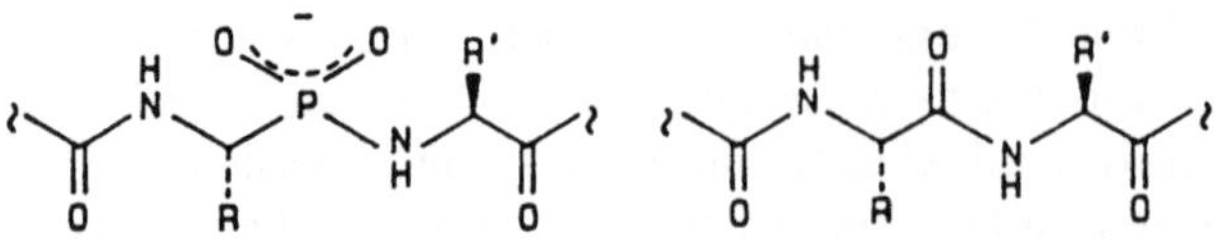

Phosphonamidate Polypeptide **Fig. 1.**

activity of certain proteases to immunoglobulins. Furthermore, the antibody's activity may be directed to any site at will by designating the amide bond to be cleaved with the phosphonamidate center in the analog used for immunization. Thus, a method would be available for the selective proteolysis of any protein whose local sequence conforms with that of the polypeptide targeted. The applications of such a method in protein chemistry, biochemistry, and medicine are without limit. For example, instead of engineering the immune system to simply bind to viruses or tumor antigens, we could aim at evoking antibodies that catalyze specific protein cleavages which inactivate viruses or kill cells. In essence, one evokes antibodies which act directly and do not depend on help from accessory factors such as complement, or complex processes like opsinization. Regardless of whether these concepts are correct in detail, one can expect antibodies of predetermined specificity to soon take on roles which may transcend their simple binding functions.

Antibody Template Directed Organic Synthesis

It is a curious paradox that although almost all biological processes are mediated by proteins, very few proteins are useful as therapeutics. There are, of course, good reasons for this. Proteins are labile, often destroyed by the acid pH of the stomach, unable to cross cell membranes or the blood-brain barrier, etc. Even a lifesaving protein such as insulin could hardly be classified as an ideal drug, in that it requires daily injection and as currently given does not prevent the many complications of diabetes. Also, important peptides such as hypothalamic releasing factors cannot easily be turned into drugs because of their limited ability to be absorbed after ingestion or cross the blood-brain barrier. So, what one really wants is to construct non-peptidyl organic mimics of proteins and peptides. One knows that this is possible because, for example, the endorphins and the opiates share the same receptor – it is just that when we look at the two structures we don't understand why the receptor "sees" them as mimics of each other. In other words, we don't know the rules by which non-peptidyl organic compounds mimic proteins. If, however, these rules could be deciphered one would see a new approach to rational design of drugs. In an interesting way, antibodies of predetermined specificity have opened a wedge which could lead to a better understanding of how to make organic mimics of proteins. The way in, of course, is to learn by making organic mimics of antigens. As long as one was dealing at the whole protein level, there was little hope for such design. But, now that the problem can be reduced to only a few amino acids, it is probably fair to say that the problem has moved from impossible to formidable. The process one can use to design organic mimics of proteins is that of antibody template directed organic synthesis. In this process one begins with a monoclonal antibody and a peptide mimic of a protein antigen (i.e., as described above for the influenza system). This is essentially the same as having, in organic chemical terms, a pure host and a pure guest. In practice, each time a chemical change is made in the peptide ligand, its ability to react with the anitbody is checked. If the ligand still reacts, the chemical step was permitted and the next step can be taken. If the antibody does not bind the ligand a different modification must be made. By continuing this process, one would hope to evolve a compound which no longer resembles the peptide but still shares immune reactivity. Indeed, non-peptidyl organic antigens would be useful, but the real goal is to learn the rules for chemical mimicking of proteins. If this could be accomplished, chemicals which mimic the functions of proteins, such as insulins and even intracellular regulators, might be a reality.

References

1. Kohler, G., and Milstein, C. (1975) Nature *256,* 495
2. Edelman, G. M. (1973) Science *180,* 830–840
3. Landsteiner, K. (1936) The specificity of serological reactivities. Charles C. Thomas
4. Marrack, J. R. (1938) The chemistry of antigens and antibodies No. 230. Medical Research Council Special Report Series. (His Majesty's Stationary Office, London, ed. 2)
5. Tiselius, A. (1937) Biochem. J. *31,* 313. (Almquist and Wiksell, Stockholm, 1937)
6. Karush, P. (1962) Advan. Immunol. *2,* 1
7. Porter, R. R. (1973) Science *180,* 713
8. Benjamin, D. C., Berzofsky, J. A., East, I. J., Gurd, F. R. N., Hannum, C., Leach, S. J., Margoliash, E., Michael, J. G., Miller, A., Prager, E., Reichlin, M., Sercarz, E. E., Smith-Gill, S. J., Todd, P. E., and Wilson, A. C. (1984) Ann. Rev. Immunol. *2,* in Press
9. Sutcliffe, J. G., Shinnick, T. M., Green, N., Liu, F.-T., Niman, H. L., and Lerner, R. A. (1980) Nature *287,* 801–805
10. Walter, G., Scheidtmann, K. H., Carbone, A. P., Laudano, A. P., and Doolittle, R. F. (1980) Proc. Natl. Acad. Sci. USA *77,* 5197–5200
11. Green, N., Alexander, H., Wilson, A., Alexander, S., Shinnick, T. M., Sutcliffe, J. G., and Lerner, R. A. (1982) Cell *28,* 477
12. Min Jou, W. M., Verhoeyen, M., Devos, R., Saman, E., Fang, R., Huylebroeck, D., Fiers, W., Threlfall, G., Barber, C., Carey, N., and Emtage, S. (1980) Cell *19,* 683–696
13. Wilson, I. A., Skekel, J. J., and Wiley, D. C. (1981) Nature *289,* 373
14. Lerner, R. A. (1982) Nature *299,* 592–596
15. Niman, H. L., Houghten, R. A., Walker, L. E., Reisfeld, R. A., Wilson, I. A., Hogle, J. M., and Lerner, R. A. (1983) Proc. Natl. Acad. Sci. USA *80,* 4949–4953
16. Jencks, W. P. (1975) Adv. Enzymology *43,* 219–410
17. Pauling, L. (1948) Amer. Scientist *36,* 58
18. Jencks, W. P., XVII International Solvay Conference, November 1983
19. Bartlett, P. A., and Marlowe, C. K. (1983) Biochemistry *22,* 4618–4624
20. Jacobsen, N. E., and Bartlett, P. A. (1981) J. Am. Chem. Soc. *103,* 654–657

Calcium Trigger Proteins

R. J. P. Williams

Inorganic Chemistry Lab., University of Oxford (UK)

I wish to make a general comment upon the nature of mechanical changes within proteins. The comment applies equally to receptors which must transmit information across a considerable distance, e.g. across a membrane, to the initiation of activity by remote control such as the interaction of ion-binding with ATP formation or hydrolysis, and even to the opening out of a large molecule so that it can be read, e.g. the loosening of histones from DNA. The motion must be directed and functional.

The general principle can be understood from the function of the calcium-trigger proteins such as calmodulin. The basic idea is that local input of energy through binding causes a movement within a cross-helical array of protein strands to an output at the distant ends of the helices. The helices act as rigid rods sliding and rotating as they move relative to one another. In the case of the calmodulins the binding of calcium is to two β-strands, which together form a small β-sheet. The β-sheet is slightly twisted (or in some cases is actually brought into being) by the calcium binding. Leading from the β-sheet are four helices which are rotated and translated by the twisting of the sheet. The far ends of these helices then link to an output connection to for example a kinase, or an ATP-ase. We know now that kinases change their shapes during activity, closing around the substrates. The kinase proteins have a hinge made from helices which roll relative to one another. It is easy to see how the rolling of the helices in the kinases can be presented or assisted by the mechanical movement of the linked calmodulin and its adjustable helices. Fortunately we have a model for exactly this transmission device in hemoglobin, where the helices of the alpha and beta chains intercommunicate to generate allosteric systems. Now we also know that there are helical structures in membranes, for example the F_0 of the ATP-synthetase. It is also the case that the F_0 protein polymer channel carries a high negative charge due to an invariant carboxylated group. There are no structural data yet but the F_0 may well respond to the uptake of protons as follows: the helical alignments of the F_0 polymer are in part controlled by the repulsion between the negative charges just mentioned. On protonation the repulsion is reduced and the helices respond by a rotational/translational notion. In so doing the F_0 molecule transduces the proton binding energy into a change in the configuration of the proton channel in the membrane and at the same time the whole protein stores energy much like the stretching of an elastic band. Now just as in the case of the calcium driven muscle contractile ATP-ase device or in the calmodulin activated kinases, so the proton channel, the F_0 or the proton-modulin can communicate with ATP-formation or hydrolysis via further helical rearrangements in the F_1 portion of the enzyme. It is also a relatively simple matter to imagine that such rearrangements of helices in a membrane could be connected to other enforced movements such as those of a molecular pump. Thus helical ion channels can be controlled.

In essence the above devices are just rearrangements of helices to open channels and/or to transmit energy. They are therefore applicable as models to all the agonist receptor systems which cause either the input of a calcium message or the formation of c-AMP from ATP. The last reaction is a special type of ATP-ase since it is a downhill use of ATP. Acetylcholine binding can be relayed to the opening of helical channels or to ATP reactions through helix movements.

In some cases reactions are triggered by the binding of anions rather than cations. A case of special interest is phosphorylation. The binding of phosphate, a dianion, can be likened to a negative calcium ion binding. We can expect that it will cause a considerable energy change in a local region of a protein. Once again this change can be transmitted through helical rods. The work of L. Johnson at Oxford makes this description very reasonable in the case of phosphorylase. Perhaps a more interesting case is that of membrane protein phosphorylation which activates the Na^+/K^+ and the Ca^{2+} pumps. My model would be that phosphorylation causes a rearrangement of helices so as to open and close membrane channels during a cycle of phosphorylation. A final example of the use of phosphorylation connected to the realignment of helices is in the removal of histones from chromatin so that DNA can be read. The histones are highly helical proteins which associate in the nucleosomes. Only the tails of the proteins remain at all exposed. These tails can be phosphorylated. On phosphorylation the tails loose positive charge and electrostatic interactions are considerably rearranged. As an inevitable consequence the tails which were pulled toward the DNA by charge attraction are now repelled in part. The repulsion will act to move the tail relative to the DNA but will also act as a force on the helices of the histones in the nucleosomes, thus generating partial unscreening of the DNA and allowing access to reading (transcribing) systems.

The above considerations have very interesting overtones as far as drug design is concerned. A drug could interact with one of the binding sites at the ends of the above devices, e.g. the calcium site on the ATPase site of muscle. This is a conventional thought. However, the long network of helical protein connections and the fact that this network changes by rotational/translational motion allows further possibilities. The drug could act on the helices at many points. It could make the motion easier or more difficult and using an interplay of thermodynamic and kinetic factors it need not block but could simply alter response times. We know that trifluoropiperazine acts like the helices of calmodulin. Again this interference with calcium messages is not very different from the effect of phosphates as allosteric effectors of hemoglobin. Using binding to remote regions of a machine rather than to either the input or output sites it is possible to describe a range of observed agonist and antagonist activities almost in terms of the friction in the machinery.

A final point concerns the selectivity of the input system. This is clearly shown in the case of calmodulin which must accept Ca^{2+} but not Mg^{2+} by a factor of 10^4. This means that the site has to be restricted in some special ways. It must not be too restrictive, however, since both the on-rate and the off-rate should be rapid. Mobile but selective sites can be deviced of course but usually they cannot have too much movement without loss of selectivity. Movement is then amplified within the receptor protein.

Glucagon Conformation in Different Environments: Implications for Molecular Recognition

K. Wüthrich

Institut für Molekularbiologie und Biophysik, ETH-Hönggerberg CH-8093 Zürich, Switzerland

Glucagon is a hormone which consists of a linear polypeptide chain of 29 amino acid residues and has a molecular weight of 3500. Early studies by circular dichroism and other physical-chemical techniques indicated a pronounced tendency of this polypeptide to adopt different spatial structures in different environments. For example, for monomeric glucagon in aqueous solution a flexible "random coil" structure was indicated [1], and for self-aggregated glucagon in aqueous solution it was shown that it could adopt either an α-helical [2] or a β-sheet [3] secondary structure and that the species formed depended critically on the peptide concentration [4]. Furthermore, interactions with lipids and detergents were also found to induce changes of the glucagon conformation [5–7]. More recently, detailed descriptions based on many-parameter techniques were obtained for glucagon conformations in three different environments, i.e. in single crystals, in aqueous solution and in the lipid-water interphase near the surface of dodecylphosphocholine (DPC) micelles. The present paper describes some implications for studies of the mechanism of action of glucagon which result from comparison of these three molecular structures.

In the glucagon single crystals which were studied by Sasaki et al. [8] by the use of X-ray methods, glucagon forms trimeric species, where the individual polypeptide chains form α-helical structures.

Monomeric glucagon in aqueous solution was studied by ^{1}H nuclear magnetic resonance (NMR). These studies confirmed earlier observations that monomeric glucagon in aqueous solution adopts a predominantly flexible "random coil" form and further revealed a structured region involving the residues – Phe 22 – Val 23 – Gln 24 – Trp 25 –, with a conformation different from an α-helix-type secondary structure [9].

The conformation of glucagon in the lipid-water interphase near the surface of perdeuterated DPC micelles was determined by NMR. Making use of nearly complete sequence-specific resonance assignments [10], a map of distance constraints between specified hydrogen atoms of the polypeptide chain was collected with two-dimensional nuclear Overhauser enhancement spectroscopy (NOESY) and subsequently used as the input for distance geometry calculations [11]. Figure 1 presents an illustration of the data on the spatial structure thus obtained. The secondary structure of micelle-bound glucagon includes three turns of irregular α-helix in the segment of residues 17–29 near the C-terminus, a stretch of extended polypeptide chain from residues 14–17, an α-helix-like turn formed by the residues 10–14 and another extended region from residues 5–10. In the N-terminal tetrapeptide H-His-Ser-Gln-Gly- the two terminal residues are highly mobile, indicating that they extend into the aqueous phase, and the mobility of the residues Gln 3 and Gly 4.

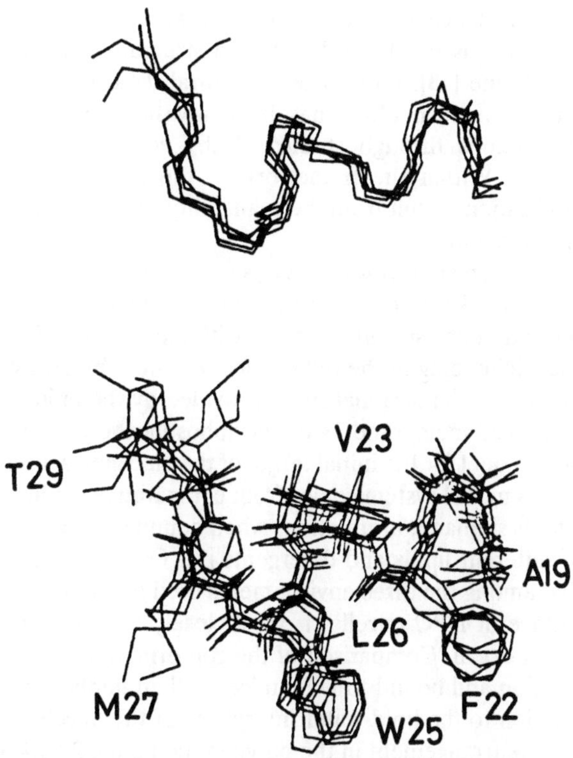

Fig. 1. Conformation of the residues 19–29 in micelle-bound glucagon. For practical reasons separate distance geometry calculations of the molecular conformation of micelle-bound glucagon were done for four distinct segments including, respectively, the residues 5–15, 10–20, 17–27 and 19–29. This figure presents the result obtained for the segment 19–29. Drawings of the backbone *(upper trace)* and of the "restrained side chain" presentation are shown. The latter includes the complete side chains for those residues for which NOE constraints were observed also for the peripheral protons; these are identified by the one-letter symbol and the sequence position. For the other residues only the backbone atoms, including the C^{α} and amide protons, and C^{β} are shown. CH_3, CH_2 and the CH groups in the aromatic rings of Phe and Tyr are represented by spherical pseudoatoms. Six structures are superimposed. Each conformer was obtained from an independent computation using the same experimental data. Each species represents a typical member of the group of similar but not identical spatial structures which are compatible with the NMR data. (Reproduced with permission from Braun et al. [11])

appears to be only partially restricted by the binding to the micelle. The experimental data showed that the polypeptide chain does not fold back on itself [7] and suggested that the overall spatial arrangement of the glucagon polypeptide chain in a lipid-water interphase is largely determined by the topology of the lipid support. The arrangement of the side chains results in formation of two hydrophobic "patches", one consisting of Phe 6, Tyr 10 and Leu 14, and the other of Ala 19, Phe 22, Val 23, Trp 25 and Leu 26.

The primary target organ for glucagon is the plasma membrane of liver and other cells, where binding to a specific receptor site mediates activation of glycogenolysis

[12]. Evidence has been presented that recognition between glucagon and its receptor depends on the ordered lipid structures surrounding the receptor site in the membrane [13]. Otherwise, structural data on the receptor system are scarce and therefore much effort has been invested to characterize the glucagon-receptor interactions through studies of the conformational properties of the hormone [8, 14]. Within the framework of the following considerations, some intriguing implications result from the availability of the spatial glucagon structures in different environments.

Fundamental considerations (see also R. Schwyzer, this volume) make it appear rather unlikely that the initial contact of glucagon with its target cell would be the formation of a specific complex with the receptor site. Rather one could expect non-specific binding to the cell surface, followed by diffusion to the specific receptor site in the two-dimensional space provided by the lipid-water interphase on the surface of the cell membrane. A dodecylphosphocholine micelle may appear to be a rather poor model for the initial target of the glucagon action, but support for the relevancy of this model system comes from comparative studies which indicated that glucagon adopts similar conformations when bound to micelles or vesicles formed by a variety of different lipids and detergents [7]. Overall, it thus seems reasonable to assume that among the three environments used for conformational studies of glucagon, the surface of DPC micelles is most closely related to the surface of the physiological target organ. Comparison of the conformations of monomeric glucagon in aqueous solution and bound to DPC micelles then implies that transfer from dilute aqueous solution to the lipid-water interphase of the micellar surface induces a major structural rearrangement in the polypeptide region 22–24, which is primarily responsible for the binding of the hormone to the receptor site [15]. Therefore, if specific binding of glucagon is determined by the conformation of the binding region, the structural features responsible for this specificity will be formed only after the hormone has been incorporated into the lipid-water interphase of the cell surface and will not be present during the transport of glucagon in an aqueous medium.

Different authors have presented evidence that while the peptide segment 20–26 is responsible for specific binding of glucagon to the receptor site, the activity depends on the presence of the complete N-terminal part of the polypeptide chain, including the N-terminal histidine [15–17]. It would thus appear that two different locations on the surface of the target organ must be recognized, respectively, by the C-terminal part and the N-terminal part of the glucagon chain in order to initiate the follow-up reactions which are controlled by the hormone. For an initial mapping of the possible relative locations of these two recognition sites it is then interesting to know the overall length of the glucagon molecule in the lipid-water interphase on the cell surface. Different conclusions result from the studies of crystalline glucagon and micelle-bound glucagon, since the overall length of the latter conformation is appreciably longer than that of the helical crystal structure [11].

Appendix: Use of Perdeuterated Lipid and Detergent Micelles for Studies of Polypeptide Conformations by ^{1}H NMR

The size and stoichiometry of the mixed micelles used for NMR studies of glucagon in a lipid-water interphase were determined by ultracentrifugation and quasi-elastic light-scattering. It was found that stoichiometrically well-defined complexes were formed between glucagon and perdeuterated dodecylphosphocholine micelles, consisting of one glucagon molecule and approx. 40 detergent molecules and with a molecular weight of ca. 17,000 [7]. Corresponding mixed micelles are formed with one molecule of melittin and ca. 40 molecules of dodecylphosphocholine [18]. Well-resolved ^{1}H NMR spectra can be obtained for the polypeptide chains in the mixed micelles, which are thus amenable to the NMR techniques which are generally used for conformational studies of proteins in solution [19]. The location and orientation of the polypeptide chain relative to the micelle surface was further investigated by NMR and electron spin resonance techniques which employed lipid nitroxide spin labels [20, 21].

With glucagon and melittin mixed micelles containing a single polypeptide chain could be obtained, and the polypeptide was found to be located predominantly in the lipid-water interphase near the surface of the micelle [7, 21]. As discussed above, studies of the micelle-bound polypeptides then provided an interesting complementation of structural data obtained in different environments, including single crystals. For other non-globular oligo- and polypeptides a micelle-bound form might turn out to be the only conformation which can be studied in detail, since this class of compounds appears to have a tendency to adopt flexible, "unstructured" forms in solution and furthermore seems to be difficult to crystallize. Since the limited mobility provides for favorable conditions for measurements of nuclear Overhauser enhancements, the micelle-bound form of a peptide is particularly suitable for structure determination by NMR [22], so that a strategy similar to that used for small globular proteins can be applied [19]. With micelles as peptide carriers, the use of these NMR techniques can actually be extended to much smaller peptides than would otherwise be possible.

A further intriguing development might result with the use of inverted, perdeuterated micelles (see P. L. Luisi, this volume) which could extend the use of NMR for conformational studies of polypeptide chains to new milieus which have not so far been accessible for detailed structural studies.

Acknowledgement

The research projects summarized here are supported by the Schweizerischer Nationalfonds (projects 3.528.79 and 3.284.82).

References

1. Panijpan, B. and Gratzer, W. B. (1974) Eur. J. Biochem. *45,* 547–553
2. Gratzer, W. B., Bailey, E. and Beaver, G. H. (1967) Biochem. Biophys. Res. Commun. *28,* 914–919
3. Epand, R. M. (1971) Can. J. Biochem. *49,* 166–169
4. Wagman, M. E., Dobson, C. M. and Karplus, M. (1980) FEBS Lett. *119,* 256–270
5. Schneider, A. B. and Edelhoch, H. (1972) J. Biol. Chem. *247,* 4986–4991
6. Epand, R. M., Jones, A. J. S. and Schreier, S. (1977) Biochim. Biophys. Acta *491,* 296–304
7. Bösch, C., Brown, L. R. and Wüthrich, K. (1980) Biochim. Biophys. Acta *603,* 298–312
8. Sasaki, K., Dockevill, S., Ackmiak, D. A., Tickle, I. J. and Blundell, T. L. (1975) Nature *257,* 751–757
9. Bösch, C., Bundi, A., Oppliger, M. and Wüthrich, K. (1978) Eur. J. Biochem. *91,* 204–214
10. Wider, G., Lee, K. H. and Wüthrich, K. (1982) J. Mol. Biol. *155,* 367–388
11. Braun, W., Wider, G., Lee, K. H. and Wüthrich, K. (1983) J. Mol. Biol. *169,* 921–948
12. Pohl, S. L., Birnbaum, L. and Rodbell, M. (1969) Science *164,* 566–569
13. Rubalcava, B. and Rodbell, M. (1973) J. Biol. Chem. *248,* 3831–3837
14. Blundell, T. L. and Wood, S. (1982) Ann. Rev. Biochem. *51,* 123–154
15. Wright, D. E. and Rodbell, M. (1979) J. Biol. Chem. *254,* 268–269
16. Rodbell, M., Birnbaumer, L., Pohl, S. L. and Sundby, F. (1971) Proc. Natl. Acad. Sci. USA *68,* 909–913
17. Carrey, E. A. and Epand, R. M. (1983) Int. J. Peptide Protein Res. *22,* 362–370
18. Brown, L. R. (1979) Biochim. Biophys. Acta *557,* 135–148
19. Wüthrich, K., Wider, G., Wagner, G. and Braun, W. (1982) J. Mol. Biol. *155,* 311–319
20. Brown, L. R., Bösch, C. and Wüthrich, K. (1981) Biochim. Biophys. Acta *642,* 296–312
21. Brown, L. R., Braun, W., Anil Kumar and Wüthrich, K. (1982) Biophys. J. *37,* 319–328
22. Wüthrich, K., Bösch, C. and Brown, L. R. (1980) Biochem. Biophys. Res. Commun. *95,* 1504–1509

Topic 2

Thermodynamics, Kinetics and Stereochemistry in Molecular Recognition

On the Economics of Binding Energies

William P. Jencks

Graduate Department of Biochemistry, Brandeis University, Waltham, MA 02254, USA

There is a widespread impression that the best, most specific small molecules that make things happen in biological processes are the molecules that bind most strongly to a specific binding site on a macromolecule. Evolution will certainly favor molecules that bind strongly enough to their receptor site that a significant fraction of these sites are occupied, under conditions when it is desirable for some event that results from the binding to occur. However, many active molecules do not bind very strongly to their receptor site and inactive molecules often bind more strongly than active ones. One reason that very strong binding is uncommon is that it is usually just as important for a molecule to be able to dissociate from its binding site as for it to bind readily; very large equilibrium constants for association lead to unacceptably slow rate constants for dissociation even if binding is diffusion controlled. However, a more fundamental reason that the most effective small molecules do not bind very strongly arises from the economics of the utilization of binding energies through interaction energies – as is well known, but often forgotten.

Virtually anything interesting that happens in biology involves interaction energies. This is because most biological events are mediated or controlled through specific binding interactions. Sometimes this binding is simple, as in the binding of biotin to avidin. More often, the binding energy is *utilized* to make something happen by changing the properties of a macromolecule. This utilization involves the economics of binding energy – you must pay for what you get; there is no free lunch. How this occurs is described by interaction energies, or linked functions [1, 2].

Why is the importance of interaction energies not more widely appreciated? They are simple and well known to most physical biochemists. They are nothing more than a way of describing the critical property of the thermodynamic box for two alternative pathways of a reaction. However, they are also subtle, poorly understood by most biochemists, and extraordinarily important. I will review here some examples of the economics for the utilization of binding energies, as described by interaction energies, and how they may be interpreted. The theme that appears repeatedly is that for systems in which large amounts of binding energy are used the observed binding is likely to be weak for the most effective ligands. This is a consequence of the economics of the system – much of the intrinsic binding energy is *used up* to provide the driving force for a process, so that the observed binding will be relatively small. Ligands that bind most strongly are often not effective mediators of biological processes; indeed, they are likely to be potent inhibitors.

A Simple Example. The ATP-driven dissociation of actomyosin to give myosin · ATP provides a well-known and relatively straightforward illustration of the utilization of an interaction energy to make a system work. The very strong binding of ATP to myosin provides the driving force for the dissociation of actomyosin in a critical step of the reaction sequence for muscle contraction (1). The

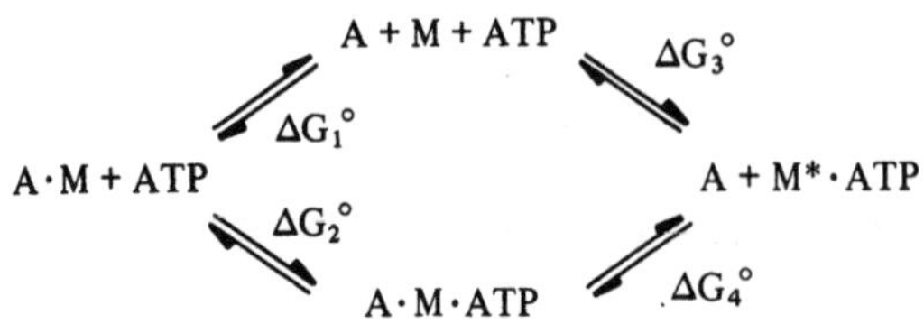

(1)

reaction proceeds because the binding of ATP to myosin, with $\Delta G_3^0 = -15.6$ kcal mol^{-1}, is slightly stronger than the binding of actin, $\Delta G_1^0 = -11.9$ kcal mol^{-1}, so that the overall reaction is favorable by some -3.7 kcal mol^{-1} at a standard state of 1 M [3, 4]. The reaction could proceed through dissociation to free myosin but it actually proceeds through the ternary complex A · M · ATP, in milliseconds rather than minutes.

The overall free energy change for the reaction is the same regardless of which path is followed, so that $\Delta G_{obsd}^0 = \Delta G_2^0 - \Delta G_4^0 = \Delta G_3^0 - \Delta G_1^0$. The interaction energy is the difference in the ways that the binding energies of myosin for actin and for ATP are expressed in the two different pathways. It is the difference in the binding energy of actin to myosin in the presence and absence of ATP, $\Delta G_I = \Delta G_4^0 - \Delta G_1^0$, which is equal to the difference in the binding energies of ATP to myosin in the presence and absence of actin, $\Delta G_I = \Delta G_2^0 - \Delta G_3^0$ [5].

This is illustrated in the bar diagram of Fig. 1, which shows the changes in Gibbs free energy for the two reaction pathways. The values in this figure are based on standard states of 1 mM for actin and 5 mM for ATP, which correspond to estimated concentrations of these species under physiological conditions ("basic Gibbs energies" [6]), and an estimated interaction factor of 10^5, corresponding to $\Delta G_I = 6.8$ kcal mol^{-1} [3, 4, 7, 8]. The path through free A, M and ATP proceeds through a high energy intermediate; the path through A · M · ATP is much lower in energy.

It is important to think of this system in two ways, which describe the two reasons that the system works. First, the stronger binding to free M of ATP than of A allows

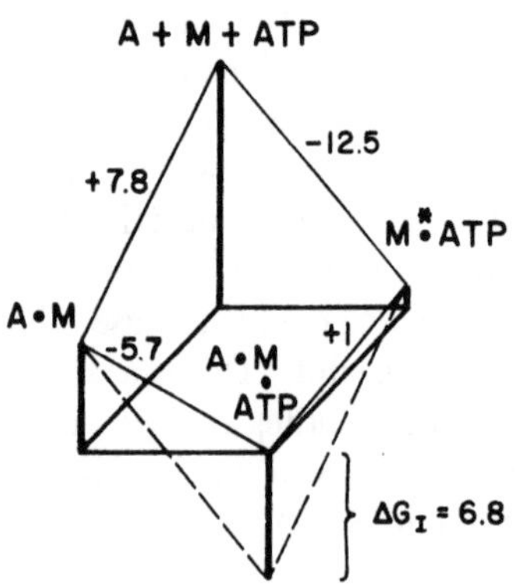

Fig. 1. Gibbs free energy diagram for the ATP-induced dissociation of actomyosin

the reaction to proceed. The large intrinsic binding energy for the binding of ATP to myosin is expressed in $M^* \cdot ATP$ and provides the driving force for the reaction. The intrinsic binding energy is the maximum binding energy that can be realized for a molecule or group under ideal conditions. Observed binding energies usually provide only limiting values for the intrinsic binding energy because ideal conditions, such as binding to an unstable conformation of a protein, generally cannot be realized experimentally. For example, the binding of ATP to M* is stronger than to M, the stable conformation of myosin in the absence of ATP.

If there were equally strong binding of ATP to actomyosin the $A \cdot M \cdot ATP$ complex would be a very low energy species, as shown by the dashed lines in Fig. 1. Such a system would not work, because the ternary complex would not dissociate to give $M^* \cdot ATP$. Thus, the second requirement of the system is that the intrinsic binding energy must *not* be expressed in this complex. Instead, a large fraction of it is used up to destabilize the system by the amount of the interaction energy, $\Delta G_I = \sim 6.8$ kcal mol^{-1}, so that ATP destabilizes the binding of myosin to A and actin can dissociate; an equal destabilization by actin facilitates the dissociation of ATP in the reverse reaction. This mutual destabilization could involve direct overlap of the binding sites for A and ATP, but is actually brought about indirectly through a conformation change that is induced by the binding of ATP. The kinetics of the reaction induced by AMPPNP are complex, but suggest that there are actually at least two states of the $A \cdot M \cdot ATP$ complex, an initial state in which the nucleotide binds weakly to induce a conformation change, and a second state, $A \cdot M^* \cdot ATP$, in which it binds more strongly and the binding to actin has been correspondingly weakened (2) [4].

$$A \cdot M \cdot ATP \rightleftharpoons A \cdot M^* \cdot ATP \qquad (2)$$

Thus, the binding energy between M and ATP is different in two states of the system. In one state, $M^* \cdot ATP$, the intrinsic binding energy is expressed directly as binding, to provide the driving force for the reaction. In the other it is not expressed; instead, much of it is utilized to produce a conformation change that is equally important in order to make the system work [9]. The amount of the difference is the interaction energy.

The Synthesis of ATP. The F_1 ATPase of mitochondria and the CF_1 ATPase of chloroplasts provide more interesting illustrations of the roles of binding and interaction energies. The synthesis of ATP during oxidative phosphorylation or photophosphorylation is mediated by coupled proton transfer through these enzyme systems. This thermodynamically unfavorable reaction might be brought about directly by the proton, following Mitchell's suggestion of a proton-driven dehydration of inorganic phosphate [10], or indirectly by the binding-change mechanism of Boyer, in which ATP is synthesized spontaneously and reversibly at the active site and the role of the proton is to bring about its dissociation from the enzyme [11]. There is precedent for the latter mechanism in the ready reversibility of ATP hydrolysis at the active site of myosin and this mechanism is consistent with a large body of experimental data. Direct support for this mechanism has recently been obtained in Penefsky's laboratory with the soluble F_1 ATPase [12]; similar results have been obtained by Feldman and Sigman [13] for the CF_1 enzyme. The problem is to understand how the hydrolysis of ATP, which is virtually irreversible in solution, can be made readily reversible at the active sites of these enzymes.

The energetics of the F_1 system, based on the data of Grubmeyer, Cross and Penefsky [12], is shown in the energy diagram of Fig. 2, which describes the reactions of (3). In solution the synthesis of ATP is unfavorable by some 7.6 kcal mol^{-1}, but at the active site of the enzyme it is spontaneous, with $\Delta G^0 = -0.4$ kcal mol^{-1}, as shown on the top and bottom of the diagram, respectively. It is apparent that the driving force for ATP synthesis is the very strong binding of ATP to F_1, with $\Delta G^0 = -16.3$ kcal mol^{-1}. This can be taken as an expression of the intrinsic binding energy between ATP and the enzyme.

$$\begin{array}{ccc} ADP + P_i & \overset{K_N}{\rightleftharpoons} & ATP \\ K_{DP} \Big\updownarrow \pm E & & K_T \Big\updownarrow \pm E \\ E{:}^{ADP}_{P_i} & \overset{K_E}{\rightleftharpoons} & E\cdot ATP \end{array} \qquad (3)$$

The other requirement for the reaction is that the energy of bound ADP and P_i must be about the same as that of bound ATP; the bound ADP and P_i must somehow be brought to a relatively high Gibbs free energy so that they can form ATP spontaneously, in contrast to the situation in solution. The ADP and P_i molecules contain the same adenine and ribose, and approximately the same number of phosphate residues and charges as ATP, so that they have the potential to have a comparable binding energy, but the observed binding is weak. It is this weak binding that makes the bound ADP and P_i relatively high in energy so that they can be converted spontaneously to ATP.

The Gibbs free energy changes for the binding of ATP and for the hydrolysis of free and bound ATP give a value of $\Delta G^0 = -8.3$ kcal mol^{-1} for the binding of ADP and P_i (Fig. 2). The difference between the binding of ADP and P_i and that of ATP is $\Delta G_I = 8$ kcal mol^{-1}, for a standard state of 1 M. It is this interaction energy that makes ATP synthesis possible; it is equal to the difference in the free energy for the hydrolysis of ATP on and off the enzyme. If there were no interaction energy, the synthesis of ATP at the active site would be just as unfavorable as it is in solution, as shown by the dashed lines in Fig. 2.

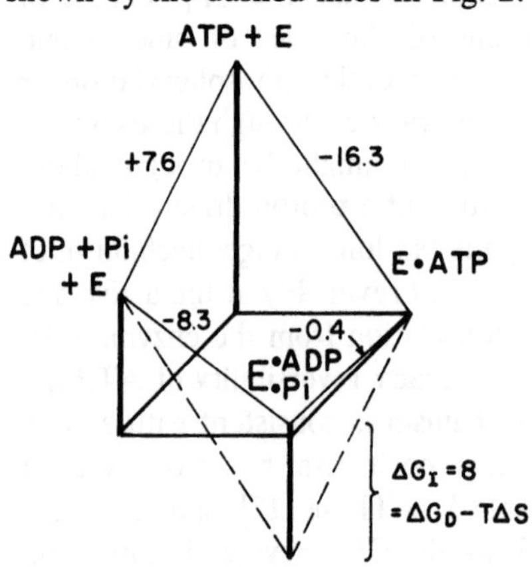

Fig. 2. Gibbs free energy diagram for the synthesis of ATP at the active site of the F_1 ATPase

The interaction energy, or increase in Gibbs energy, represents loss of entropy, $-T\Delta S$, and strain or destabilization, ΔG_D, of the bound ADP and P_i. The obvious, and probably the most important difference between the nonenzymic and enzymic reactions is that the bimolecular nonenzymic reaction is very improbable because it requires a large loss of translational and rotational entropy of the reactants in order to form a new covalent bond between them, whereas the intramolecular reaction at the active site of the enzyme requires a much smaller loss of entropy and is correspondingly more probable. There are a number of intramolecular chemical reactions that are faster by factors of $10^5 - 10^8$ compared with corresponding bimolecular reactions at a concentration of 1 M, and factors of up to 10^8 M in favor of the intramolecular reaction are reasonable on theoretical grounds [14–16]. The loss of entropy, $-T\Delta S$, upon binding ADP and P_i corresponds to an increase in Gibbs free energy, ΔG, which is paid for by utilization of the binding energy of the reactants. The entire advantage of the enzymic reaction could be accounted for by this advantage from induced intramolecularity, in principle, but it is virtually certain that there is also destabilization of the bound reactants by strain, desolvation and/or electrostatic effects. One reason for this is that a large entropy loss requires very tight binding of the reactants next to each other [16]. Thus, the interaction energy, ΔG_I, represents the sum of the contributions of entropy loss and destabilization of the bound ADP and P_i (4) [9].

$$\Delta G_I = \Delta G_D - T\Delta S \tag{4}$$

One consequence of the interaction energy is that the binding of P_i to the $E \cdot ADP$ complex is weak. The equilibrium constants [12] for the steps give a value of $K'_P = 3$ M for the dissociation of P_i from the $E \cdot ADP \cdot P_i$ complex and kinetic measurements give a value of K'_P of the order of 0.03 M [17]. The difference in these values may reflect uncertainties in the other constants, but they both represent weak binding. Phosphate binds to a site on the free enzyme with a dissociation constant of $K_P = 2.9 \times 10^{-4}$ M at pH 8.1 and is displaced from this site by nucleotides and other reagents that bind to the active site [18]. If this site is the productive binding site for P_i, the presence of ADP reduces the binding of P_i by a factor of 100-fold or more. This reduction represents an interaction energy, or a mutual destabilization, between the bound ADP and P_i, ΔG_{12}, that can account for a significant fraction of the overall interaction energy, ΔG_I [19]. Phosphate also binds very weakly to the $CF_1 \cdot ADP$ complex, with an apparent dissociation constant of 0.2 M for the synthesis of $CF_1 \cdot ATP$ at pH 7 [13].

The reversible synthesis and hydrolysis of ATP at the active site of myosin is very similar, as shown in the energy diagram of Fig. 3. The strong binding of ATP reflects the expression of some -15.6 kcal mol^{-1} of intrinsic binding energy and the interaction energy for the binding of ADP and P_i is $\Delta G_I = 6.4$ kcal mol^{-1} for a standard state of 1 M, so that ATP synthesis on the enzyme is only slightly unfavorable, with $\Delta G^0 = 1.4$ kcal mol^{-1} [3, 19, 20]. For standard states of 5 mM for ATP, 0.05 mM for ADP and 1 mM for P_i [7] ("basic Gibbs energies" [6]), the interaction energy is 13.2 kcal mol^{-1}. The interaction energy includes the loss of entropy that is required for the synthesis of ATP from dilute reactants, so that it is larger for physiological concentrations of reactants.

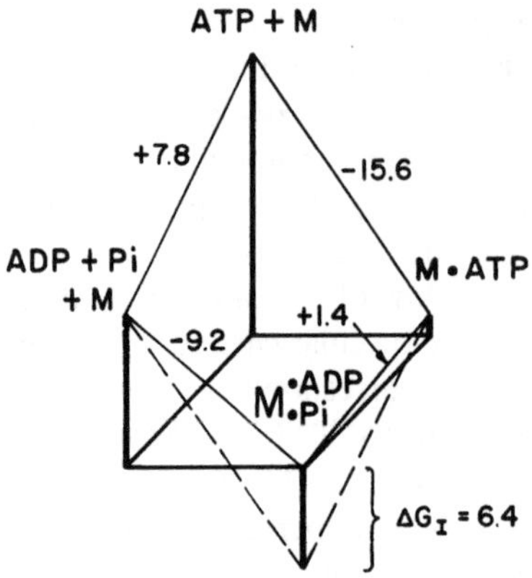

Fig. 3. Gibbs free energy diagram for the synthesis of ATP at the active site of myosin

In this reaction it is possible to estimate the interaction energy that is brought about by the binding of ADP and P_i individually to myosin and that which is brought about by interaction between ADP and P_i when they are bound together. The part that is brought about directly by the protein arises from the loss of entropy upon binding the *individual* molecules of ADP and P_i and any destabilization of these bound molecules that is relieved upon ATP formation, ΔG^s. A value of ΔG^s = 3.6 kcal mol^{-1} is given by comparison of ΔG_T^0 for the binding of ATP with the sum of ΔG_D^0 and ΔG_P^0 for the binding of ADP and P_i individually (5, 6) [8, 19, 20]. The part that is brought about by interaction between bound ADP and P_i is given by the difference between the binding of ADP to free M and to $M \cdot P_i$, or the difference between the binding of P_i to M and to $M \cdot ADP$ to form the ternary complex; both

$$\begin{array}{ccc} M & \underset{\pm P_i}{\overset{K_P}{\rightleftharpoons}} & M \cdot P_i \\ K_D \updownarrow \pm ADP & & K_D' \updownarrow \pm ADP \\ M \cdot ADP & \underset{\pm P_i}{\overset{K_P'}{\rightleftharpoons}} & M{\cdot ADP \atop \cdot P_i} \end{array} \tag{5}$$

$$M + ATP \overset{K_T}{\rightleftharpoons} M^* \cdot ATP \tag{6}$$

of these differences correspond to the same interaction energy, $\Delta G_{12} = \Delta G_P' - \Delta G_P = \Delta G_D' - \Delta G_D$ (5). This interaction energy arises from additional loss of freedom of movement and entropy, and from direct mutual destabilization by compression, desolvation and electrostatic interaction of the bound molecules in the ternary complex. For myosin the value of ΔG_{12} is 2.8 kcal mol^{-1} [8, 19, 20]. If, as appears likely, these unfavorable interactions are relieved when ATP is formed [21], the overall driving force for ATP synthesis that arises from ΔG_1 can be divided into approximately equal contributions from the direct binding interactions of ADP and P_i that arise when each binds to the protein, ΔG_s = 3.6 kcal mol^{-1}, and from the interaction between the ADP and P_i when they are both bound, $\Delta G_{12} = 2.8$ kcal mol $^{-1}$ [19].

In this system actin plays the role of protons and other nucleotide molecules with the F_1 ATPase, by decreasing the affinity of the enzyme for bound nucleotides and accelerating their dissociation from the enzyme.

Coupling. The purpose of the utilization of binding energy is primarily kinetic. The different expression of binding energies in different states of the system is used to bring the different states to similar Gibbs energy levels under physiological conditions so that turnover can occur rapidly. This serves to avoid the formation of very high energy or very low energy intermediates that would increase the kinetic barriers along the reaction path and require very large rate constants for the individual steps in order to permit turnover to occur at a useful rate.

The binding energies do not account for coupling. The mechanism of coupling may be described by a set of rules, which represent enzyme specificities that change in the different states through which the enzyme passes during the reaction cycle. The coupling rules are independent of the binding energies; they may or may not be related to the mechanisms through which the binding energies are expressed. A system that obeyed the coupling rules would still bring about a coupled reaction if there were no interaction energies, but would usually do so at too slow a rate to serve any physiological function [9].

The net synthesis of ATP in oxidative phosphorylation or photophosphorylation requires that its intrinsic binding energy be expressed to give very strong binding, as described above, but also requires that the binding must then become weak so that the ATP can dissociate to give free ATP. This is presumably mediated by the proton gradient, according to the binding change mechanism [11]. However, when the binding of ATP becomes weak the equilibrium constant for its synthesis becomes unfavorable, so that a requirement of the mechanism is that the enzyme must have a way of preventing the reversion of ATP back to ADP and P_i before it dissociates. One mechanism for meeting this requirement is that the activity of the enzyme for catalysis of the synthesis and hydrolysis of ATP is turned off in a low-affinity state of the enzyme.

A model for oxidative phosphorylation that illustrates the kind of rules and reaction sequence that could give such a coupled process is shown in (7) [9]. This

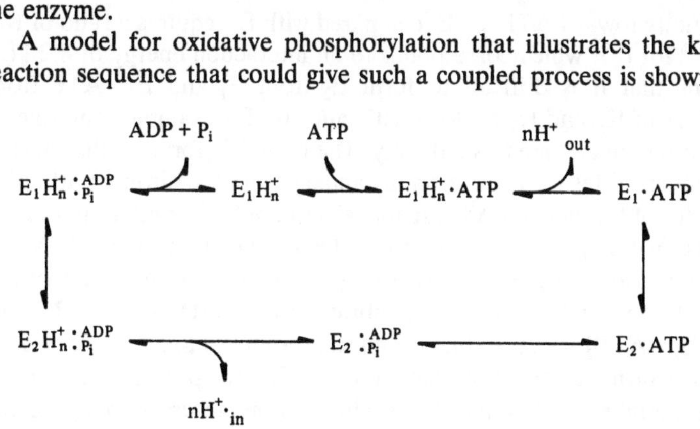

(7)

scheme is a modification of the model originally proposed by Boyer [22, 23]. It is unlikely to be correct in the form shown; it is only meant to illustrate the kind of properties that would give a workable system. The rules that define the coupling in this model are:

1) The binding and dissociation of ATP, ADP, and P_i occur only with the E_1 state of the enzyme, whereas catalysis of the cleavage and synthesis of ATP occurs only with the E_2 state of the enzyme.
2) Proton transfer from one side of the membrane to the other occurs only with the species $EH_n^+ \cdot ADP \cdot P_i$, whereas interconversion of the unprotonated E_1 and E_2 forms occurs only with the species $E \cdot ATP$.

In principle, the coupling mechanism does not require a change in the affinity of the binding sites for the proton in different states of the enzyme. In this model, the larger concentration of the unprotonated species that results from the higher pH inside mitochondria will increase the rate of synthesis of ATP in this state. The larger concentration of the protonated form at the lower pH outside will increase the rate constant for the dissociation of ATP. These differences influence the rate of observed ATP synthesis, but are not directly responsible for the coupling process itself.

Different affinities for ATP in the E_1 and the E_2 forms of the enzyme introduce another interaction energy into the system. This may be illustrated by the oversimplified scheme of (8), which involves transfer of only a single proton, omits $E \cdot ADP \cdot P_i$, and includes the equilibrium constants for steps that do not occur directly in the coupled system. If we consider only the binding of ATP, a 10^6-fold

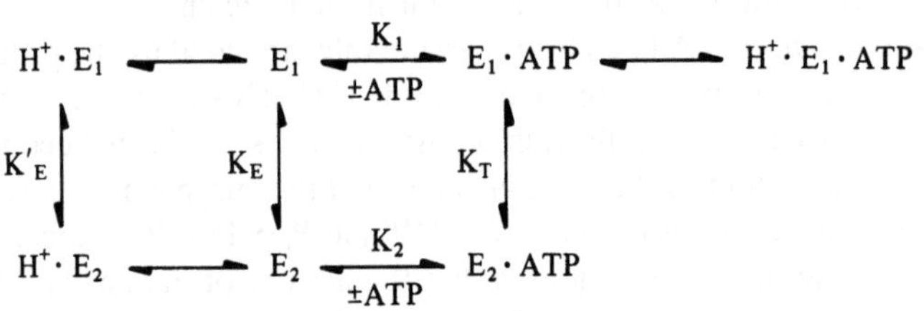

(8)

decrease in the affinity toward ATP for E_1 compared with E_2 requires a ratio of $K_E/K_T = K_1/K_2 = 10^{-6}$ for (8), which corresponds to an interaction energy of 8.2 kcal mol^{-1}. This means that it is difficult to form E_2 from E_1 and $E_1 \cdot ATP$ from $E_2 \cdot ATP$; the values of K_E and K_T could be 10^{-3} and 10^3, for example. The role of the proton gradient is to overcome this difficulty. The E_1 and E_2 forms of the enzyme could have different affinities for the proton, by analogy with the different affinities of the different states of the calcium ATPase for calcium, but this is not required. If, for example, we assume $pK_A = 7$ for each protonated enzyme form, pH values of 5 outside and 9 inside, and transport of only a single proton, then the ratios of the concentrations of the relevant species at equilibrium are $[E_2]/[H^+ \cdot E_1] = 0.1$ and $[H^+ \cdot E_1 \cdot ATP]/[E_2 \cdot ATP] = 0.1$, so that the concentrations of these species would be adequate to allow turnover at a reasonable rate. With three protons transported instead of one the equilibria will be more favorable and the necessary pH gradient smaller.

Note that not all the reactions shown in (8) actually occur through the pathways that are shown in the equation; some of them are forbidden by the specificity rules.

The purpose of the equation is only to illustrate the economics of the energy balance in this kind of system. There is no direct interconversion of E_1 and E_2, for example. This is required for coupling, according to rule 2, and is also required in order that the proton gradient can be used to produce useful concentrations of the $E_2 \cdot ADP \cdot P_i$ form of the enzyme, as just described. The E_1 and E_2 species are not directly at equilibrium with each other; direct interconversion would produce a proton leak. Their ratio is shifted from the equilibrium ratio of $E_2/E_1 = 10^{-3}$ to a ratio of 0.1 by the proton gradient because their interconversion occurs only through $H^+ \cdot E_1$ and $H^+ \cdot E_2$.

The real situation is, of course, much more complex, but the same principles are applicable. An important additional property of the system is that the binding of nucleotides to a second binding site, probably on an adjacent β subunit of the ATPase, causes a dramatic increase of ca. 10^6 in the rate and equilibrium constants for dissociation of nucleotides from the active site [24]. This cannot occur during ATP synthesis because the synthesis of ATP requires that it binds very strongly (Fig. 2). The proton may act as a key to regulate this allosteric action of nucleotides. It is not unreasonable that the binding and dissociation of nucleotides should require the protonated state of the enzyme for the allosteric effect as well as for catalysis. Thus, the proton may act as a key that permits other nucleotides to provide a driving force that facilitates the dissociation of ATP.

Nucleotide binding to a second site increases the rate of catalysis of ATP hydrolysis as well as dissociation [24, 25]. An alternative rule for avoiding hydrolysis is simply that binding of the proton and nucleotide bring about dissociation of ATP faster than it reverts to ADP and P_i; it is possible that the dissociation of ATP is brought about specifically by ADP and P_i [11]. There is precedent for such a mechanism in the actin-driven dissociation of ATP in good yield from the equilibrium mixture of $M \cdot ATP \rightleftharpoons M \cdot ADP \cdot P_i$, which is the analogous reaction in the myosin system [26].

The E-P ATPases. The calcium and sodium-transporting ATPases must solve similar but more difficult thermodynamic problems in order to undergo turnover at a useful rate:

1) They must synthesize an acyl phosphate at the active site that is unfavorable by $\Delta G^0 = 10$ kcal mol^{-1} in solution.
2) They bring about a reversible interconversion of the enzyme-phosphate intermediate between a low energy form, which is at equilibrium with inorganic phosphate, and a high energy form, at equilibrium with ATP.
3) They must bring about a reversible interconversion between a strong-binding species for the transported ions, which can pick up the ion from dilute solution, and a weak-binding species, from which the ion has a high escaping tendency so that it can dissociate into a concentrated solution [27, 28].

These remarkable properties are all illustrated by the classic experiment in which ATP is synthesized spontaneously from ADP and P_i in a system with no ion gradients or intact membranes upon addition of the transported ion, as shown in (9) for the calcium enzyme. This result was demonstrated by Taniguchi and Post with the sodium enzyme [29] and by Knowles and Racker with the calcium enzyme [30]. The synthesis of ATP, which seems to be magic at first glance, is driven by the chemical

$$\mathrm{E + P_i \rightleftharpoons E\cdot P_i \rightleftharpoons E{-}P \xrightarrow[\text{ADP}]{\text{Ca}^{2+}} ATP + E\cdot Ca_2} \quad (9)$$

potential of the added calcium or sodium ions and by the strong binding of these ions to the free enzyme.

The spontaneous synthesis of the acyl phosphate, E-P, is analogous to the synthesis of ATP from ADP and P_i by the F_1 ATPase and by myosin, except that in this case one component of the reaction becomes covalently bound to the enzyme. The energetics of the system at pH 7 in the presence of 5 mM Mg^{2+} and 0.1 M KCl is illustrated by Fig. 4 [31]. Experimentally, the binding of P_i to the enzyme is downhill by -1.9 kcal mol^{-1} and the synthesis of E-P is slightly more downhill by -0.3 kcal mol^{-1}. The synthesis of an acyl phosphate from inorganic phosphate and the carboxylate group of acetate (10) is uphill by 10.3 kcal mol^{-1} at pH 7. Formation of the acyl phosphate at the active site will be uphill by the same amount if the carboxyl

$$\mathrm{RCOO^- + P_i \rightleftharpoons R\overset{O}{\overset{\|}{C}}OPO_3^{2-}} \quad (10)$$

group has the same properties as acetate and there are no noncovalent interactions with the binding site, as shown at the top of the diagram. Formation of the observed E-P complex is made possible when these noncovalent binding interactions develop, as shown on the right side of the diagram. The driving force for the reaction arises from the expression of some -12.5 kcal mol^{-1} of intrinsic binding energy between the covalently bound phosphate group and the active site, which includes a bound magnesium ion. Since this binding is intramolecular and requires little loss of entropy, it can express close to the maximal possible amount of the intrinsic binding energy. In contrast, the binding of ATP to the F_1 ATPase and myosin is bimolecular and requires loss of translational and rotational entropy.

The same binding interactions should be available for noncovalently bound P_i, but strong binding of P_i is not observed; the observed binding energy is only $\Delta G^0 = -1.9$ kcal mol^{-1}. Again, this bound E · P_i species must be kept at a high energy

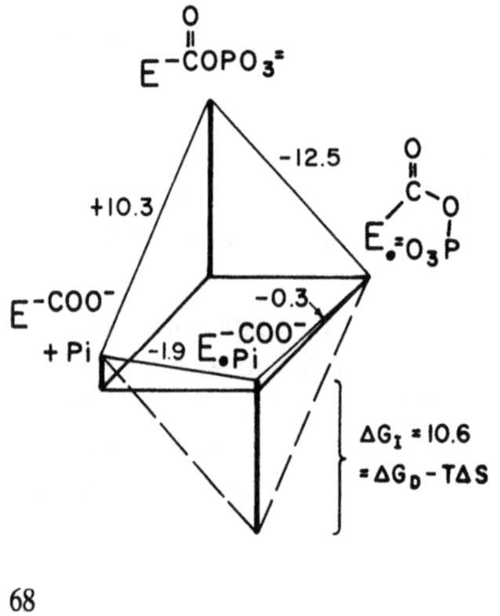

Fig. 4. Gibbs free energy diagram for the spontaneous formation of an acyl phosphate at the active site of the calcium-transporting ATPase [31]

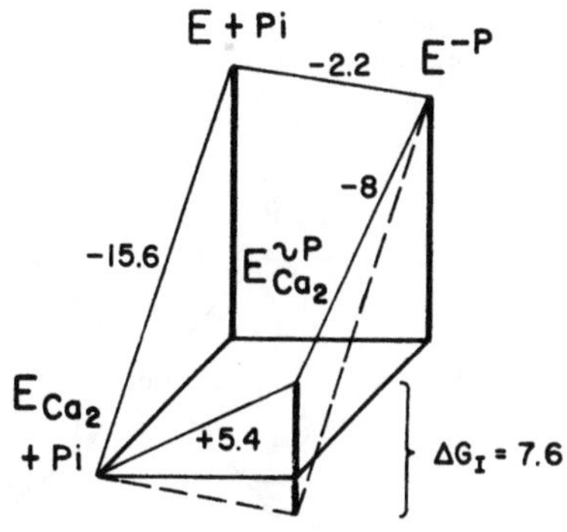

Fig. 5. Gibbs free energy diagram to show the mutual destabilization of bound calcium and phosphate in E-P · Ca_2 through the interaction energy ΔG_I [31]

level so that it can form the acyl-phosphate bond spontaneously. If there were no interaction energy the synthesis of E-P would not occur, as shown by the dashed lines in the figure. Part of the interaction energy that is responsible for the high Gibbs energy of E · P comes from the loss of entropy on binding, which makes the conversion of E · P_i to E-P an intramolecular reaction, and part from destabilization mechanisms that can involve strain, desolvation and electrostatic effects. Model reactions suggest that intramolecularity alone could account for the favorable equilibrium constant for the synthesis of E-P, but there is little doubt that destabilization mechanisms are also involved. The situation with the sodium enzyme is very similar.

The changes in the escaping tendency of bound phosphate and calcium are described by Fig. 5. Calcium binds very strongly to the free enzyme and phosphate utilizes a large amount of binding energy to form E-P, as just described (Fig. 4). However, the binding of calcium ions to E-P to give E-P · Ca_2 is much weaker. Consequently, the hydrolysis of E-P · Ca_2 is strongly downhill with $\Delta G^0 = -5.4$ kcal mol^{-1}. There is a mutual destabilization of bound phosphate by calcium and of bound calcium by phosphate that amounts to an interaction energy of $\Delta G_I = 7.6$ kcal mol^{-1}. This activates the phosphate group so that it can be transferred reversibly to ADP and activates calcium so that it can dissociate into the high concentration of calcium inside the sarcoplasmic reticulum vesicle. It is not known whether the mechanism of this destabilization is direct or is mediated through a conformation change, nor whether it can be divided into several steps [31].

The final result of this utilization of binding energy is that the free energies of the intermediate states of the enzyme along the reaction sequence are nearly balanced under physiological conditions; they proceed slightly downhill along the reaction path for calcium transport [31]. This is consistent with the primary kinetic role of binding energies that serves to avoid very high or low energy intermediates along the reaction path.

The enzyme maintains the energy of the activated phosphate and calcium of E-P · Ca_2 in a simple manner by utilizing an ordered kinetic mechanism for the reactions of this key intermediate, as shown in (11). The ordered mechanism requires that in order for calcium to dissociate to the outside the phosphate must first be transferred to ADP to give ATP. If calcium dissociated first and the interaction energy was lost, the phosphate would no longer be activated and could not be transferred to ADP. On the inside calcium dissociates first, against the high chemi-

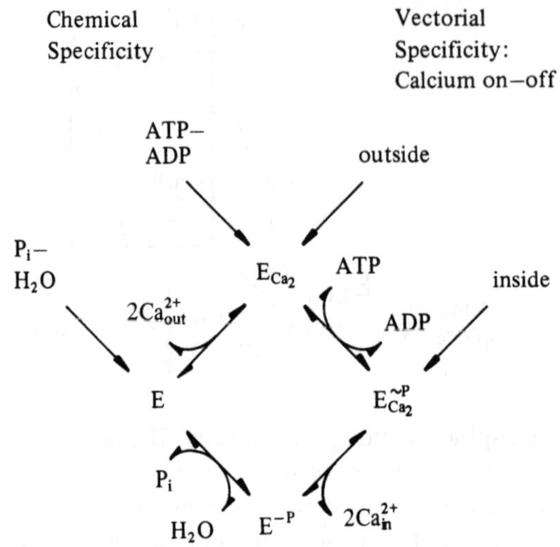

(11)

cal potential of the concentrated calcium inside the vesicle, before the phosphate is transferred to water to give P_i. Again, transfer of the phosphate group first with loss of the interaction energy would cause a loss of the activation of calcium, so that it could not dissociate to the inside of the vesicle [31, 32].

Thus, the ordered kinetic mechanism plays an essential role in the coupling process of this system. The coupling process may be described by a simple set of rules that represent changes in chemical and vectorial specificity of the different chemical states of the system (11).

1) The free enzyme exhibits a chemical specificity for catalysis of phosphorylation by P_i and phosphate transfer from E-P to water, whereas E · Ca is specific for catalysis of phosphate transfer to ADP and from ATP.
2) The vectorial specificity involves binding and dissociation of calcium for transport, which occurs only on the outside of the vesicle with free E (or E · ATP) and only on the inside of the vesicle with E-P [31, 32].

These rules avoid the designations E_1 and E_2, for which there is not a universally accepted definition. The different chemical and vectorial specificities appear to be mediated by the different chemical forms of the enzyme – by E and E · Ca_2 for the chemical specificity and by E and E-P for the vectorial specificity. The rules and the ordered kinetic mechanism of the enzyme are consistent with well known chemical properties of the enzyme, including the "occluded" calcium of the key intermediate E-P · Ca_2 that is exposed to neither side of the membrane. This intermediate behaves as if the calcium ion is simply covered up by covalently bound phosphate, or as if phosphorylation induces a conformation change that covers it up and prevents calcium dissociation to the outside; it also makes possible a further conformation change that allows its dissociation inside. This dissociation occurs in a slow step that is affected by the properties of the membrane and is rate determining under most conditions.

The sodium-potassium ATPase has very similar properties, including ordered mechanisms for the binding and release of both of the transported ions [27, 32].

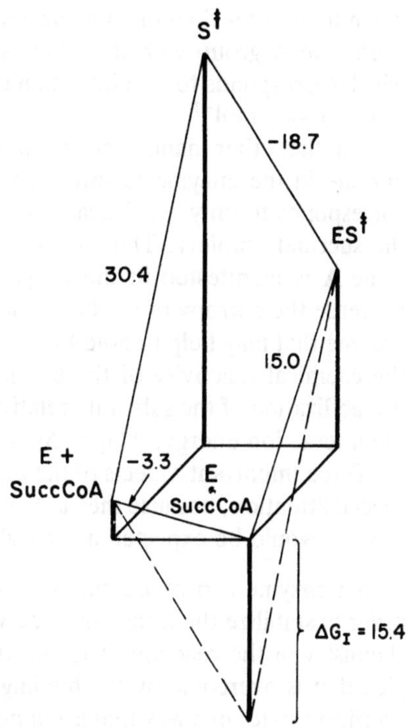

Fig. 6. Gibbs free energy diagram to show the rate acceleration brought about by CoA transferase in the reaction with succinyl CoA

Catalysis by Enzymes. The principal difference between enzymic catalysis and ordinary chemical catalysis is that enzymes utilize the intrinsic binding energy of specific substrates to stabilize the transition state, whereas most chemical catalysts do not. This difference is well known, but its importance is not always appreciated. In order for the intrinsic binding energy to make a large contribution to catalysis, or to bring about an increase in the reaction rate of the ES complex, it is necessary that it be expressed as stabilization, or strong binding, of the transition state. It is equally important that it *not* be expressed as strong binding in the ES and EP complexes. Instead, it can be utilized to increase the reaction probability by bringing about loss of entropy of the bound substrate and by overcoming the unfavorable ΔG_D of destabilization mechanisms that involve desolvation, electrostatic repulsion and mechanical strain [16, 33].

This is illustrated in Fig. 6 for the reaction of succinyl CoA: acetoacetate coenzyme A transferase with its specific substrate succinyl CoA, compared with the corresponding nonenzymic reaction through the transition state $S^{\ddagger}$ [34]. Both reactions involve thiol interchange to form a new thiol ester of coenzyme A. Catalysis by the enzyme increases the reaction rate by a factor of 5×10^{13}, which corresponds to a stabilization of the transition state by 18.7 kcal mol^{-1}. Most of this stabilization arises from the utilization of noncovalent binding interactions between the coenzyme A moiety of the substrate and the enzyme. A short chain thiol ester of succinate with methyl mercaptopropionate has the same chemical properties as succinyl CoA but lacks the specific binding groups of the normal substrate; this compound reacts with the enzyme some 10^{12} more slowly than succinyl CoA, only an order of magnitude faster than the nonenzymic reaction. The noncovalent interactions of the coenzyme A group with the active site are responsible for a rate increase of 10^{12}, which corresponds to a stabilization of the transition rate of the enzymic reaction by some 17 kcal mol^{-1}.

On the other hand, the coenzyme A group interacts very weakly with the enzyme in the enzyme-substrate complex. The observed binding of succinyl-CoA corresponds to only -3.3 kcal mol^{-1} and some of this binding is brought about by the succinate moiety. Thus, only a small fraction of the binding energy of coenzyme A is manifested in the enzyme-substrate complex; most of it is utilized to decrease the entropy of the bound substrate, to drive a conformation change of the enzyme that may help to hold the reacting groups firmly in position, and to increase the chemical reactivity of the bound substrate by a mechanism that may involve destabilization of the substrate relative to the transition state. These factors make up the interaction energy, $\Delta G_I = \Delta G_D - T\Delta S = 15.4$ kcal mol^{-1} (Fig. 6).

Three important aspects of the utilization of binding energy for catalysis deserve special attention, because they are not immediately obvious and may be the opposite of what would be expected at first glance [35].

1) An enzyme can utilize the intrinsic binding energy of parts of the substrate in order to stabilize the transition state when these groups are not at all involved in the chemistry of the reaction. The enzyme can create a destabilization in the ES complex that is overcome by the binding energy of the substrate and is relieved in the transition state, in a way that is not possible for ordinary chemical catalysts in water.

Consider, for example, the postulated stabilization by lysozyme of the carbocation-like transition state of a glycoside substrate by the aspartate β-carboxylate group at the active site, I. This stabilization would be small in water because of the

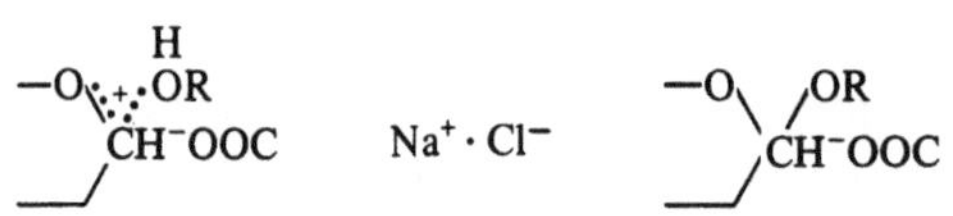

I II III

high polarity of water; there is virtually no stability of the ion pair of sodium chloride, II, in water. However, insertion of the uncharged substrate next to this carboxylate group at the active site can cause a large *destabilization* by desolvation, III. The solvation energy of the carboxylate ion from the gas phase is enormous and a nonaqueous environment can destabilize a carboxylate group by a factor of at least 10^8 [36]. Binding of the substrate is observed because the binding energy of the sugar residues is strong enough to overcome this destabilization. When the transition state is reached, I, the destabilization disappears. The binding energy of the nonreacting parts of the substrate is now fully expressed and serves to stabilize the transition state, even though these residues are not involved directly in the chemistry of the reaction [35, 37, 38].

2) It is not necessary that entropy loss and substrate destabilization mechanisms, such as strain, are actually manifested directly in the enzyme-substrate complex; it is only necessary that they exist so that the substrate cannot bind strongly. The requirements for optimal binding may be so stringent that parts of the substrate do

not bind correctly or do not bind in the active site at all in the observed enzyme-substrate complex. It may be useful to think of a second, metastable enzyme-substrate complex, ES*, in which these groups do bind correctly before the transition state is reached (12).

$$E + S \underset{K_{S(obsd)}}{\rightleftharpoons} ES \overset{K_2}{\rightleftharpoons} ES^* \rightleftharpoons ES^{\neq} \tag{12}$$

This possibility, which is not at all unlikely, has the disconcerting consequence that the observed enzyme-substrate complex, ES, does not exhibit the destabilization mechanisms and firm fixation of the bound substrate that are important in bringing about the observed rate acceleration. It is not important for pure transition state theory, which requires consideration of only the ground state and the transition state, but it raises a disturbing experimental problem if the structure of a complex that is determined by X-ray diffraction differs significantly from that of the productive ES* complex.

3) There is a semantic problem that arises from two very different meanings of words such as "tight" and "loose". This can cause confusion if the economics for the utilization of binding energy are not taken into account.

The tight, firm, exact binding of a substrate with a large loss of entropy can take place only by the expenditure of a large amount of binding energy. Thus, the tight, firm, exact binding will be relatively *weak* because much of the available intrinsic binding energy, ΔG_{int}, will be used up to pay for the loss of entropy (13). A loss of entropy corresponds to a positive Gibbs energy and makes the observed binding

$$\Delta G_{obsd} = \Delta G_{int} + \Delta G_D - T\Delta S \tag{13}$$

weaker (Fig. 6). Conversely, the loose, imprecise, binding of a substrate does not require a large expenditure of binding energy, so that the observed binding of this substrate can be *strong*. It is important to realize that tight binding can be weak and loose binding can be strong.

This can be most easily understood in terms of probability. This tight, exact binding may have only one position for binding (or a small number of positions). Suppose that loose binding forms the same hydrophobic and hydrogen bonds, but has 10^3 slightly different positions in which these bonds can be formed. The loose binding will then be stronger by a factor of 10^3, because it is more probable. In order to obtain the maximal loss of entropy it is necessary to restrict the number of available positions or states very strongly indeed; even a crystalline organic solid has a large entropy.

These examples require a careful consideration of the economics for the utilization of binding energy for catalysis; they illustrate the importance of balancing the books.

Two other points deserve mention. First, we can ask why destabilization of the ES complex should be important for an enzyme that operates at substrate concentrations below K_m under physiological conditions, so that the observed rate is controlled by k_{cat}/K_m. Destabilization of the ES complex does not affect k_{cat}/K_m directly, because this second-order rate constant depends only on the relative energies of the free reactants, E and S, and the transition state, $ES^{\neq}$. Most enzymes do operate at $[S] < K_m$.

The answer is that destabilization of ES is not important for small differences in rate. An increase in the intrinsic binding energy of a good substrate can stabilize the transition state and give a rate increase even if ES is also stabilized. However, any large rate increase from binding interactions requires that the binding must not be expressed in the ES complex. The difference between physiological substrate concentrations and K_s is not very large and a stronger binding in the ES complex will soon lead to a situation in which ES has a lower Gibbs energy than E under physiological conditions. Additional stabilization energy that is expressed in ES will not give a rate increase under these conditions because it will stabilize the ground state, ES, and the transition state equally. In the limit, equal stabilization of ES and $ES^{\neq}$ can give no catalysis because the rate constant for reaction of ES is the same as that for S. This corresponds to the situation in which there is no interaction energy, ΔG_I, as shown by the dashed lines in Fig. 6.

Thus, catalysis by this mechanism cannot be separated from the concentrations of substrate that are of interest. In other words, the standard state that is chosen for consideration is important to the conclusion that is reached. This is essentially an entropy effect [16, 35].

The second point concerns the question of whether it is essential that the substrate be destabilized or "pushed" in such a way as to resemble the transition state in the ES complex. At first sight it might seem that all that is needed is that the substrate must not bind strongly to E, so that any kind of poor fit or destabilization would be satisfactory for catalysis.

The answer to this question arises directly from the requirement that the binding energy must stabilize the transition state in order to give catalysis. This stabilization means that the active site must be complementary to the transition state, so that destabilization of the bound substrate by geometric strain or other mechanisms will tend to drive the substrate toward the structure of the transition state. Any destabilization of the bound substrate that is not relieved in the transition state does not give catalysis.

Receptors. The binding of hormones, neurotransmitters, drugs, or other effectors to the appropriate receptors makes something happen. For most receptors the driving force for these events is the intrinsic binding energy of the effector, as discussed recently by Franklin [39]. However, there is no necessary relationship between the observed binding strength and the size of the response, or between the affinity and the efficacy of a series of effectors. The most active effectors may show relatively weak binding.

The equilibrium constants for the binding of an effector are described by (14) [40, 41]. Figure 7A shows the Gibbs free energy balance of the system. Initially the receptor is in an inactive state, R, with an unfavorable equilibrium constant, K_1, and free energy change, ΔG_1^0, for conversion to the active state, R^*. The magnitude of K_1 and ΔG_1^0 determine the discrimination factor of the receptor, i.e. the extent to which it is turned off in the ground state. When an effector binds to the receptor the more favorable equilibrium constant, K_4, and free energy change, ΔG_4^0, result in the conversion of some of the effector to the active state, $R^* \cdot E$, and some biochemical or physiological event happens.

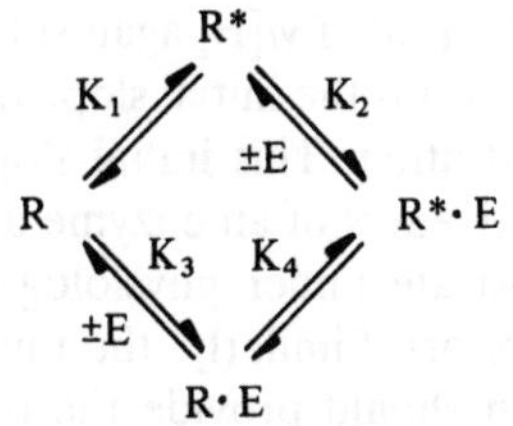

(14)

The binding energy of the effector is expressed maximally to give strong binding to the active form of the receptor, K_2 and ΔG_2^0, whereas the observed binding to the inactive receptor, K_3 and ΔG_3^0, is much weaker. This is apparent in Fig. 7A and follows from the relationship $K_2/K_3 = K_4/K_1$ for (14). The system works because an effective effector is complementary to the active form, R*, not the inactive form, R, of the receptor. Strong binding of the effector to the active state pulls the reaction toward that form, while the relatively weak binding to the inactive state means that it has a relatively high Gibbs energy so that it can form the active state readily. The intrinsic binding energy is expressed in the active $R^* \cdot E$ complex, but is not well expressed in the inactive $R \cdot E$ complex. The difference is the interaction energy, ΔG_I, which describes the imbalance in the expression of binding energy in $R \cdot E$ and $R^* \cdot E$.

The situation is exactly analogous to an enzyme. The active site of the enzyme is complementary to the transition state, so that the intrinsic binding energy stabilizes the $ES^{\ddagger}$ complex, but is not precisely complementary to the substrate, so that ES is high in energy and can reach the transition state easily. At concentrations below saturation the most active effectors are those that have the most favorable intrinsic binding energy toward the active form of the receptor so that they produce a high concentration of RE^*, just as the most effective substrates for enzymes have the most favorable binding to the enzyme in the transition state, $ES^{\ddagger}$. For an enzyme

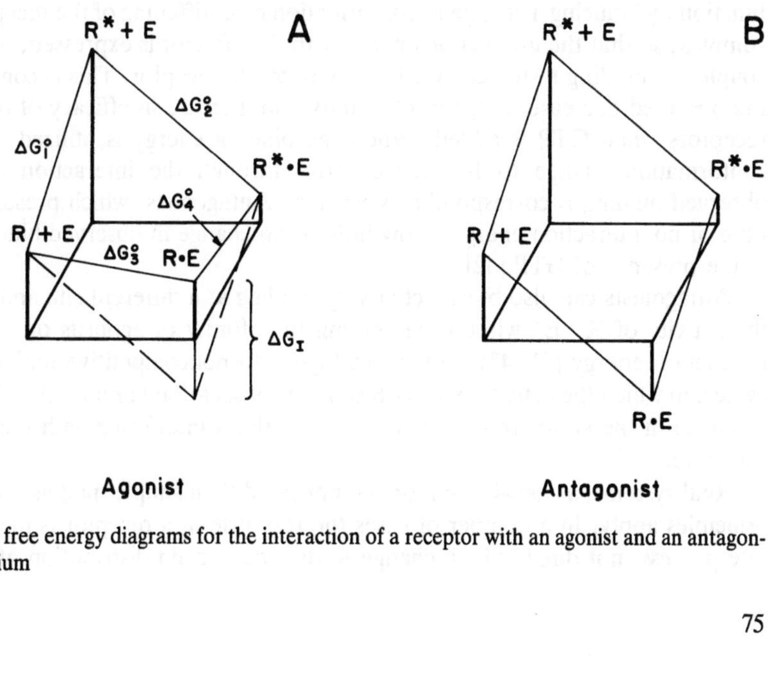

Fig. 7. Gibbs free energy diagrams for the interaction of a receptor with an agonist and an antagonist at equilibrium

this measure of activity corresponds to the initial slope of a plot of v/[E] against [S], i.e. to the second-order rate constant k_{cat}/K_m; for a receptor it is the initial slope of a (nonlogarithmic) plot of activity against effector concentration. This initial slope, k_{cat}/K_m, is probably the best overall measure of the effectiveness of an enzyme as a catalyst because it describes the activity toward a substrate under physiological conditions or in the presence of other substrates or inhibitors. Similarly, the initial slope of a plot of activity against effector concentration should provide the best measure of effector activity at concentrations below saturation or in the presence of competing effectors or inhibitors.

It is not necessary that all or most of the receptor be converted to the active state ($K_4 > 1$, $\Delta G_4^0 < 0$); in fact, the most efficient utilization of binding energy will occur when less than half of R · E is converted to R* · E at saturation. Most of the binding energy that gives a larger conversion is wasted when the receptor is saturated – for example, the amount of binding energy that is required to increase K_4 tenfold, from 0.01 to 0.1, will increase the fraction of active R* · E ninefold, from 1% to 9%, but will only increase it 1.8 fold, from 50% to 91%, when K_4 increases from 1.0 to 10.

If the binding of an effector to the inactive receptor, R, were as strong as to the active form, R*, there would be no interaction energy and conversion to the active form would be just as unfavorable for R · E as for free R, as shown by the dashed lines in Fig. 7A. If the interaction energy is small, there will be a correspondingly small activation of the receptor. The efficacy of an effector at saturation of the receptor increases as the interaction energy increases as long as $K_4 < 1.0$; i.e. $RT \ln K_4 = -\Delta G_4^0 = \Delta G_I^0 - \Delta G_1^0$.

Thus, when ΔG_I is small or zero nothing will happen when the effector binds to the receptor, and the binding of a different, active receptor will be inhibited, sometimes very strongly. Such an inactive effector is an antagonist (Fig. 7B). An agonist is an effective effector, with a large ΔG_I. A partial agonist has an intermediate value of ΔG_I.

Activator molecules that increase the efficacy of effectors, such as GTP, may function by inducing a change in conformation or a stiffening of the receptor protein complex, so that the interaction energy with the effector is expressed; they serve to couple the binding to the change in activity of the receptor. This is consistent with the observed decrease in apparent affinity and increase in efficacy of β adrenergic receptors when GTP is added. When the binding energy is utilized to drive the conformation change to the active form, through the interaction energy, the observed binding is correspondingly weaker. Antagonists, which presumably have little or no interaction energy, show little or no change in observed binding affinity in the presence of GTP [42].

Antagonists can also block activity by binding at a different site and decreasing the activity of R · E* without decreasing the affinity of agonists or changing the interaction energy [41, 43]. This is analogous to noncompetitive inhibition of enzymes, in which the activity of the ES complex is decreased or abolished by inducing a change in the structure of the active site without interfering with the binding of substrate.

Real systems, as usual, are more complicated than simple models, but the same principles apply. In a number of cases the response of a receptor is mediated by a rate process, not directly by a change in the equilibrium distribution of active and inactive receptor. In this situation the interaction energy of an agonist serves to increase the rate by stabilizing the transition state, just as in enzyme catalysis [39]. According to transition state theory, the pseudo-equilibrium constant $K^{\ddagger}$ that determines the concentration of the rate-limiting transition state relative to the $R \cdot E$ complex is more favorable when the interaction energy is larger. The situation is illustrated by Fig. 7A if the transition states $R^{\ddagger}$ and $R \cdot E^{\ddagger}$ are substituted for R^* and $R \cdot E^*$. Binding of the effector will usually bring about a change in the overall equilibrium constant as well, in order to make the formation of activated receptor thermodynamically favorable.

A special case of this situation occurs when an agonist increases the rate constant k_2 for formation of the active state of the receptor to the point that binding of the agonist is irreversible and becomes the rate-limiting step; this corresponds to $k_2 > k_{-1}$ in (15). An antagonist that does not increase k_2 may bind reversibly, which corresponds to $k_2 < k_{-1}$ in Eq. (15) [41].

$$R + E \underset{k_{-1}}{\overset{k_1}{\rightleftharpoons}} R \cdot E \xrightarrow{k_2} R^* \cdot E \qquad (15)$$

Coupling and Directionality. There are a number of processes in which the free energy of hydrolysis of an energy-rich compound is utilized to provide a directionality that is more temporal than physical; it provides control rather than performing work.

A well-known example is the utilization of two molecules of ATP to drive the synthesis of carbamyl phosphate toward completion in the urea cycle (16) [44]. There is no chemical reason to use a second molecule of ATP to activate bicarbonate, because carbon dioxide itself is much more reactive toward nucleophilic attack

$$HOCO_2^- + NH_3 + 2ATP \longrightarrow H_2NCO_2PO_3^{=} + P_i + 2ADP \qquad (16)$$

than any addition derivative, XCO_2^-, but the hydrolysis of the ATP pulls the overall reaction toward completion and prevents the accumulation of toxic levels of ammonia. Similarly, the coupling of carboxylation reactions to the hydrolysis of ATP through the intermediate synthesis of carboxybiotin serves to make these reactions thermodynamically favorable under physiological conditions so that they proceed in the forward direction, but does not produce an intermediate with a higher chemical reactivity than carbon dioxide. In both of these systems the activation also serves a kinetic function by attaching a chemical handle to carbon dioxide so that it can be delivered to the active site of the enzyme at a high effective concentration; carbon dioxide itself has very little binding energy to utilize for catalysis [16].

The hydrolysis of GTP is coupled to the operation of many of the factors that control initiation, elongation and termination of protein synthesis in order to provide directionality and specificity to these processes [45]. The free energy of hydrolysis of ATP is utilized indirectly to provide specificity in proofreading processes, such as the cleavage of incorrectly activated amino acids on aminoacyl-tRNA or aminoacyl-AMP molecules [46]. GTP hydrolysis in the reaction cycle initiated by the binding of hormones or drugs provides a directionality and time dependence to the action of certain receptors that are coupled to adenylate cyclase and lead to release of cyclic AMP; inhibition of this hydrolysis leads to a continued activity of the receptor, as in the action of cholera toxin [47]. Again, the free energy of ATP hydrolysis is used indirectly, by driving the synthesis of acetylcholine, to provide a short burst of activity of the acetylcholine receptor at a synapse through rapid and irreversible hydrolysis of the effector by acetylcholinesterase [48].

The changes in specificity and activity in this kind of coupling are analogous to those in the more complex systems described above. In some cases, such as the hydrolysis of acetylcholine or incorrectly activated amino acids, the mechanism is obvious. In others, such as the factors that control protein synthesis, there is a more complex series of events which represent the operation of rules and changes in specificities in different states of the system that are analogous to those in more conventional coupled vectorial processes. The identification of these rules and the mechanisms through which they are expressed provides a promising area for increasing our understanding of how cells work.

References

1. Wyman, J., Jr. (1964) Adv. Prot. Chem. *19,* 223–286. Weber, G. (1975) Adv. Prot. Chem. *29,* 1–83. Hill, T. C. "Free Energy Transduction in Biology", Academic Press, New York, 1977
2. These terms are preferred to "coupling energy", in order to avoid the implication that the interaction energy is responsible for coupling. The mechanism of coupling is described by a set of rules that are not necessarily related directly to the interaction energies of a coupled vectorial system, as pointed out below
3. Goody, R. S., Hofmann, W. and Mannherz, H. G. (1977) Eur. J. Biochem. *78,* 317–324
4. Konrad, M. and Goody, R. S. (1982) Eur. J. Biochem. *128,* 547–555
5. Botts, J. and Morales, M. (1953) Trans. Faraday Soc. *49,* 696–707. Highsmith, S. (1976) J. Biol. Chem. *251,* 6170–6172
6. Simmons, R. M. and Hill, T. L. (1976) Nature *263,* 615–618
7. Sleep, J. A. and Smith, S. J. (1981) Curr. Topics Bioenergetics *11,* 239–286
8. Holmes, K. C. and Goody, R. S. (1983) Biochim. Biophys. Acta *726,* 13–39
9. Jencks, W. P. (1980) Adv. Enzymol. *51,* 75–106
10. Mitchell, P. (1974) FEBS Lett. *43,* 189–194
11. Kayalar, C., Rosing, J. and Boyer, P. D. (1977) J. Biol. Chem. *252,* 2486–2491. Rosen, G., Gresser, M., Vinkler, C. and Boyer, P. D. (1979) J. Biol. Chem. *254,* 10654–10661
12. Grubmeyer, C., Cross, R. L. and Penefsky, H. S. (1982) J. Biol. Chem. *257,* 12092–12100
13. Feldman, R. I. and Sigman, D. S. (1982) J. Biol. Chem. *257,* 1676–1683
14. Page, M. I. and Jencks, W. P. (1971) Proc. Natl. Acad. Sci. USA *68,* 1678–1683
15. Page, M. I. (1973) Chem. Soc. Rev. *2,* 295–323
16. Jencks, W. P. (1975) Adv. Enzymol. *43,* 219–410
17. Penefsky, H. Personal communication
18. Kasahara, M. and Penefsky, H. S. (1978) J. Biol. Chem. *253,* 4180–4187
19. Jencks, W. P. (1981) Proc. Natl. Acad. Sci. USA *78,* 4046–4050
20. Bagshaw, C. R. and Trentham, D. R. (1974) Biochem. J. *141,* 331–349. Trentham, D. R., Eccleston, J. F. and Bagshaw, C. R. (1976) Q. Rev. Biophysics *9,* 217–281
21. Any mutual destabilization of bound ADP and P_i that is not relieved upon ATP formation will decrease the observed binding of both and will not affect ΔG_I; it will increase the fraction of ΔG_I that arises from binding ADP and P_i individually. If P_i binds to a non-productive site on M (such as part of the ADP site), the productive binding site must have weaker binding. Therefore the contribution of ΔG_s will be underestimated and that of ΔG_{12} overestimated in this situation also

22. Boyer, P. D. (1975) FEBS Lett. *58,* 1–6
23. Boyer's model [22] clearly describes a set of properties that can account for the coupling of ATP synthesis to proton transport. The model involves the reversible synthesis of ATP from bound ADP and P_i when the unprotonated binding site moves from one side of the membrane to the other. Recent work has directly demonstrated the reversible synthesis of ATP at the active site without proton transport [12], so that the rules for a complete model should allow these processes to occur in separate steps
24. Cross, R. L., Grubmeyer, C. and Penefsky, H. S. (1982) J. Biol. Chem. *257,* 12101–12105
25. Hutton, R. L. and Boyer, P. D. (1979) J. Biol. Chem. *254,* 9990–9993
26. Sleep, J. A. and Hutton, R. L. (1978) Biochemistry *17,* 5423–5430
27. Cantley, L. C. (1981) Curr. Topics Bioenerg. *11,* 201–237
28. de Meis, L. and Vianna, A. L. (1979) Ann. Rev. Biochem. *48,* 275–292
29. Taniguchi, K. and Post, R. L. (1975) J. Biol. Chem. *250,* 3010–3018
30. Knowles, A. F. and Racker, E. (1975) J. Biol. Chem. *250,* 1949–1951
31. Pickart, C. M. and Jencks, W. P. (1984) J. Biol. Chem. *259,* 1629–1643
32. Jencks, W. P. (1983) Current Topics in Membranes and Transport *19,* 1–19
33. Pauling, L. (1948) Amer. Scientist *36,* 58
34. Moore, S. A. and Jencks, W. P. (1982) J. Biol. Chem. *257,* 10893–10907
35. Jencks, W. P. In "Molecular Biology, Biochemistry and Biophysics", Vol. 32, Chapeville, F. and Haenni, A.-L., eds., Springer-Verlag: New York, 1980, pp. 3–25
36. Kemp, D. S. and Paul, K. (1970) J. Am. Chem. Soc. *92,* 2553–2554
37. Vernon, C. A. (1967) Proc. R. Soc. London, Ser. B *167,* 389–401
38. Levitt, M. Quoted in Ford, L. O., Johnson, L. N., Machin, P. A., Phillips, D. C. and Tjian, R. (1974) J. Mol. Biol. *88,* 349–371
39. Franklin, T. J. (1980) Biochem. Pharmacol. *29,* 853–856. Franklin, T. J. (1980) Trends in Pharmacol. Sci. *1,* 430–433
40. Katz, B. and Thesleff, S. (1957) J. Physiol. *138,* 63–80
41. Goldstein, A., Aronow, L. and Kalman, S. M. (1974) "Principles of Drug Action: The Basis of Pharmacology", 2nd ed., John Wiley & Sons: New York, pp. 82–111. Ariens, E. J. (1979) Trends in Pharmacol. Sci. *1,* 11–15
42. Howlett, A. C., Arsdale, P. M. V. and Gilman, A. G. (1978) Mol. Pharmacol. *14,* 531–539
43. U'Prichard, D. C., Greenberg, D. A. and Snyder, S. H. (1977) Mol. Pharmacol. *13,* 454–473
44. Metzenberg, R. L., Hall, L. M., Marshall, M. and Cohen, P. P. (1957) J. Biol. Chem. *229,* 1019–1025
45. Kaziro, Y. (1978) Biochim. Biophys. Acta *505,* 95–127
46. Fersht, A. R. (1981) Proc. R. Soc. London, Ser. B *212,* 351–379
47. Cassel, D. and Selinger, Z. (1977) Proc. Natl. Acad. Sci. USA *74,* 3307–3311
48. Boyd, N. D. and Cohen, J. B. (1980) Biochemistry *19,* 5353–5358 and 5344–5353

Discussion

List of interventions

1 Intervention of Scheraga

In your model, you use binding energy to overcome the entropy loss (from translation and rotation) upon association. However, in some association reactions, e.g. the association of positively and negatively charged groups, the entropy of dehydration outweighs the entropy due to translation and rotation, so that there is an overall *increase* in entropy upon association. How does this affect your model?

1.1 Jencks

The two effects are independent. The entropy change from dehydration, and from other changes involving the solvent, is included in the observed association constant and binding free energy. The important loss of translational and rotational entropy refers to the loss of the quantities needed to reach the transition state. The observed entropy of activation includes both of these contributions, which are not easily distinguishable. The contribution of translational and rotational entropy is estimated by comparison of the rate constants for bimolecular and intramolecular reactions.

2 Intervention of Franklin

You propose that in order to account for the inability of the Ca^{++}-dependent ATP-ase to act as an ATP-synthesizing system when the vesicles are charged with Ca^{++}, the enzyme on the inner face of the membrane must change its specificity. However, is it not possible that the Ca^{++} inside the vesicles is not at a high free concentration but is in fact largely sequestered by binding proteins? The ATP-ase would then not need to change its specificity since the enzyme would not in fact be exposed to a high concentration of Ca^{++} inside the vesicles.

2.1 Jencks

Some of the calcium is sequestered inside by binding proteins, but there is also a high activity and concentration of free calcium ions. This is perhaps most directly shown by the observation that the calcium gradient from inside to outside *can* be utilized to synthesize ATP from ADP and phosphate.

3 Intervention of Williams

Why are $E \sim P$ intermediates used in some pumps while only ATP is used in others? May I suggest that $E \sim P$ is used when the system needs to be irreversible so that outer membrane pumps work further from equilibrium than the mitochondrial and chloroplast pumps.

3.1 Jencks

The calcium ATPase of sarcoplasmic reticulum is an $E \sim P$ pump and is readily reversible.

4 Intervention of Lehn

Is it known whether the calcium and phosphate binding sites of Ca-ATP-ase are close together or far part? This would determine whether the mutual influence between Ca^{++} and phosphate is of a conformational nature or is due to more direct interactions between the binding sites of each species.

4.1 Jencks

No, it is not known.

Dynamics of Ligand Binding to Proteins

Martin Karplus

Department of Chemistry, Harvard University, Cambridge, Massachusetts 02138, USA

Dynamics is of importance to ligand binding, if for no other reason than that ligand binding is a reaction in which motion as a function of time is the essential element. In what I shall discuss my focus will be on certain detailed aspects of the dynamics of ligand binding, both in relation to the kinetics of the reaction and the thermodynamic properties of the system. "Ligand" in the present context is interpreted very broadly; it might involve oxygen binding to myoglobin, substrates or inhibitors binding to enzymes and hormones binding to receptors. Thus, the ligand can be as small as an electron, an atom or diatomic molecule and as large as a protein. To set the stage, I have indicated diagrammatically in Fig. 1 some of the dynamical aspects of the ligand binding problem. There is first of all the relative motion of the ligand and the protein to which it binds. This is expected to be governed by the diffusion equation appropriately modified to take account of attractive interactions and activation barriers that are present in many cases. It has been suggested, for example, that in the binding of O_2^- to superoxide dismutase, attractive and even channeling interaction between the ligand and the enzyme are involved. A second aspect of the binding problem involves internal motions of the ligand. For small ligands like

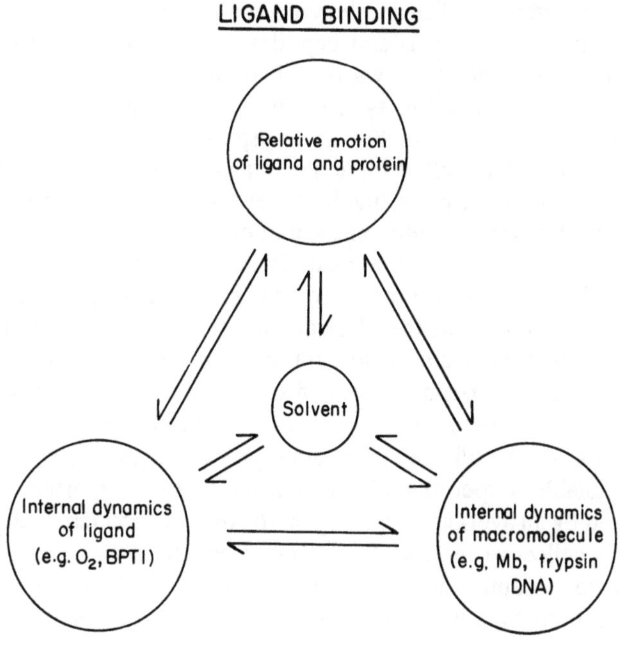

Fig. 1. Dynamic elements in ligand binding

oxygen, where only the bond stretching vibration is present, internal motions are expected to play a negligible role. By contrast, in the case of the bovine pancreatic trypsin inhibitor, a ligand for trypsin, the motions of the side chains are likely to be involved in the binding. The internal dynamics of the macromolecule to which the ligand binds are also expected to be of great importance. Finally, the center of the diagram is occupied by the solvent which affects all of the motions. It is obviously dominant in the relative diffusion of ligand and macromolecule, but the solvent also plays a role in the motion of external side chains and, as Hans Frauenfelder has emphasized, may be significant for the internal dynamics as well.

Overview

It may be useful to review briefly some of the different types of macromolecules where internal dynamics is likely to be important. For hemoglobin there is more evidence concerning the role of motion than for any other protein. The tertiary and quaternary structural changes that occur on ligand binding and their relation to the allosteric mechanism are well documented. An important role of the quaternary structural change is to transmit information over a longer distance than could take place by tertiary structural changes alone; the latter are generally damped out over rather short distances unless amplified by the displacement of secondary structural elements or domains. In both myoglobin and hemoglobin, fluctuations in the globin appear to be essential to the binding process; that is, the protein matrix in the X-ray structure is so tightly packed that there is no sufficiently low energy path for the ligand to enter or leave the heme pocket. Only through structural fluctuations in certain bottleneck regions can the barriers be lowered sufficiently to obtain the observed rates of ligand binding and release.

In many proteins and peptides the transport of substances occurs through the molecule rather than via overall translation as in hemoglobin. The most obvious cases are membrane systems, in which fluctuations are likely to be essential in determining the kinetics of transport. For channels that open and close (e.g. gramicidin) as well as for active transport involving enzymes (e.g. ATPases), fluctuations, in some cases highly correlated ones, must be involved. At present, structural details and studies of the motions are lacking, but this is an area where dynamic analyses are likely to be made in the near future.

For electron-transport proteins, such as cytochrome c, protein flexibility is likely to be involved in the electron transfer. Evidence now favors a vibronic-coupled tunneling mechanism for transfer between cytochrome c and other proteins, although outer-sphere mechanisms are not fully excluded. In the vibronic-coupled tunneling theory, processes which would be energetically forbidden for rigid proteins become allowed if the appropriate energies for conformational distortions are available. Experimental data indicate that the important fluctuations are characterized by an average frequency on the order of 250 cm^{-1}, close to that associated with the collective modes of proteins. Also, the transfer rate is a sensitive function of donor-acceptor distance and may be greatly increased by surface side chain displacements that allow for the closer approach of the interacting residues or ligands.

For proteins that participate in binding, flexibility and fluctuations enter into both the thermodynamics and the kinetics of the reactions. For the rate of binding of two macromolecules (protein-antigen and antibody, protein-inhibitor and enzyme), as well as for smaller multisite ligands, structural fluctuations of side chains, hydrogen-bonding groups, etc., can lead to lowering of the free energy barriers. The rate of the binding process can be increased by dividing it into successive steps for which flexibility in the ligand may be required. The required fluctuations are likely to be sufficiently small and local that they will be fast relative to the binding and therefore not rate limiting.

The relative flexibility of the free and bound ligand, as well as changes in the binding protein, must be considered in the overall thermodynamics of the binding reaction. If the free species has considerable flexibility and fluctuations are involved in the binding step as described above, it is likely that the bound species will be less flexible and a significant entropic destabilization will result. Thus, for strong binding in cases where the rate is not important, relatively rigid species are desirable. This would reduce the conformational entropy decrease and could lead to a very favorable enthalpy of binding if there is high complementarity in the binding site. However, some flexibility and an increase in the conformational space available to the bound species has a stabilizing effect that partly compensates for the loss of translational and rotational entropy on binding. Conversely, the entropy loss of binding a flexible substrate or the rigidification of a protein on substrate binding can be used to modulate the binding constant even when strong, highly specific enthalpic interactions are present. The required balance between flexibility and rigidity will be determined by the function of the binding in each case.

In the role of proteins as catalysts, there is the greatest possibility of contributions from motional phenomena. The role of flexibility per se has often been discussed, particularly from the viewpoint of structural changes induced by the binding of the substrate. In addition to cooperative effects caused by quaternary alterations, a variety of results can arise from the perturbation of the tertiary structure. One example is the ordered binding of several substrates (or effectors and substrates), with the first molecule to bind altering the local conformation so as to increase or decrease the subsequent binding of other molecules. The occurrence of large-scale changes, such as the closing of active-site clefts by substrate binding, as in certain kinases, has been interpreted in terms of catalytic specificity, alteration of the solvent environment of the substrate, and exclusion of water that could compete with the enzymatic reaction. In large enzymes with more than one catalytic site or in coupled enzyme systems, conformational freedom may be important in moving the substrate along its route from one site to the next.

The flexibility of the substrate-binding site in enzymes can result in effects corresponding to those already considered in receptor binding. In the enzyme case there exists the often-discussed possibility of enhanced binding of a substrate with its geometry and electron distribution close to the transition state; for this to occur, conformation fluctuations are essential. Entropic effects also are likely to be of significance, both with respect to solvent release on substrate binding and possible changes in vibrational frequencies that alter the entropy of the bound system in the enzyme-substrate complex or in the transition state. There are also indications that the inactivity of enzyme precursors can result from the presence of conformational freedom in residues involved in the active site. The entropic cost of constraining them in the proper geometry for interacting with the substrate may be so high that the activity is significantly reduced relative to that of the normal enzyme where the same residues are held in place more rigidly. Such a control mechanism has been suggested for the trypsin, trypsinogen system, and for other proteins.

As to the time dependence of fluctuations and structural alternations, there are a variety of possibilities to be considered. In the binding of reactants and release of products, the time course of fluctuations in the enzyme could interact with the motion of the substrate. The opening and closing fluctuations of active-site clefts may be modified by interactions with the substrate as it enters or leaves the binding site.

Hemoglobin as an Example

As already mentioned, the hemoglobin system is the one in which the structural changes that occur on ligand binding have been studied in most detail. Figure 2 provides a schematic outline of the various processes involved in the binding of O_2 or CO to myoglobin and hemoglobin; in myoglobin, which is a monomer, there are of course no quaternary structural changes (g). In what follows I shall discuss two aspects of ligand binding – the first concerns the passage of the ligand through the protein into the heme pocket (i.e. aspect b) and c)) and the second concerns the structural changes in the heme group and the subunit induced by ligand binding (aspects d) and e)).

The high-resolution X-ray structure of myoglobin [1–4] does not reveal any path by which ligands such as O_2 or CO can move between the heme-binding site and the outside of the protein. Since this holds true both for the unliganded and liganded protein, i.e. myoglobin [3] and oxymyoglobin [4], structural fluctuation must be

Ligand Binding in Hemoglobin

a) Approach of ligand to protein
b) Passage of ligand through protein
c) Entrance of ligand into heme pocket
d) Formation of heme-ligand bond
e) Structural changes in heme group
f) Tertiary structural changes induced in subunit
g) Quaternary structural changes in tetramer

Fig. 2. Steps in ligand binding to hemoglobin

involved in the entrance and exit of the ligands. Empirical energy function calculations [5] show that the rigid protein would have barriers on the order of 100 kcal/mol; such high barriers would make the transitions infinitely long on a biological time scale. Figure 3, panel I, gives the nonbonded potential contour lines seen by a test particle representing an O_2 molecule in a plane (xy) parallel to the heme and displaced 3.2 Å from it in the direction of the distal histidine; the coordinate system in this and related figures has the iron at the origin and the z-axis normal to the heme plane. The low potential-energy minimum corresponds to the observed position of the distal O atom of an O_2 molecule forming a bent Fe–O–O bond [4]. The shortest path for a ligand from the heme pocket to the exterior (the low energy region in the

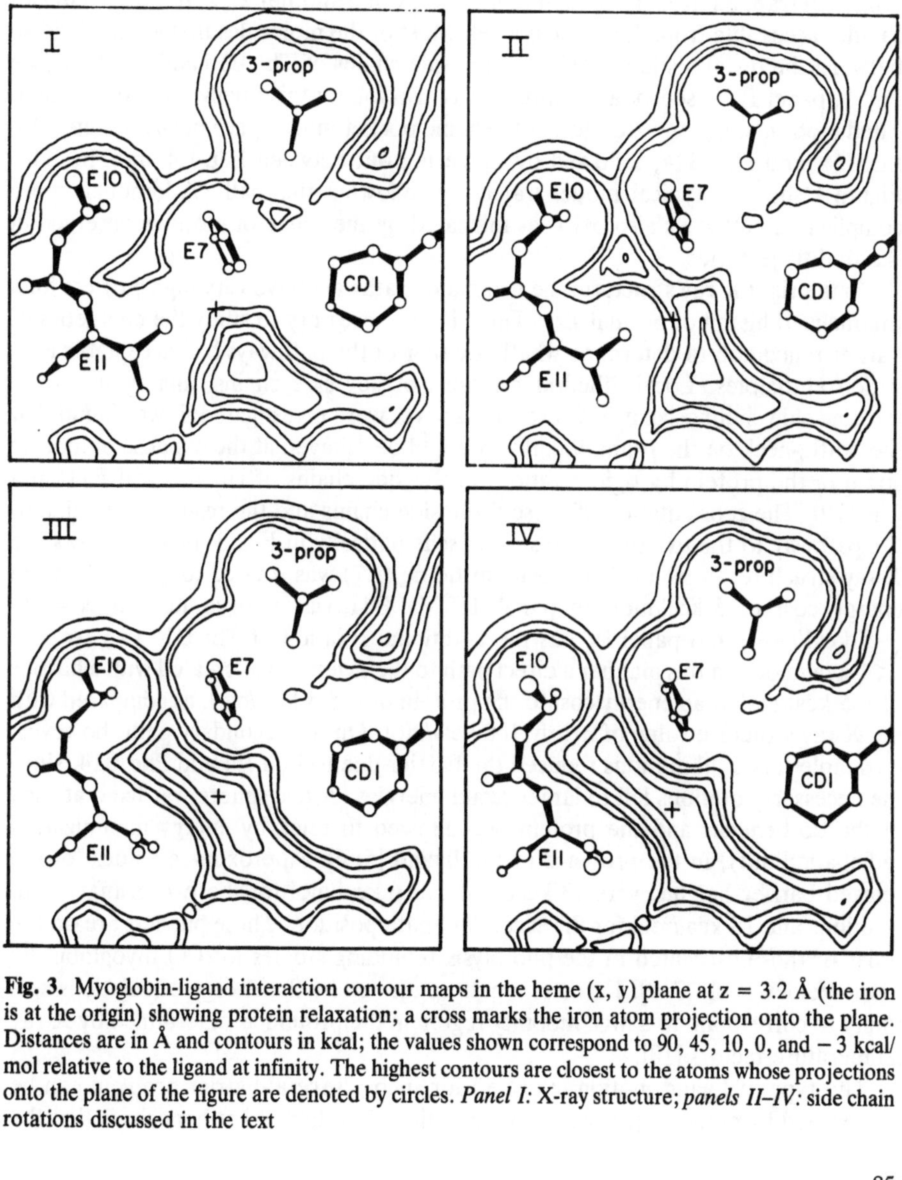

Fig. 3. Myoglobin-ligand interaction contour maps in the heme (x, y) plane at z = 3.2 Å (the iron is at the origin) showing protein relaxation; a cross marks the iron atom projection onto the plane. Distances are in Å and contours in kcal; the values shown correspond to 90, 45, 10, 0, and − 3 kcal/mol relative to the ligand at infinity. The highest contours are closest to the atoms whose projections onto the plane of the figure are denoted by circles. *Panel I:* X-ray structure; *panels II–IV:* side chain rotations discussed in the text

upper left of the figure) is between His E7 and Val E11. However, this path is not open in the X-ray geometry because the energy barriers due to the surrounding residues indicated in the figure are greater than 90 kcal/mol.

To analyse pathways available in the thermally fluctuating protein, ligand trajectories were calculated with a test molecule of reduced effective diameter to compensate for the use of the rigid protein structure. A trajectory was determined by releasing the test molecule with substantial kinetic energy (15 kcal/mol) in the heme pocket and following its classical motion for a suitable length of time. A total of 80 such trajectories were computed; a given trajectory was terminated after 3.75 ps if the test molecule had not escaped from the protein. Slightly more than half of the test molecules failed to escape from the protein in the allowed time; 25 molecules remained trapped near the heme-binding site, while another 21 were trapped in two cavities accessible from the heme pocket. Most of the molecules that escaped did so between the distal histidine (E7) and the side chains of Thr E10 and Val E11 (see Fig. 3, panel I). A secondary pathway was also found; this involves a more complicated motion along an extension of the heme pocket into a space between Leu B10, Leu E4, and Phe B14, followed by squeezing out between Leu E4 and Phe B14. Figure 4 shows a typical model trajectory following this path. Additional, more complicated pathways also exist, as indicated by the range of motions observed in the trial trajectories.

In the rigid X-ray structure, the two major pathways have very high barriers for a thermalized ligand of normal size. Thus, it was necessary to study the energetics of barrier relaxation to determine whether either of the pathways had acceptable activation enthalpies. Local dihedral rotations of key side chains, analogous to the tyrosine side chain oscillations described above, were investigated; it was found that the bottle-neck on the primary pathway could be relieved at the expense of modest strain in the protein by rigid rotations of the side chains of His E7, Val E11, and Thr E10. The reorientation of these three side chains and the resultant opening of the pathway to the exterior is illustrated schematically in Fig. 3; panel I shows the X-ray structure; in panel II the distal histidine (E7) was rotated to $\chi_1 = 220°$ at an energy cost of 3 kcal/mol; in panel III, Val E11 was also rotated to $\chi_1 = 60°$ ($\sim$5 kcal/mol); and panel IV has the additional rotation of Thr E10 to $\chi_1 \cong 305°$ ($<$1 kcal/mol). In this manner a direct path to the exterior was created with a barrier of $\sim$5 kcal/mol at an energy cost to the protein of $\sim$8.5 kcal/mol, as compared with the X-ray structure value of nearly 100 kcal/mol. On the secondary path, however, no simple torsional motions reduced the barrier due to Leu E4 and Phe B14, since the necessary rotations led to larger strain energies. A test sphere was fixed at each of the bottlenecks and the protein was allowed to relax by energy minimization (adiabatic limit), in the presence of the ligand [5, 6]. Approximate values for the relaxed barrier heights were 13 kcal/mol and 6 kcal/mol for the two primary path positions and 18 kcal/mol for the secondary path position. These barriers are on the order of those estimated in the photolysis rebinding studies for CO myoglobin by Frauenfelder et al. [7–9]. Further, a path suggested by the energy calculations was found to correspond to a high mobility region in the protein as determined by X-ray temperature factors [10].

The type of ligand motion expected for such a several-barrier problem can be determined from the trajectory studies mentioned earlier. What happens is that the

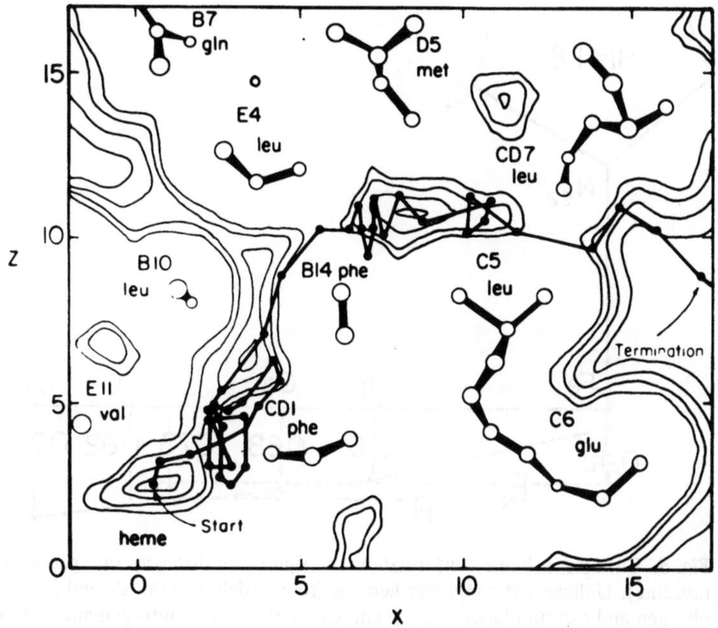

Fig. 4. Diabatic-ligand trajectory following a secondary pathway (see text); a projection of the trajectory on the plane of the figure [(x, z) plane at y = 0.5 Å] is shown with the dots at 0.15 ps intervals. The start of the trajectory at the heme iron and the termination point exterior to the protein are indicated by arrows

ligand spends a long time in a given well, moving around in and undergoing collisions with the protein walls of the well (see Fig. 4). When a protein fluctuation is sufficient to significantly lower the barrier, or the ligand gains sufficient excess energy from collisions with the protein, or more likely both at the same time, the ligand moves rapidly over the barrier and into the next well where the process is repeated. In a completely realistic trajectory involving a fluctuating protein and ligand-protein energy exchange, the time spent in the wells would be much longer than that found in the diabatic model calculations (Fig. 4). Further, from the complexity of the range of pathways in the protein interior, it is likely that the motion of the ligand will have a diffusive character.

The analysis of myoglobin suggests that the native structure of a protein is often such that the small molecules that interact with the protein cannot enter or leave if the atoms are constrained to their average positions. Consequently, side chain and other fluctuations may be required for ligand binding by proteins and for the entrance of substrates and exit of products from enzymes.

I shall now briefly describe the changes in structure that are produced in the heme group and the surrounding protein by ligand binding. Figure 5 shows a simplified cross section of a "domed" heme group prior to ligand binding and the "undoming" that is induced by the ligand in accord with model compounds results and calculations. As the ligand binds, the heme group approaches a much more nearly planar geometry. The source for the heme doming in the five-coordinate systems used to be ascribed entirely to the larger effective radius of the high-spin

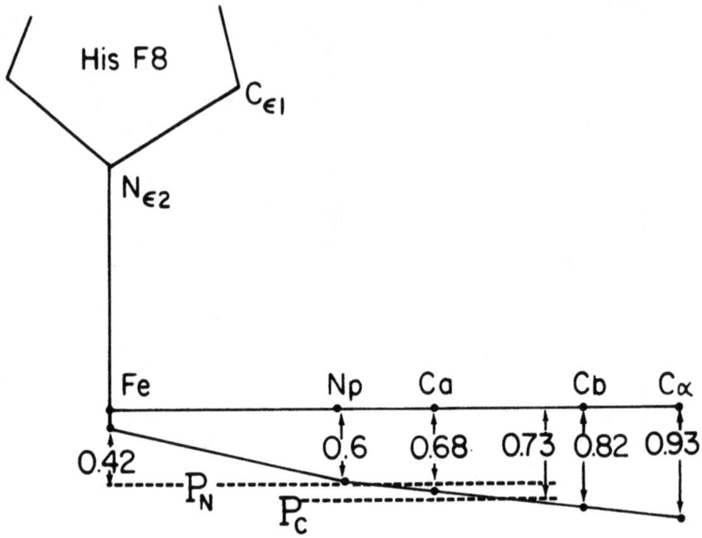

Fig. 5. Changes in heme conformation and significant distances on ligation (relative to fixed axial histidine). Unliganded nonplanar heme is from 2-MelmFeTPP. P_N and P_C are the mean porphyrin nitrogen and carbon planes; N_p, C_a, and C_b are the pyrrole nitrogen and carbons in standard order; C_α is the first carbon of a substituent on C_b

ferrous iron with an electron in the in-plane d_{xy} orbital, which results in a repulsive interaction with the (N_p) pyrrole nitrogen σ orbitals. It now appears that this is only one factor involved. Experiments and calculations have shown that size of the central porphyrin hole (i.e., the pyrrole nitrogen N_p-N_p distance), which obviously would play the dominant role in the iron d_{xy} orbital nitrogen repulsion, is considerably more variable than had been thought, making it possible for the pyrrole nitrogens to move somewhat farther apart and reduce the iron-N_P repulsion. Ab initio quantum-mechanical calculation by Olafson and Goddard [11] and by Dedieu *et al.* [12] and a semiempirical calculation by Warshel [13] suggest that an additional important factor is the steric interaction between the axial nitrogens of the imidazole (N_ε) and the pyrrole nitrogens (N_p). This is repulsive for an in-plane iron and is clearly decreased by an out-of-plane iron position, which is not destabilizing in the absence of a sixth ligand.

In going from the five-coordinate high-spin to the six-coordinate low-spin state, the combination of an empty d_{xy} orbital and the pull on the iron due to the second axial ligand leads to a balance of forces in which an in-plane geometry for the iron is generally favored in model systems. Simultaneously, all the iron-nitrogen distances become somewhat shorter, the heme undomes, and a geometry change of the type illustrated in Fig. 5 results.

All the local structural changes just described are expected to occur in model compounds, as well as in heme proteins, in going from a five- to a six-coordinate structure and so are not per se related to cooperativity in hemoglobin. The cooperative mechanism must therefore involve interactions between the ligand-initiated structural changes and the globin. In the unliganded deoxy-hemoglobin structure (Fig. 6), the heme group is in an essentially unstrained position [14, 15]. Calcula-

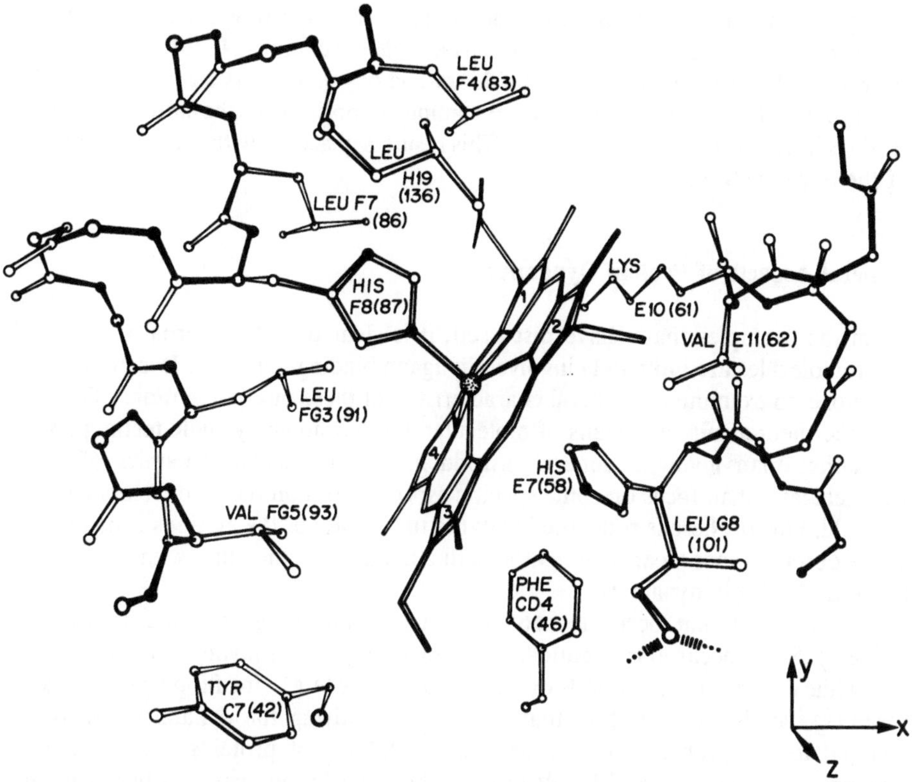

Fig. 6. Heme group and its environment in the unliganded α-chain. Solid bonds denote polypeptide backbone. Only selected side chains are shown; the heme 4-propionate is omitted for clarity

tions show that the heme group is in a broad minimum and that any globin "pull" on the iron through the proximal histidine is very small on the scale of the energy of cooperativity. The reduced oxygen affinity of deoxy-hemoglobin results from the strain on the liganded subunit in a tetramer with the deoxy quaternary structure; i.e., the tertiary structural changes induced by ligation require more energy if the subunit is in the deoxy than in the oxy tetramer. An essential factor in producing the strain is the proximal histidine F8, which is asymmetrically positioned relative to the heme (see Fig. 6); i.e., the distance between the imidazole $C_{\varepsilon 1}$ and pyrrole 1 nitrogen is significantly shorter than the distance between the imidazole $C_{\delta 2}$ and pyrrole 3 nitrogen in the deoxy structure [14–16]. Although there is no repulsion between the imidazole carbons and domed-heme group in the unliganded structure, significant repulsion sets in from the $C_{\varepsilon 1}$-1 N contact when the heme group becomes planar on ligand binding; as is evident from Fig. 5, the $C_{\varepsilon 1}$-1 N contact is much smaller after ligation. The eclipsed O_2, imidazole position may be a contributing factor in this strain mechanism [4, 15]. The quaternary constraints in the deoxy structure on the F helix, of which His-F8 is a part, appear to constrain the imidazole from taking a more symmetric position. This prevents the local strain from being dissipated and contributes to the reduced ligand affinity. In addition, the repulsive contacts result in a tilting of the heme, whose displacement initiates other tertiary structural changes of the globin that form the reaction path in transmitting the effect of ligand binding from the heme group to the surface of its subunit [14, 15]. As has been shown in a detailed comparison of the deoxy and oxy quaternary structures [17], the imidazole of His-F8 has a much more symmetric position in the latter so that no strain is introduced on ligand binding. This result is in accord with the higher affinity of the oxy structure.

General Aspects of Protein Dynamics

From the examples that I have discussed, it is clear that the internal motions of macromolecules are intimately involved in ligand binding reactions. It is of interest, therefore, to examine the general characteristics of these internal motions. This can be done theoretically by means of molecular and harmonic dynamic techniques.

Rather than giving a general introduction to molecular dynamics of macromolecules I shall focus on some specific aspects based on recent developments in the field. One of these is concerned with the time scales of the motions contributing to the observed root mean square atomic fluctuations and the other with a harmonic analysis of protein dynamics.

The motions that occur in proteins cover a vast range of time scales (see Table 1). Some occur in the femtosecond range (e.g., those involving bond stretching) while others go over periods of seconds or longer (e.g., ring flipping of aromatic residues), so that the method that is used for studying the dynamics has to be adapted to the problem. The folding and unfolding of proteins is a dynamical phenomenon that is most difficult to treat theoretically because it is both complicated and slow. Other problems such as enzyme kinetics, which are also slow macroscopically, involve individual events (i.e., barrier crossing trajectories) that are fast so that they can be studied by specialized molecular dynamics simulation methods [18]. Present "brute force" simulations extend up to several hundred picoseconds. With supercomputers an increase of one or two orders of magnitude is possible but still longer simulations will have to wait for new, perhaps specialized computers.

Table 1. Classification of internal motions of globular proteins

Scales of motions (300°K)	
Amplitude	0.01 to 100 Å
Energy	0.1 to 100 Kcal
Time	10^{-15} to 10^{3} sec
Types of motions	
Local	Atom fluctuations, side chain oscillations, loop and "arm" displacements
Rigid body	Helices, domains, subunits
Large-scale	Opening fluctuation, folding and unfolding
Collective	Elastic-body modes, coupled atom fluctuations, soliton and other non-linear motional contributions

Although considerable experimental information from X-ray crystallography is available on the magnitude of atomic fluctuations in proteins [19], much less is known about their time scale. Molecular dynamics simulations have now been employed in an analysis of the time dependence of the fluctuations [20]. A 25 ps trajectory [21] for BPTI was used to obtain sub-averaged root-mean-square (rms) fluctuations; that is, the entire trajectory was divided into a series of intervals of a given length, the mean-square displacements relative to the mean for each interval determined and the results averaged for the entire trajectory. Figure 7 shows the results for the C^{α} atoms. There is a significant contribution to the rms fluctuations from the subpicosecond motion. At 0.2 ps, the C^{α} rms averaged over all residues is already 0.13 Å, about 40% of the value at 5.0 ps. The average rms value increases slowly with time during the trajectory. Even at 25 ps, the length of the simulation, the asymptotic value for the average rms fluctuation may not have been reached for all atoms, though the results for most atoms have converged.

Comparing the relative values of the C^{α} fluctuations, we see that at 0.2 ps the results are rather uniform. Since local high-frequency oscillations are making the main contribution to this sub-average, it appears that the effective potential does not vary significantly throughout the protein. This is in accord with expectations, if the dominant factor for the high frequency oscillations is the librational potential associated with torsional motion of the backbone atoms. It should be noted, however, that

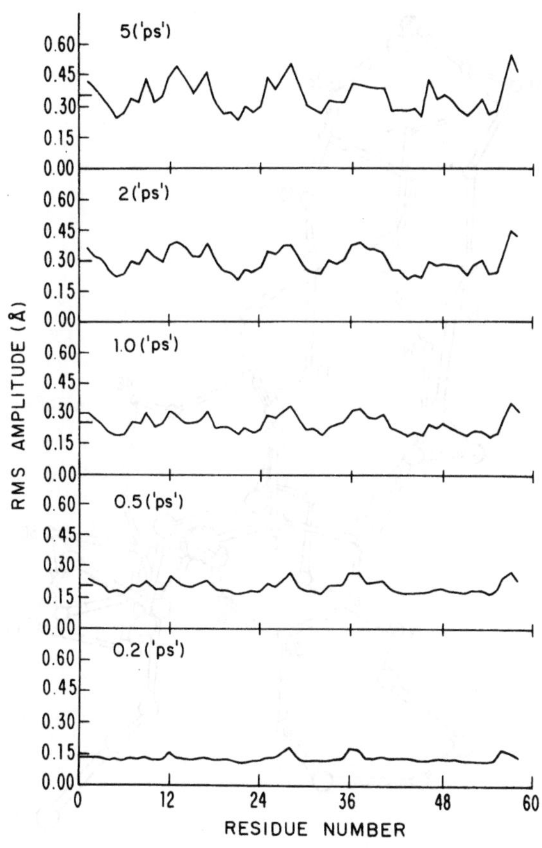

Fig. 7. Root mean square displacement sub-averages as a function of residue number for the C^{α} atoms (T_{Av} = 0.2, 0.5, 1, 2 and 5 ps)

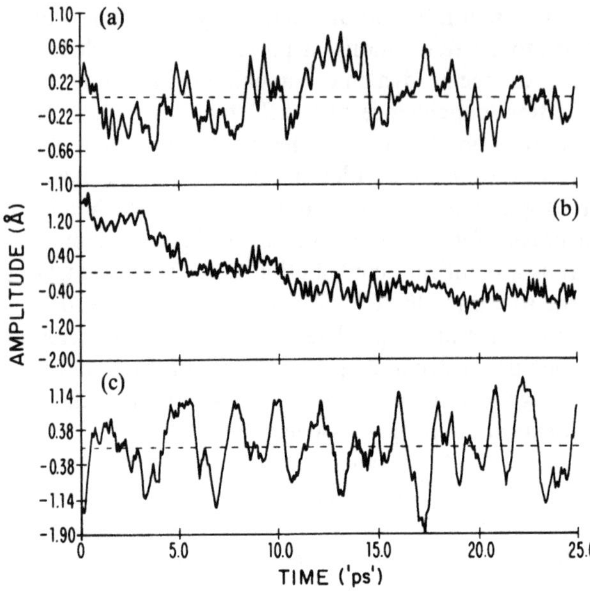

Fig. 8 a–c. Time series for largest Cartesian component in principal axis system of thermal ellipsoid: **a** Tyr 21 $C^{\delta 2}$, **b** Asp 50 C^{β}, **c** Lys 15 C^{δ}

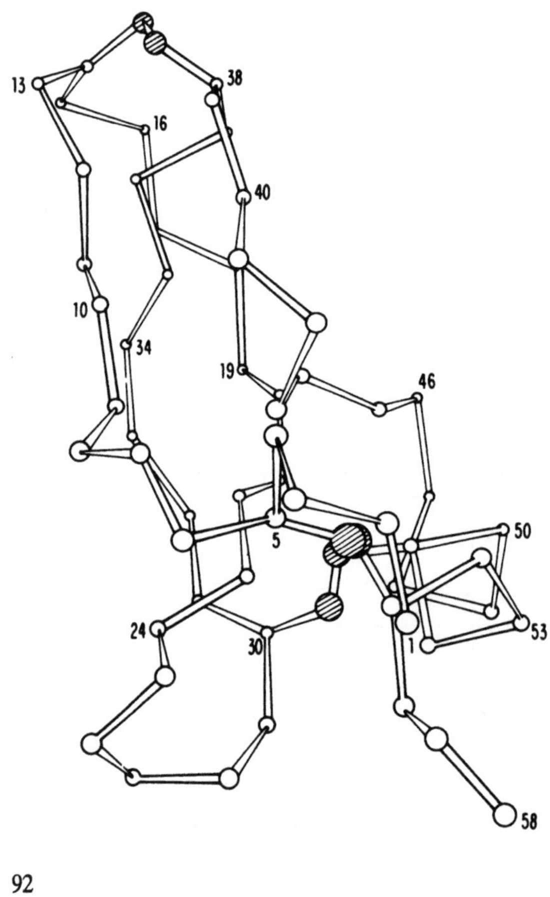

Fig. 9. X-ray structure of BPTI [22]; only α carbons and disulphide bridges are shown

even on the 0.2 ps time-scale there is some suggestion of inhomogeneity, in that in the neighbourhood of atoms 12, 28, and 36, as well as of the N and C terminal ends, slightly larger fluctuations are found; these regions all have greater than average fluctuations when the low frequency modes are included (as can be seen from a comparison of the 0.2 and 5.0 ps values). From the time series results (Fig. 8), the 0.5 ps averages include all the high frequency contributions and the lower frequency motions are becoming more important. This is clearly indicated in Fig. 7 by the greater variation in the fluctuations along the polypeptide chain. As longer time subaverages are examined, we see that the magnitudes of the fluctuation continue to increase in certain regions, in accord with the longer relaxation times and lower frequency collective character of the modes involved; examples (see Fig. 9) are the loop region (residues 24–28) and the region at the top of the molecule (around residues 14 and 38) [15]. By contrast, for other parts of the protein (e.g. the β-sheet region, 18–24), the C^α fluctuations have already reached their asymptotic value by 2 ps.

Time series and time-development of the mean-square displacements, as well as correlation functions for the atoms of BPTI [20], suggest that the fluctuations generally involve the superposition of two types of motions. One is a high frequency oscillation of relatively small magnitude, while the other has a considerably lower frequency and larger amplitude. From the characteristics of the individual atom fluctuations, and from the relation between the displacements of different atoms, we can draw qualitative conclusions concerning the nature of the two types of motional contributions. The high frequency oscillations are local, in the sense that they correspond to the librational motion of the individual atoms in an effective potential. This potential is a summation of dihedral angle terms in the potential function for the backbone or side chain of which the atom is a part and of non-bonded interactions with the atoms of the surrounding protein matrix. By contrast, the lower frequency components have a non-local, more collective origin, in that they involve the correlated motion of groups of atoms, ranging from a small number of atoms next to each other along the backbone and/or along a single side chain to a much larger number in certain regions of the protein. The frequencies for the collective mode contributions to the correlation functions vary from $\sim 1 \times 10^{11}\ s^{-1}$ to $\sim 1 \times 10^{12}\ s^{-1}$; in wave numbers this corresponds to values between 3 and 30 cm^{-1}. This range is in accord with the lowest normal mode frequencies obtained for BPTI (see below).

The results of this analysis, which suggests that protein atom fluctuations have two types of components, are of general interest. There are local high frequency oscillations, which are relatively small and rather uniform throughout the protein, and collective, lower frequency modes, which lead to larger displacements and have a more inhomogeneous character. The latter may well be of particular importance in the biological function of the protein; e.g. they may be involved in the displacements of side chains, loops or other structural units that are required for the transition from an inactive to the active configuration of a globular protein and in the correlated fluctuations that may play a direct role in enzyme catalysis. Further, the extended nature of these motions makes them more sensitive to the environment (e.g. the observed differences in simulation results between a vacuum and a solvent run). Because they involve sizeable portions of the surface of the protein, the collective motions are expected to be quenched at low temperature by freezing of the solvent. Their contribution to the mean-square fluctuations may explain the transition observed near 200 K in the temperature dependence of fluctuations in proteins such as myoglobin [23].

Harmonic Model for Motions

An alternative approach to the dynamics of a protein or one of its constituent elements (e.g., an α-helix) is to assume that the harmonic approximation is valid. Early attempts to examine dynamical properties of proteins or their fragments used the harmonic approximation. They were motivated by vibrational spectroscopic studies [24], where the calculation of normal mode frequencies from empirical potential functions has long been a standard step in the assignment of infrared spectra [25]. In calculating the normal vibrational modes of a molecule, one assumes that the vibrational displacements of the atoms from their equilibrium positions are

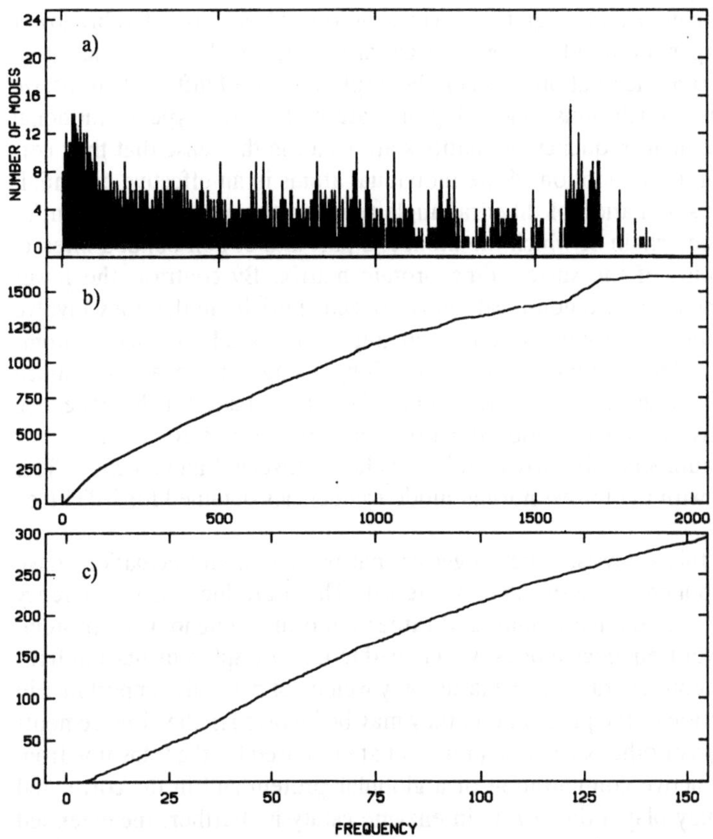

Fig. 10a–c. Normal frequencies: **a)** histogram of the number of normal modes per 5 cm^{-1} interval (hydrogen stretches are not shown); **b)** number of modes below a given frequency; **c)** expanded version of **b)** in the low frequency region

small enough that the potential energy can be approximated as a sum of terms which are quadratic in the displacements. The coefficients of these quadratic terms form a matrix of force constants which, together with the atomic masses, can be used to set up a matrix equation for the vibrational modes of the molecule [25]. Solution of this equation generally requires the diagonalization of a 3N dimensional matrix, where N is the number of atoms in the molecule. The result is obtained as a set of 3N eigenvalues (vibrational frequencies) and 3N eigenvectors (normal modes). Six of these eigenvectors are associated with eigenvalues of zero, corresponding to overall translation and rotation of the molecule. The remaining 3N – 6 eigenvalues are the internal vibrational frequencies of the molecule; the associated eigenvectors give the directions and relative amplitudes of the atomic displacements in each normal mode.

Although the harmonic model does not provide a complete description for the motional properties of a protein because of the contribution of anharmonic terms to the potential energy, it is nevertheless of considerable importance because it does serve as a first approximation for which the theory is highly developed. Further, the harmonic model is essential for quantum mechanical treatments of vibrational contributions to the heat capacity and free energy [26, 27].

A harmonic dynamic analysis of the BPTI using the CHARMM program potential function has recently been performed in the full conformational space of the molecule [28]; that is, all bond lengths and angles, as well as dihedral angles, are included for the 580 atom system consisting of all heavy atoms and polar hydrogens.

The histogram in Fig. 10 shows the calculated normal spectrum: a) lists all the frequencies up to 2000 cm^{-1} (hydrogen stretching frequencies are not shown), b) gives the cumulative distribution for the number of modes below a given frequency, and c) shows an expanded cumulative distribution for the lowest 300 modes of primary interest. There is an essentially continuous, though not completely uniform, distribution of frequencies between 3.1 cm^{-1} and 1200 cm^{-1}. Between 1200 cm^{-1} and 1800 cm^{-1}, the frequencies tend to come in groups, many of which are dominated by bond stretching vibrations. There are twenty modes between 3.1 and 13 cm^{-1} and there is a peak in the frequency distribution near 50 cm^{-1}. Because the structure used was not an absolute minimum, seven negative modes were found; energy searches along these modes, which are all local in character, indicated that their correct frequencies are in the range 20 to 40 cm^{-1}.

The root mean square (rms) atom fluctuations, $\langle \Delta r_k^2 \rangle$, were calculated from the normal modes by evaluation of the classical expression [26]

$$\langle \Delta r_k^2 \rangle = k_B T \sum_i \frac{|\vec{a}_{ik}|^2}{\omega_i^2}$$

where $\vec{a}_{ik}$ is the vector of the projections of the ith normal mode with frequency ω_i on the Cartesian components of the displacement vector for the kth atom, k_B is the Boltzmann constant and T is the absolute temperature; quantum corrections are negligible above 50 °K [26]. Figure 11 shows the normal mode rms fluctuations calculated at 300 °K and compares them with the results of a molecular dynamics simulation of PTI in a van der Waals solvent [21]; this simulation was used because its average structure is closest to that employed for the normal mode analysis. The results for main chain and side chain averages as a function of residue number are

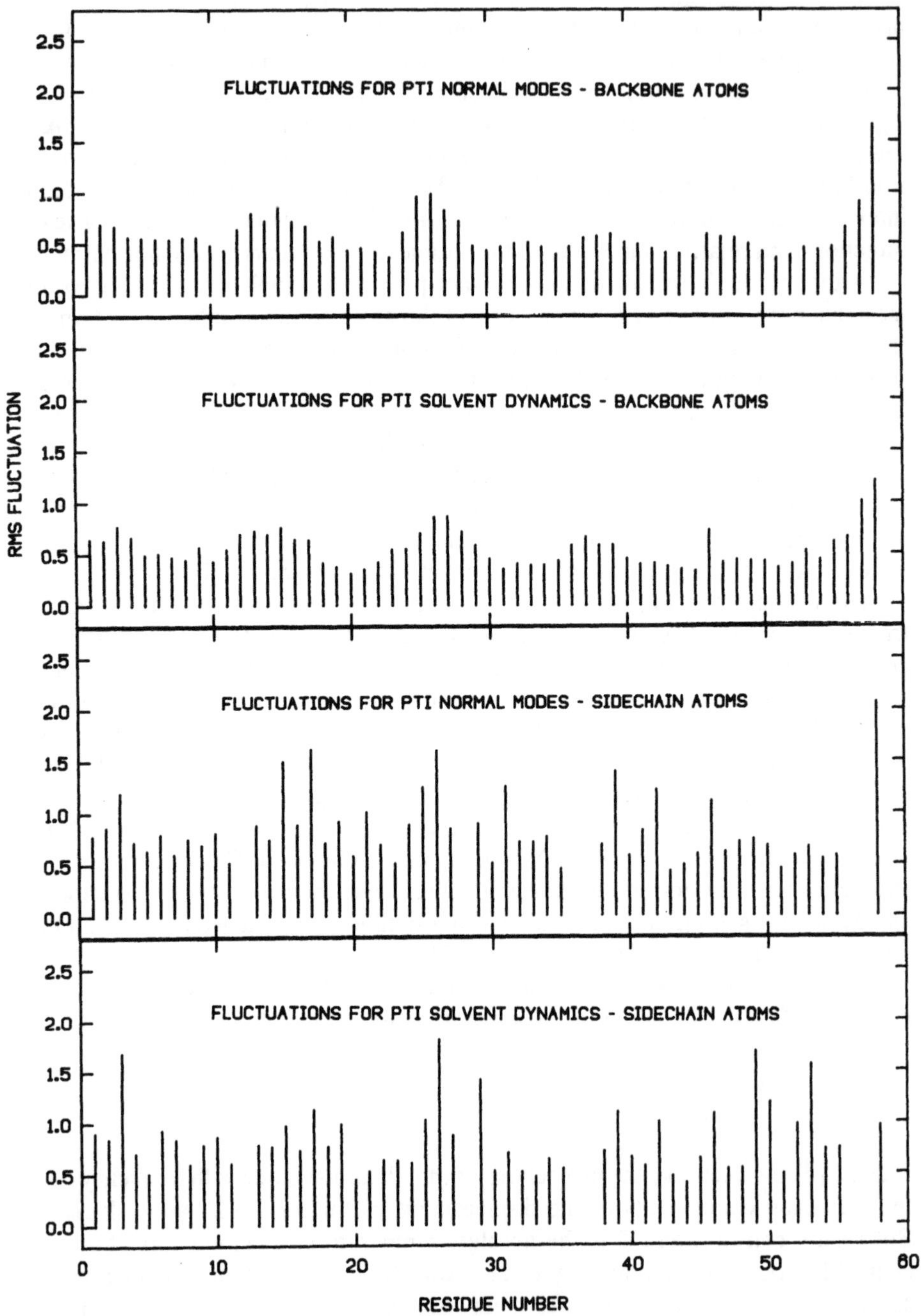

Fig. 11. Root mean square atomic fluctuations (Å) at 300 °K averaged over each residue from normal mode and molecular dynamics; separate plots are given for main chain (N, C^{α}, C) and side chain heavy atoms

Table 2. Root mean square fluctuations (Å)

Type	Normal Modes	Dynamics[a]
all atoms	0.776	0.714
main chain	0.598	0.582
side chain	0.905	0.846
C^{α}	0.611	0.536
C^{β}	0.712	0.662
γ position	0.736	0.768
δ position	0.841	0.912
ε position	0.996	0.970

[a] From reference 28

given (see also Table 2). For the main chain fluctuations, the molecular dynamics and normal mode values are very similar; for the side chains, there is also a correspondence though the differences are more pronounced. The main chain values show that the C terminal end has large fluctuations, as does the loop region at the bottom of the molecule (residues 25–29) and the binding site in the neighborhood of residues 14 and 38 at the top of the molecule. By contrast the β sheet residues [18–24, 29–35] show smaller fluctuations; the α-helices [3–7, 47–56] are intermediate (see Fig. 9). These results are to be compared with the molecular dynamics evaluation of the time development (see above).

The origin of the differences between the molecular dynamics and normal mode results is likely to have contributions from anharmonic and solvent effects and from the difference between the average dynamics structure and that used for the normal mode analysis. The main chain atoms apparently experience a potential of mean force that is closer to the harmonic potential than do the side chains; since the dynamics simulation was done in a van der Waals solvent, the exterior side chains are expected to be most perturbed. To investigate the nature of the potential in the neighborhood of the structure used for the harmonic model, we have done an energy search along the low-frequency normal-mode displacements. The energy dependence showed significant anharmonic contributions. Fitting the resulting energies to a parabola generally led to a small increase in the effective frequencies ("adjusted" frequencies) which would tend to reduce the rms fluctuations; the relative values of the fluctuations would be essentially unchanged.

Analysis of the time scale as well as the magnitude of the fluctuations has been made for the molecular dynamics results. From the calculated time series and correlation functions (see above), it has been found that the atomic motions contributing to the rms displacements generally have a small local high frequency component (~ 0.2 ps or $\sim 150\ \mathrm{cm}^{-1}$) on which are superposed motions of a more collective character with time scales ranging from 1 to 10 ps or 30 to 3 cm^{-1}. The present normal mode study essentially confirms the dynamics results. Figure 12 shows the contributions of the different normal modes to the displacements of some of the atoms whose motions where analysed in the molecular dynamics simulations; also included is the fluctuation of the radius of gyration for the molecule. In most cases, the dominant contributions come from low frequency modes in the range 3 to 50 cm^{-1}, although non-negligible contributions come from higher frequencies up to

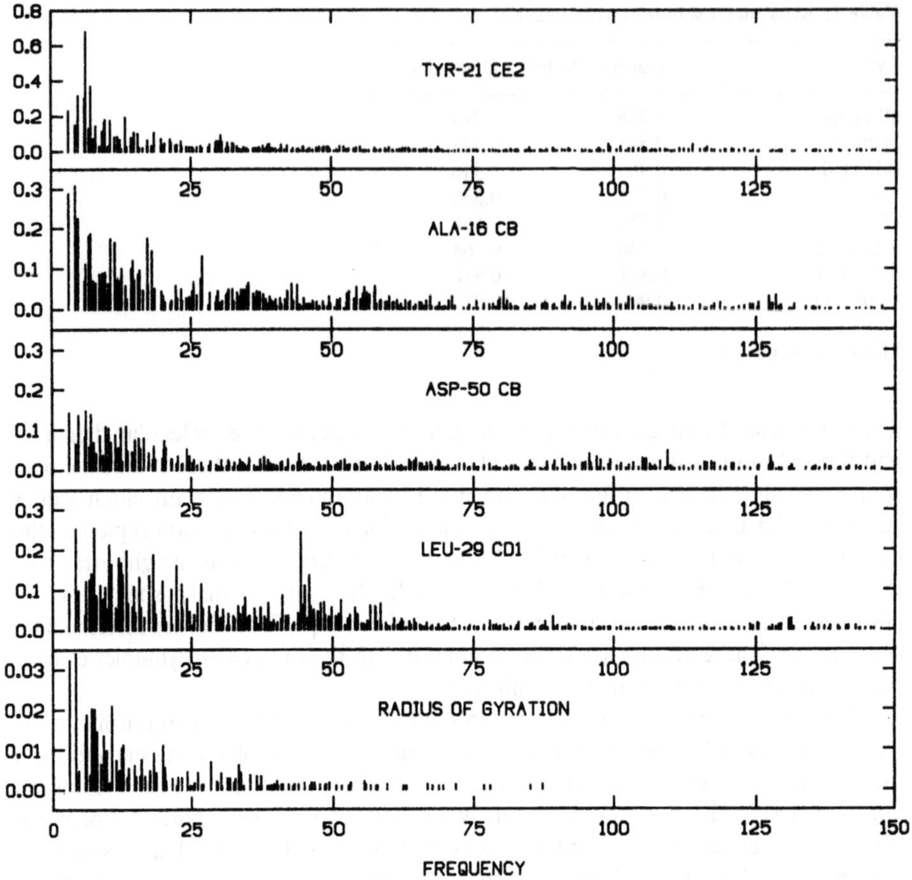

Fig. 12. Contribution of normal modes to rms fluctuations (Å) as a function of frequency; selected atoms and the radius of gyration are included

130 cm^{-1}. It is evident that for certain atoms (e.g. Tyr-21, $C^{\varepsilon 2}$), only a very small number of modes are important, while for other atoms (e.g. Ala 16 C^{β}, Asp 50 C^{β}) a range of frequencies are involved; for Leu 29 $C^{\delta 1}$, a mode at 44.5 cm^{-1} makes a large contribution.

It is of considerable interest to examine the form of the normal modes themselves. This is of particular importance for the evaluation of the correlation between the motions of different atoms and different groups of atoms. Analysis of the dynamics results have indicated that the larger scale motions have a collective character that may involve a few neighboring atoms, a residue, or groups of many atoms in a given region of a protein. We show in Fig. 13 the distribution of the displacements over residues of some of the low frequency modes. Also included is one of the translation modes, which clearly demonstrates the purity of this mode, a pictorial verification of the accuracy of the normal mode determination procedure. Most of the 300 lowest modes are highly delocalized; they are generally distributed over the entire molecule. A striking exception is one of the "negative" modes that is

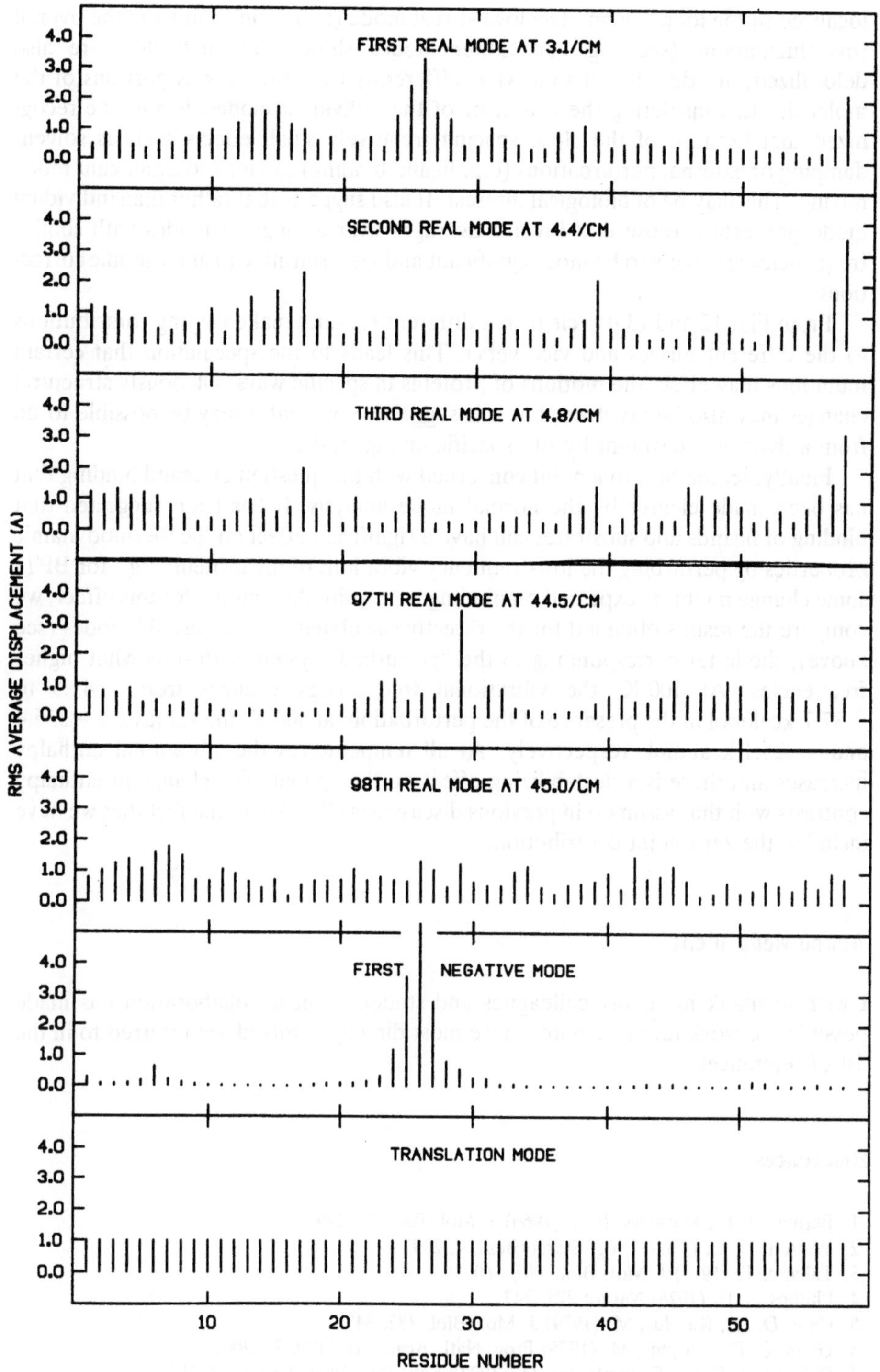

Fig. 13. Normal mode distribution: rms average displacement of atoms within a residue for a 1 Å rms displacement along selected modes

localized in the loop region. The lowest real mode (at 3.1 cm^{-1}) mirrors the overall rms fluctuations (see Fig. 11). Other modes shown, although they are also delocalized, are distributed somewhat differently over the various portions of the molecule. In considering the character of the individual modes, it must be recognized that because of the close spacing, relatively small effects such as solvent damping or external perturbations (e.g. ligand binding) can lead to significant mode mixing. This may be of biological interest. It also suggests that rather than individual mode properties, those that involve averages over a range of modes with similar frequencies are likely to be most significant and least sensitive to anharmonic corrections.

From Fig. 12 and 13 it is clear that different residues make varying contributions to the different modes and vice versa. This leads to the speculation that certain mutations may effect the motions of proteins in specific ways. Obviously structural changes may also be involved, but it is suggestive of what it may be possible to do from a dynamic viewpoint by site-specific mutagenesis.

Finally, let me turn to a point concerned with the question of ligand binding that has been made clearer by the normal mode analysis. It has been suggested that binding of ligands and substrates can have a significant effect on the thermodynamic properties by perturbing the low-frequency vibrations of the protein: e.g., for BPTI some change might be expected on binding to trypsin. As a model for this effect, we compare the results obtained for the directly calculated and "adjusted" modes (see above), the latter corresponding to the "perturbed" system with somewhat higher frequencies. At 100 K, the vibrational free energy changes from −41.5 to −37.7 kcal/mol in the presence of the perturbation; at 300 K, the values are −336.4 and −325.1 kcal/mol, respectively. At all temperatures the vibrational enthalpy increases and there is a destabilizing effect on the system. The change in enthalpy contrasts with that assumed in previous discussions [29], due to the fact that we have included the zero-point contribution.

Acknowledgement

I wish to thank my many colleagues and students whose collaboration has made possible the work reported here; those most directly involved are referred to in the list of references.

References

1. Perutz, M. F., Mathews, F. S. (1966) J. Mol. Biol. *21,* 199
2. Watson, H. C. (1969) Prog. Stereochem. *4,* 299
3. Takano, T. (1977) J. Mol. Biol. *110,* 569
4. Phillips, S. E. (1978) Nature *273,* 247
5. Case, D. A., Karplus, M. (1979) J. Mol. Biol. *132,* 343
6. Gelin, B. R., Karplus, M. (1975) Proc. Natl. Acad. Sci. USA *72,* 2002
7. Debrunner, P. G., Frauenfelder, H. (1982) Ann. Rev. Phys. Chem. *33,* 283
8. Austin, R. H., Beeson, K. W., Eisenstein, L., Frauenfelder, H., Gunsalus, I. C. (1975) Biochemistry *14,* 5355

9. Beece, D., Eisenstein, L., Frauenfelder, H., Good, D., Marden, M. C. et al. (1980) Biochemistry *19,* 5147
10. Frauenfelder, H., Petsko, G. A., Tsernoglou, D. (1979) Nature *280,* 558
11. Olafson, D. B., Goddard, III W. A. (1977) Proc. Natl. Acad. Sci. USA *74,* 1315
12. Dedieu, A., Rohmer, M.-M., Viellard, H., Viellard, A. (1979) Nouv. J. Chimie *3,* 653
13. Warshel, A. (1977) Proc. Natl. Acad. Sci. USA *74,* 1789
14. Gelin, B. R., Karplus, M. (1977) Proc. Natl. Acad. Sci. USA *74,* 801
15. Gelin, B. R., Lee, A. W.-M., Karplus, M. (1983) J. Mol. Biol. *171,* 489
16. Fermi, G. (1975) J. Mol. Biol. *97,* 237
17. Baldwin, J., Chothia, C. (1979) J. Mol. Biol. *129,* 175
18. Northrup, S. H., Pear, M. R., Lee, C.-Y., McCammon, J. A., Karplus, M. (1982) Proc. Natl. Acad. Sci. *79,* 4035
19. Petsko, G. A., Ringe, D. (1984) Ann. Rev. of Biophysics *13,* 331
20. Swaminathan, S., Ichiye, T., van Gunsteren, W. F., Karplus, M. (1982) Biochemistry *21,* 5230
21. van Gunsteren, W. F., Karplus, M. (1982) Biochemistry *21,* 2259
22. Deisenhofer, J., Steigemann, W. (1975) Acta Crystallogr. Sect. B *31,* 238
23. Parak, F., Frolow, E. N., Mossbauer, R. L., Goldanskii, V. I. (1981) J. Mol. Biol. *145,* 825
24. Miyazawa, T. (1967) in *Poly-α-Amino Acids,* ed. Fasman, G. D., Marcel Dekker, New York
25. Wilson, E. B., Decius, J. C., Cross, P. C. (1955) *Molecular Vibrations,* McGraw-Hill, New York
26. Levy, R. M., Perahia, D., Karplus, M. (1982) Proc. Natl. Acad. Sci. USA *79,* 1346
27. Go, N., Scheraga, H. (1976) Macromolecules *9,* 535
28. Brooks, B., Karplus, M. (1983) Proc. Natl. Acad. Sci. USA *80,* 6571
29. Sturtevant, J. M. (1977) Proc. Natl. Acad. Sci. USA *74,* 2236

Discussion

List of interventions

1 Intervention of Schulten

You describe the dynamics of the atoms in proteins by integration of the Newtonian equations of motion, i.e. by deterministic equations. One may except, however, that a protein as a system of many strong interacting atoms with a large number of degrees of freedom should behave much like a liquid in which the motion along a single degree of freedom, e.g. a side group rotation, after a short time span of about a picosecond should appear like a Brownian motion. I wonder if this behavior, if true, may not allow one to describe proteins also by stochastic differential equations or by Fokker-Planck-equations. Such a description may allow one to extend the time periods over which one can computationally monitor the protein dynamics to longer and biochemically more relevant times.

1.1 Karplus

At the limit of diffusive motion, the velocity relaxes so rapidly that it does not contribute to the displacement. That does not appear to be true in general for the individual atoms of a protein, though it is likely to be true for groups of atoms (e.g., the heme group in myoglobin and hemoglobin).

2 Intervention of Huber

In the calculation of the pathway of a ligand to its binding site on a protein the simulation remains close to the starting conformation. It may be, however, that the actual pathway of the diffusing ligand is based on a structure substantially different, at least with respect to the conformation of the side chains.

3 Intervention of Lerner

Could you tell us more about what you mean by "incidence" of denaturation of proteins in solvents which we usually think of as non-denaturing.

3.1 Karplus

Perhaps I could comment on some of the work done on hydrogen exchanges. Lysozyme is perhaps the clearest case. If you look at hydrogen exchange, at low temperature, in general the individual hydrogen atoms have different activation energies, different behaviors. If the temperature is increased they all seem to reach the same activation energy which is close to that for denaturation. Thus at certain temperatures, denaturation fluctuations are dominant in exchange. These large fluctuations still may be infrequent, but they are important for hydrogen exchange because all the other ways for solvent to enter the protein are less likely.

Binding Energy and the Stimulation of Hormone Receptors

T. J. Franklin

Imperial Chemical Industries PLC, Pharmaceuticals Division, Alderley Park, Macclesfield, Cheshire, UK

Essential Features of the Agonist-Receptor Interaction

To fulfill their physiological roles a hormonal agonist and its receptor must first recognize each other and proceed to a highly specific binding interaction. It is generally believed that this interaction leads to some degree of conformational change in the receptor (receptor stimulation) which in many cases is coupled to an effector system that regulates the intracellular levels of a second messenger such as cyclic AMP or calcium ions. Amplification of the signal arising from the agonist-receptor interaction usually depends upon the initiation of a cascade by the second messenger which culminates in the appropriate biological end response. In other instances the receptor directly controls ion fluxes across the cell membranes (e.g. nicotinic receptors). Steroid receptors after stimulation translocate to the cell nucleus and activate transcription of specific regions of the genome. The amplification systems, which are such a striking feature of hormonal and neuroendocrine action, have been the subject of intense research activity for the past two decades. The situation is very different with regard to the nature of the primary interaction between agonists and their receptors and the conformational changes that ensue. There have been a few reports of efforts to analyze the basic thermodynamics of the binding phenomenon in receptor systems which will be discussed later but little or no effort has been directed to the study of the influence of ligand structure on essential conformational changes in the receptor.

There is ample indirect evidence that the initial binding interactions between agonists and their receptors do indeed lead to conformational changes in the receptors. For example, β receptors [1], α_2 receptors [2], nicotinic receptors [3, 4], muscarinic receptors [5, 6] and hydroxytryptamine receptors [7] all exhibit striking changes in affinities for their agonists after the initial interaction suggesting a change in receptor conformation. The cytoplasmic receptors for estrogens undergo a profound alteration in sedimentation coefficient and a change in agonist affinity following estrogen binding [8]. These agonist-induced changes are almost certainly due to conformational rearrangements of the receptor molecules. So far, however, there are no details concerning the intramolecular nature of these rearrangements or their thermodynamics and there is little understanding of the contributions of the binding interactions between agonist and receptor to the subsequent conformation movements in the agonist-receptor complexes. It seems to me that the nature of these events will be essential to our eventual understanding of the relationship between the chemical structures of natural hormones and their synthetic variants and their activities as agonists, partial agonists and pure antagonists. It is my purpose in this paper to examine some recent studies into the nature of ligand-protein interactions and protein conformational changes that may be relevant to the general phenomenon of hormone-receptor interactions. Hopefully it will be possible to envisage a number of experimental approaches that may appeal to biochemical pharmacologists.

Energetics of the Primary Interaction between Agonists, Antagonists and Their Receptors

Weiland, Minneman and Molinoff [9, 10] described thermodynamic studies on the interaction of β-adrenergic agents with the β-receptors of turkey erythrocytes and rat tissues. They examined the effects of temperature on the binding constants of agonists, partial agonists and antagonists. Using the Van't Hoff relationship, $\ln K_A = -\Delta H°/RT + \Delta S°/R$, where K_A is the association constant, both the enthalpy, $\Delta H°$ and the entropy, $\Delta S°$, of binding were determined. It was found that the binding of full agonists resulted in an unfavorable decrease in entropy that was compensated by a larger, favorable decrease in enthalpy. The situation was reversed for antagonist binding, i.e. a large increase in entropy was the driving force for stabilization of the ligand-receptor complex. In general a good correlation was found between the efficacy of agonists and partial agonists and the decreases in entropy although two antagonists with partial agonist effects on the heart *in vivo,* which were classified by Weiland et al. as pure antagonists, exhibited entropy-driven binding. Under the conditions of their study the "good" agonists activated the β-adrenergic receptor. It seems very likely therefore that the enthalpic and entropic changes were not simply due to binding but also reflected conformational changes occurring in the receptor and associated proteins. The antagonists studied by Weiland et al. were in general much more hydrophobic molecules than the agonists. It is not surprising that the binding of an antagonist like propranolol is entropy-driven since the naphthalene moiety of this compound would be expected to participate in hydrophobic interactions which are known to be associated with favorable entropy changes. The entropy-driven character of the binding of the β-adrenergic antagonists studied by Weiland et al. is therefore more likely to be the result of the physicochemical nature of these molecules rather than due to the phenomenon of antagonism *per se.*

Möhler and Richards [11] compared the interaction of agonist and antagonist benzodiazepine compounds with synaptic membrane preparations. At 37 °C the binding of the agonist, clonazepam, was enthalpy-driven since there was an unfavorable decrease in entropy. The antagonist compound Ro15-1788 bound with favorable enthalpic and entropic changes. Enthalpy-driven binding of benzodiazepine agents was also observed by Quast, Mahlmann and Vollmer [12]. Barlow and his colleagues [13, 14] found that the binding of most muscarinic antagonists to the receptor was entropy-driven.

Almost certainly, therefore, it will not be possible to make general statements about the thermodynamics of the binding of agonists and antagonists. In the case of glucocorticoid agonists, Wolff et al. [15] have shown that the binding is apparently entropy driven although again the binding thermodynamics were not distinguished from any agonist-induced changes in the receptors. To avoid this problem it may be necessary to carry out binding studies over much shorter time courses under non-equilibrium conditions. A further complication in the interpretation of ligand binding studies is the frequent involvement of additional ligands such as GTP and metal ions. Such accessory ligands, which are believed to play an important regulatory role *in vivo,* affect the thermodynamic parameters measured *in vitro* since they almost certainly alter receptor conformation themselves. Unless *in vitro* conditions are made as near physiological as present knowledge allows, thermodynamic analysis of the binding of hormonal and pharmacological ligands could be very misleading. Nevertheless, as will be seen later, the application of radioligand displacement studies, carried out under carefully defined conditions, to closely related series of ligands can be a useful aid to deriving quantitative information about the contribution of structural features of the ligands to the binding interactions.

Conformational Changes in Proteins

The ability of proteins to undergo specific conformational changes in response to interactions with other molecules and ions is at the heart of enzymic activity, membrane transport processes, muscle contraction, cell motility, vision, oxygen delivery by hemoglobin and receptor stimulation by hormones and neuroendocrine effectors. The mechanisms involved in initiating and carrying through these changes are therefore of crucial importance in biology. Even in the absence of binding ligands the structure of proteins is never motionless. The individual atoms vibrate on a subpicosecond time scale while at a much more leisurely pace there are slow structural fluctuations, often referred to as protein breathing movements, in the range of 1 to 1000 s. These intrinsic movements of protein molecules are relevant to their interactions with specific ligands. For example, it has been noted that the equilibrium structures of hemoglobin and myoglobin provide no access for oxygen to reach the buried heme groups. In reality, of course, the thermal motions of the polypeptide chain allow access to occur quite readily [16].

Conformational changes in proteins that are consequent upon ligand binding also depend upon the inherent flexibility and mobility of protein structures. For example, the remarkable ability of calmodulin to bind to and activate an astonishingly wide range of enzymes and proteins with other functions is thought to be due to the high level of internal flexibility in the calmodulin molecule [17] permitting a range of conformational possibilities. For many other proteins with more limited conformational opportunities intramolecular flexibility and motion is restricted by disulphide bridges and multiple non-covalent interactions. Delineation of the pathways of specific conformational changes in protein is clearly a long way off but it seems most unlikely that a protein samples all possible pathways when a shift in conformation follows ligand binding [18]. It has been calculated that a "typical" protein with a single polypeptide chain of 100 amino acids would take 10^{50} years to arrive at a new equilibrium position between two major conformers if it sampled all possible routes [19]. There is little doubt that the amino acid sequence codes for protein folding although amino acid substitutions seem to be possible at almost all residues without major changes in the folding pattern [20]. X-ray structures of globins only distantly related through evolution reveal a strikingly uniform "globin fold" despite little commonality in amino acids [20]. However, the importance of *specific* amino acids in determining efficient allosteric control is indicated, for example, by the adverse action of mutations affecting the α_1-β_2 chain contacts in hemoglobin. These contacts are known to have a key role in transmitting allosteric interactions since there is considerable movement at these interfaces on oxygenation. Mutant hemoglobins with abnormalities in the α_1-β_2 contacts exhibit diminished cooperativity which adversely affects their oxygen delivery function [21].

Ligand-Mediated Conformational Changes

Conformationally labile proteins establish an equilibrium between their energetically distinct conformers. In the case of proteins that respond to specific ligand binding with a major shift in the conformational equilibrium, the ligand-free equilibrium favors the uncomplexed, thermodynamically more stable ground state. A common assumption has been that a shift in equilibrium to a conformationally altered, less stable form of a protein is caused by the preferential binding of the ligand to the latter conformer to yield a thermodynamically more stable ligand-protein complex. The induced fit theory of Koshland and Neet [22] for enzyme-substrate interactions proposed that the shift in equilibrium towards the substrate-activated state of an enzyme would not be rate-limiting for catalysis. Although this is now known not to be generally true [23] it engendered the belief that although the ground state, E, and its substrate activated form, E*, are thermodynamically distinct they must be in fast exchange with each other if the binding of substrate, S, to E* rapidly establishes a new equilibrium favoring E* (or rather E*S).

$$E \rightleftharpoons E^* + S \rightleftharpoons E^*S \tag{1}$$

The schematic energetics of this type of reaction are illustrated in Fig. 1a. Clearly there could only be a rapid shift in the conformational equilibrium by this route if the activation energy barrier separating E and E* is low. No doubt this mechanism for ligand-mediated conformational change is adequate when the energy difference between E and E* is small. Where major conformational changes are involved, however, the equilibrium displacement mechanism might cause the conformational change to be rate-limiting in enzyme catalysis. Indeed, this could underlie the slow transitions from the inactive to the active forms of chymotrypsin [24, 25], glyceraldehyde 3-phosphate dehydrogenase, glutamate dehydrogenase and alkaline phosphatase [26].

There is however, another possible mechanism by which ligands can mediate conformational changes in proteins which is represented by the equation:

$$E + S \rightleftharpoons ES \rightleftharpoons E^*S \tag{2}$$

The ligand binds to the ground state of the protein forming a Michaelis complex which then isomerizes to the active form (the "inductive" model). This model for substrate activation of enzymes was advocated forcefully by Citri [26] as a plausible alternative to the mass action model described by Eq. (1).

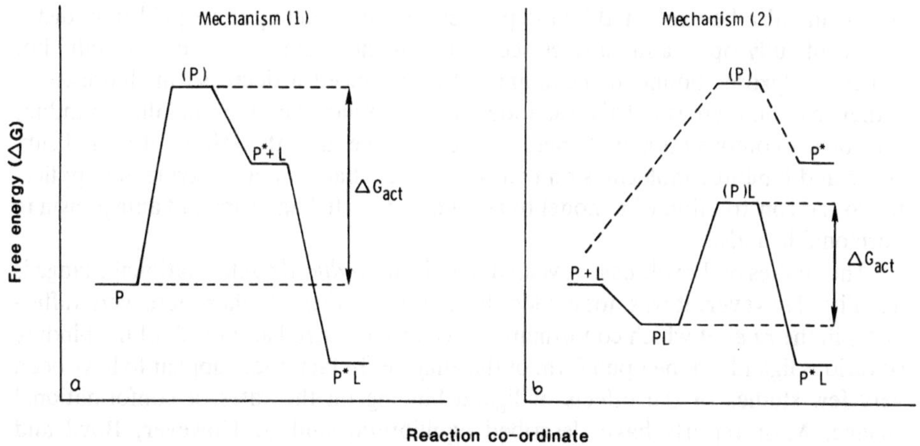

Fig. 1a, b. Free energy diagrams for alternative mechanisms of agonist-mediated excitation of receptors. **a** In mechanism (1) the agonist L displaces the equilibrium between the ground and excited states of the receptor, P and P*, by binding specifically to P*. **b** In mechanism (2) the agonist actively induces the conformational change in the receptor by binding to the ground state P. In both cases intrinsic binding energy is used to "pay" for the free energy difference between P and P* but in mechanism (2) it also contributes to a reduction in the activation energy requirement to achieve the transition state intermediate (P) between P and P*. (Reproduced by permission of Elsevier Publications (Cambridge) from T. J. Franklin, *Trends in Pharmacological Sciences,* November 1980, pp. 430–3)

As Morawetz [27] pointed out an important implication of Eq. (2) is that: "association with the substrate may not only modify the relative thermodynamic stability of different conformations (of the enzyme) but *may conceivably lower the energy barrier to be surmounted in their interconversion*" (Fig. 1b).

Thus in contrast to the mass action model [13], in the inductive model the ligand has the potential to increase the *rate* at which the active state of the enzyme is formed. In principle, this could be an important consideration when major, and presumably therefore energetically demanding, conformational rearrangements are involved in the shift from the ground state to the active form of the protein.

Good evidence that a ligand can indeed *induce* conformational change has been provided by Steitz and his colleagues [28, 29]. They have shown that when glucose binds inside the active site cleft of the hexokinase molecule it induces a closing of the cleft through a hinge-like motion. This closing of the cleft brings the potential catalytic groups into the active site and provides a binding site for the three phosphate groups of the cosubstrate ATP. It was found that o-toluylglucosamine, which binds to hexokinase in exactly the same way as glucose, is not a substrate for the enzyme because the o-toluyl group prevents the essential conformational change from occurring [29]. Ohning and Neet [30] used a fluorescent probe to monitor the conformational changes in hexokinase. They found that the binding of good substrates, like glucose and fructose, led to maximal changes in fluorescence which the authors equated with the optimum closing of the substrate cleft. Poor substrates, like mannosamine and 5-thioglucose, caused a smaller fluorescence change which was interpreted as a partial closing of the cleft, although it was conceded that the data could also be explained by the poor substrates affecting the equilibrium distribution of fully open and fully closed forms of the enzyme. A 5-carbon inhibitor molecule, lyxose, bound to hexokinase but produced a decrease in fluorescence rather than an increase. This was thought to be indicative of an inability to induce the correct conformational change. As we shall see later the effects of partial substrate and inhibitor molecules on conformational change in an enzyme are applicable to the construction of a model to account for partial agonism and antagonism in hormonal ligands.

The studies on hexokinase revealed that ligand-*induced* conformational change is a reality. However, it was not evident from those studies whether there were differences in the *rates* at which conformational change occurred as a result of the binding of various ligands to the open form of the enzyme. In fact, there appear to have been very few studies of the effects of ligand binding on the rates of conformational change. Most reports have described equilibrium studies. However, Boyd and Cohen [3, 4] used a rapid mixing technique to follow the rate of changes in the affinity of the *Torpedo* cholinergic receptors for acetylcholine following the interaction of the receptor with its agonists. They concluded that a proportion (20%) of the receptors pre-exist in a conformation that binds acetylcholine with high affinity. The remainder of the receptors bind the agonist with low affinity and the resulting low affinity complexes isomerize slowly to the high affinity state which is believed to represent the desensitized form of the cholinergic (nicotinic) receptor. Measurement of the effect of agonists on this conformational transition revealed that acetylcholine increased the isomerization rate constant by 100 times compared with the transition of the unliganded receptor. Carbamoylcholine increased the rate by some 250-fold. It would be of considerable interest to extend this type of investigation to a wider range of cholinergic agonists to establish whether there might be a general relationship between the efficacy of agonists and their ability to enhance the rate of receptor conformational change as suggested by Franklin [31].

The complexities inherent in studies on agonist-mediated receptor transformations are well illustrated by the studies of Notides and his group on the estrogen receptor. The transformation of this cytoplasmic receptor to its active form, which is capable of movement into the nucleus and subsequent genomic activation, involves a conversion of the inactive 4 S form of the receptor to a 5 S species following the binding of an agonist to the 4 S molecule. The 4 S–5 S transition is a second order, dimerization reaction which is strongly temperature-dependent. Weichman and Notides [32] followed the activation of the receptor (in a crude cytosolic fraction from uterus) by the full agonist estradiol and the partial agonists estriol and estrone. At 25 °C the second order rate constants for receptor transformation were similar for all three agonists. However, more recently Sasson and Notides [8] have concluded that although estriol and estrone occupy the same site on the estrogen receptor as estradiol, the two partial agonists are less effective than estradiol at 30 °C in inducing the essential conformational changes in the receptor. It is worth noting that there is an agonist-mediated change in the estrogen receptor which precedes the 4 S–5 S transformation [33]. It is possible, therefore, that the rate constant for this primary step is sensitive to the degree of agonist efficacy while the 4 S–5 S change is independent of this property. Detailed studies with purified estrogen receptors should provide additional information on this point.

Energetics of Ligand-Mediated Conformational Change

In addition to the free energy difference between the ground and ligand transformed states of a protein which determines the equilibrium position between these two states, the activation energy requirement to drive the protein through one or more transition states must also be considered. Activation is used here in the thermodynamic sense and should be distinguished from the biological meaning of activation which, of course, refers to the conversion of a protein from the ground state to its physiologically active form. The Arrhenius activation energy requirement (which is an enthalpic parameter) for major conformational change can be substantial. The agonist-mediated transformation of the estrogen receptor from 4 S to 5 S involves approximately 20 kcal mol^{-1} [34] and the substrate activation of chymotrypsin requires 21.7 kcal mol^{-1} [25]. It would not be surprising if the activation phenomena of many hormonal receptors demanded similar Arrhenius energy inputs. The enthalpic energy requirement to achieve the transition state in the conformational change may, of course, be partially or wholly offset by a favorable entropy contribution if there is a decrease in order in the transition state. However, at present, knowledge of the entropy content of transitional states of receptor proteins is nonexistent and is likely to remain so until an improved availability of pure receptors permits detailed monitoring of their conformational changes.

The Role of the Ligand

The energy requirement to drive the protein through its transition state and to stabilize it in its active conformation is largely derived from the energy made available when the ligand interacts with its specific binding site. In the ligand-induced model of conformational change (Eq. (2)) some of the ligand-binding energy can be used to lower the requirement for activation energy to reach the transition state. Jencks [35] proposed that binding energy is used in enzyme catalysis to lower the activation energy requirement to achieve the transition state. In enzyme catalysis the transition state separates the chemically different species of substrate and product whereas in a conformational shift there is, of course, no true chemical change. However, the suggestion that binding energy can be used to lower activation energy is still applicable. Jencks's analysis of the contribution of binding energy to enzyme catalysis provided major new insights into the process. For example, it revealed that observed binding energies, i.e. those given by the Km values, indicate only a fraction of the total energy released when substrate and enzyme interact. A substantial amount of this *intrinsic* binding energy is used to facilitate the catalytic reaction. This is achieved in several ways:

1) by lowering the requirement for the reduction in rotational and translational entropy for the formation of the enzyme-substrate complex;
2) by destabilizing the substrate through geometric distortion, desolvation and electrostatic mechanisms;
3) by inducing conformational changes in the enzyme that are necessary to achieve the active state.

$$\text{Thus } \Delta G_{obs} = \Delta G_{int} + \Delta G_D + \Delta G_c - T\Delta S_{int} \quad (3)$$

where

ΔG_{obs} = observed free energy of binding ($-$ RTln $1/K_m$)
ΔG_{int} = intrinsic free energy of binding
ΔG_D = free energy of substrate destabilization
ΔG_c = free energy of conformational change in enzyme
$T\Delta S_{int}$ = intrinsic entropy changes exclusive of solvation effects associated with the formation of ES.

Note that ΔG_D, ΔG_c and $T\Delta S_{int}$ are unfavorable terms although in many instances they will be partially offset by the favorable gain in entropy associated with the displacement of structured water from the substrate binding site and possibly due to conformational changes in the enzyme. Nevertheless, this examination of the composition of the observed binding values goes some way to explaining the apparent paradox between the exquisite specificity of many enzymes for their substrates and the often surprisingly high K_m values, i.e. between 0.1 and 0.01 mM, which suggest weak binding.

Jencks [35] distinguished between the productive component of intrinsic binding energy which is used to facilitate catalysis and the remainder which is observed as the binding parameter (Fig. 2). "Good" substrates with a high turnover rate are

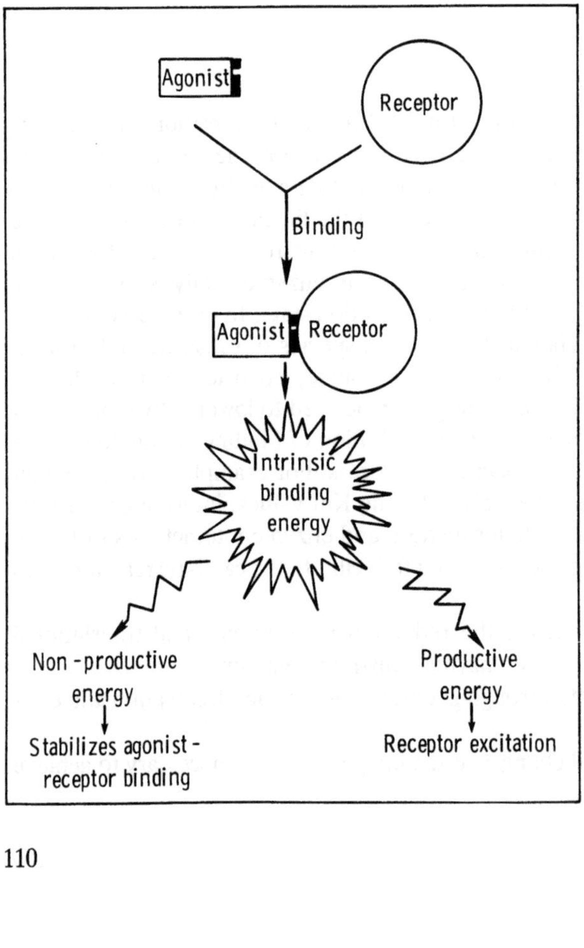

Fig. 2. Diagrammatic representation of the concept of intrinsic binding energy and its partitioning into the energy from "productive" interactions between the agonist and its receptor, which is used to transform the receptor to its excited state, and "non-productive" energy which stabilizes the binding between the agonist and receptor. (Reproduced by permission of Elsevier Publications (Cambridge) from T. J. Franklin, *Trends in Pharmacological Sciences*, November 1980, pp. 430–3)

characterized as molecules which optimize productive binding energy when they interact with their enzymes. "Poor" substrates, on the other hand, because of suboptimal interactions with their enzymes, generate less productive binding energy. This results in lower values for V_{max}, frequently accompanied by lower K_m values because non-productive binding energy is used to stabilize substrate binding. A substrate analog with a V_{max} approaching zero and a low K_m would of course be an excellent competitive inhibitor. Such a molecule is seen as releasing little or no productive energy on binding to an enzyme.

In attempting to relate Jencks's binding energy hypothesis to hormone receptor-ligand interactions we can see that the intrinsic binding energy of agonists is used in part to "pay" for the intrinsic entropy loss involved in forming the agonist-receptor complex and for the essential conformational change in the receptor that leads to the characteristic agonist response. Remaining binding energy is expressed as the $K_{dissociation}$ (K_D), which is therefore an underestimate of the total energy released on interaction. The generation of productive binding energy probably underlies the pharmacological phenomenon of efficacy. Partial agonism in pharmacological ligands has been something of a mystery since its discovery more than 40 years ago. Molecules with this property may saturate the receptors but fail to elicit a full biological response. There are numerous possible explanations for this important phenomenon but in terms of Jencks's concept of productive binding energy, parallels can be seen between poor substrates and competitive inhibitors on the one hand and between partial agonists and pure antagonists on the other.

Partial agonists would be expected to generate insufficient productive binding energy to ensure stimulation of an adequate fraction of the receptor pool for a maximum biological response. Thus the equilibrium constant given by

$$\frac{[R^*L]}{[RL]}$$

(where R is the ground state and R* the active state of the receptor and L is the stimulating ligand) would be lower for a partial agonist than for a full agonist. In the case of a pure antagonist the constant is insignificant. We should also expect that the *rate* at which R*L is formed would be less for a partial agonist than for a full agonist since less binding energy is available to achieve the transition state (RL) in mechanism 2, although this would not apply if the conformational change is achieved by mechanism 1. The free energy diagrams for the interaction of agonists with receptors are given in Fig. 1a, b.

Partial Conformational Changes

Another mechanism for partial agonism must also be considered. As previously discussed, Ohning and Neet [30] suggested that poor sugar substrates for hexokinase may induce a partial closing of the substrate cleft in this enzyme, resulting in a less than optimal catalytic system. It is conceivable that some partial agonists may induce incomplete conformational changes in receptors, either because of an inadequate energetic contribution or perhaps because of some steric interference by the ligand that prevents a full change. A partially transformed receptor molecule might be able to elicit an incomplete agonist response. Future experiments with purified receptors should facilitate the direct investigation of this possibility.

Application of the Binding Energy Concept to Structure-Activity Relationships

A complete understanding of the molecular details of physiologically important conformational changes in proteins and their relationships to the structural features of initiating ligands is a distant dream. In the meantime, however, there are some practical possibilities for assessing the contributions of the individual constituent groups of ligands to receptor binding and to the mediation of conformational changes. As we have seen, the intrinsic energy that is released when two interacting molecules bind together is not discernible from the observed binding parameter. However, Jencks [35] pointed out that it is possible to obtain an estimate of the contribution of individual structural features of ligand molecules if a series of related analogs is available. In the simplest case, if we consider a ligand A and its substituted derivative A-B which both bind to the same receptor protein, an approximation to the intrinsic free binding energy contribution of B is given by the difference in the observed binding energies $\Delta G^{B} = \Delta G^{A\text{-}B} - \Delta G^{A}$

$$\text{i.e.}\quad \Delta G^{B} = (-RT\ln K^{A\text{-}B}) - (-RT\ln K^{A}) \tag{4}$$

where K^{A} and $K^{A\text{-}B}$ are the respective association constants. This relationship is approximately correct in certain circumstances because the intrinsic entropy change (exclusive of solvation effects) attendant upon binding is essentially the same for A and A-B. Such a simple approach to the estimation of intrinsic binding energy is well suited to the technique of radioligand displacement by a series of closely related compounds provided that the displacement is strictly competitive. Some authors (e.g. [6, 32]) have attempted to use this approach to derive the total binding energies of complex molecules (steroids and triiodothyronine) by the addition of the individual binding energies of the constituent parts of the molecules. Although the values obtained were reasonable approximations to the observed binding energies of the complete molecules, there can be no real basis for such a procedure [35]. For example, the addition of binding energies is equivalent to the multiplication of equilibrium constants, which gives numbers with units of M^{-2} that have no simple relationship to an equilibrium constant with a unit of M^{-1}.

The most that can be expected from a measure of the difference in binding energies of A and A-B is therefore an estimate of the intrinsic binding energy of B. This, however, is subject to a number of important limitations:

1) If the substituent B exerts an electronic effect on A this could well affect the interaction of other regions of the ligand with its receptor and lead to an erroneous estimate of the contribution of B.
2) If B is a flexible substituent group its intrinsic contribution to the binding of A-B will be partially obscured by an additional unfavorable entropy requirement to restrict its mobility in the binding site.

3) Similarly, if the accommodation of B at the receptor binding site involves strain in A-B which does not occur with the binding of A alone the unfavorable enthalpic requirement will diminish the apparent intrinsic energy contribution of B.
4) If the binding of A-B causes a conformational change in the receptor that does not result from the binding of A, the energy requirement for the change will of course have to be "paid for" out of the binding energy released from the interaction of A-B with the receptor.

The last point is especially relevant to a comparison of the binding of agonists and antagonists or of agonists with different levels of efficacy. We may expect binding to give rise to a range of values for the isomerization constant [R*L]/[RL] and perhaps to different rates at which the equilibria are established. Under these circumstances it may be difficult to assess the intrinsic binding energy contributions of substituent groups. An indication of the nature of this problem is given by some typical results obtained by my colleague Dr. A. E. Wakeling on the displacement of the tritiated steroidal androgen, methyltrienolone, from the cytosolic androgen receptor by a series of anilide androgen antagonists (Table 1). The parent compound, hydroxy-flutamide, appears to be a pure antagonist as judged by *in vivo* studies whereas the closely related analogs A and B have significant androgenic activity. The substitution of a CH_3 group in hydroxy-flutamide by a CF_3 in analog A and by a CHF_2 group in B confers increases of binding energy of 0.63 and 1.07 kcal mol^{-1} respectively. However, since analogs A and B are partial agonists the observed binding energies (9.10 and 9.54 kcal mol^{-1}) may be less than their intrinsic

Table 1. Binding energies in anti-androgens. Free energies of binding derived from competitive displacement of [3 H]-methyltrienolone from cytosolic androgen receptors. $\Delta\Delta G^0$ values are vs. hydroxy-flutamide

		Observed binding energies ($k \cdot cal \cdot mol^{-1}$) ΔG^0	$\Delta\Delta G^0$	Agonism in vivo
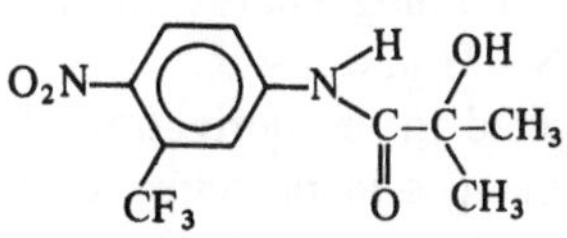 Hydroxy-flutamide		−8.47	–	0
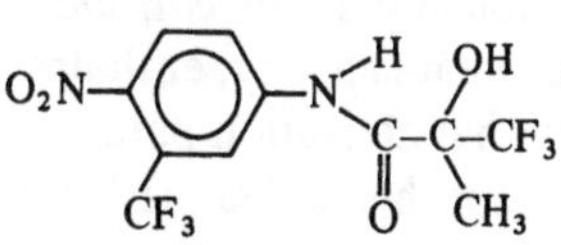	A	−9.10	−0.63	++
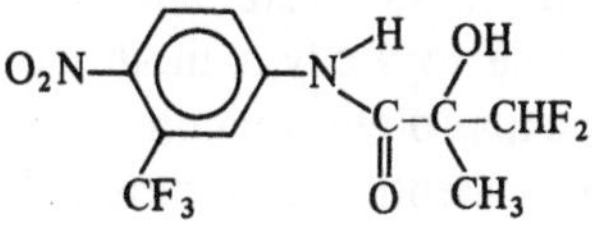	B	−9.54	−1.07	+++

energies because of the energy required to transform the androgen receptor. A further complication is that the agonist-induced conformational change may increase the affinity of the receptor for the ligand. An alternative strategy is therefore needed in order to permit a more accurate assessment of the contributions of individual groups when conformational changes in the receptor are involved. The following section suggests a possible method based on an elementary application of transition state theory to protein conformational changes.

Transition States and Conformational Change

Fersht [36] recognized that a comparison of the dissociation constants of a substrate S-H and its substituted derivative S-R would probably lead to an underestimate of the binding contribution of R because of the likely use of the binding energy to lower the activation energy of the catalytic reaction. He suggested that it would be better to compare the ratio of k_{cat}/K_m for the two substrates. The parameter k_{cat} is defined as the rate constant for the conversion of the enzyme-substrate complex to the enzyme-product complex:

$$E + S \underset{k_{-1}}{\overset{k_1}{\rightleftharpoons}} ES \xrightarrow{k_{cat}} EP \tag{5}$$

When $k_{-1} \gg k_{cat}$ then K_m is equivalent to k_{-1}/k_1, which is the dissociation constant for the Michaelis complex.

Transition state theory yields the following relationship [36]:

$$\ln(k_{cat}/K_m) = \ln(kT/h) - (\Delta G_a - \Delta G_b)/RT \tag{6}$$

where k = Boltzman constant, h = Planck constant, ΔG_a = inherent activation energy due to the electronic effects of bond making and breaking, and ΔG_b is the intrinsic binding energy of the substrate to the enzyme. Provided that differences in ΔG_a for similar substrates can be ignored, the intrinsic binding energy of the group R can be obtained from the difference in ΔG_b for S – H and S – R.

It would be very helpful if this principle could be applied to the derivation of binding energies for ligands that cause conformational changes in proteins. Can transition state theory, which was originally developed to deal with the making and breaking of covalent bonds between reactant molecules, be applied to conformational change in proteins? Certainly the concept has been implicit in many considerations of protein dynamics. However, the dependence of the heat capacity, enthalpy and entropy of proteins on temperature [37] makes it likely that activation parameters will vary with temperature. In addition, Frauenfelder and his colleagues [38] have revealed a complex involvement of the viscosity of the solvent with internal protein dynamics that further complicates the application of transition state theory.

However, despite these difficulties it should be possible to apply a modified version of Eq. (6) to ligand-induced conformational change in proteins for comparative purposes, i.e. by deriving ΔG_b for related ligands under identical experimental conditions and at a temperature as close to physiological as possible. The k_{cat} term of

Eq. (6) can be replaced by k_{act} which is the rate constant for the transformation of the ground state of the receptor-agonist complex to the biologically active form R*L via the transition state (RL). Km is replaced by K_D, the dissociation constant given by

$$\frac{[R][L]}{[RL]}$$

Fersht [36] indicated that the application of Eq. (6) to enzyme substrates requires the assumption that differences in the free energy of activation to the transition state for similar substrates can be ignored. A similar assumption must be made for agonist-mediated conformational change in receptors, although of course it is unknown at present whether or not this is valid.

Accepting the various caveats surrounding the application of transition state theory to conformational changes in proteins, it would be instructive to investigate the effects of related agonists and partial agonists on the rates of conformational change in their specific receptors. An interesting test case of current practical interest is provided by the estrogen receptor. The previously mentioned indirect approach of Notides and his colleagues, which follows the 4S-5S transformation, could be misleading if an essential agonist-induced change in the 4S protein precedes the dimerization step. The anticipated availability of purified 4S receptor should facilitate a direct investigation of agonist-induced conformational changes. Furthermore, it seems likely that because of rapid developments in DNA-recombinant biology, presently elusive receptor proteins should eventually become available for this type of study. Reconstitution of membrane-localized receptors in appropriate environments may pose additional technical problems. Nevertheless there is now an exciting prospect of gaining a greater understanding of the energetics and kinetics of the interactions of hormonal and neuroendocrine agents with their receptors. It should then be possible to relate these parameters to the individual molecular features of the ligands. This in turn will assist the design of pharmacological agents with improved efficacy and selectivity of action. The study of conformational changes in hormone and neuroendocrine receptors and the role of the ligands that mediate them has lagged well behind that of other proteins. It seems a fairly safe prediction that this unsatisfactory situation should improve dramatically in the next few years.

References

1. Kent, R. S., De Lean, A. and Lefkowitz, R. J. (1980) Molec. Pharmacol. *17,* 14–23
2. Hoffman, B. B., Kilpatrick, D. M. and Lefkowitz, R. J. (1980) J. Biol. Chem. *255,* 4645–4652
3. Boyd, N. D. and Cohen, J. B. (1980) Biochemistry *19,* 5344–5353
4. Boyd, N. D. and Cohen, J. B. (1980b) Biochemistry *19,* 5353–5358
5. Roeske, W. R., Ehlert, F. J., Barritt, D. S., Yamanaka, K., Rosenberger, L. B., Yamada, S., Yamamura, S. and Yamamura, H. I. (1983) Adv. Biochemical Psychopharmacol. *36,* 15–30
6. Bolger, M. B. and Jorgensen, E. C. (1980) J. Biol. Chem. *255,* 10271–10278
7. Fillion, G. (1983) Adv. Biochemical Psychopharmacol. *36,* 115–123
8. Sasson, S. and Notides, A. (1983) J. Biol. Chem. *258,* 8113–8117
9. Weiland, G. A., Minneman, K. D. and Molinoff, P. B. (1979) Nature *281,* 114–117
10. Weiland, G. A., Minneman, K. D. and Molinoff, P. B. (1980) Molec. Pharmacol. *18,* 341–347

11. Möhler, H. and Richards, J. G. (1981) Nature *294,* 763–765
12. Quast, U., Mahlmann, H. and Vollmer, K-O. (1982) Molec. Pharmacol. *22,* 20–25
13. Barlow, R. B. and Burston, K. N. (1979) Br. J. Pharmacol. *66,* 581–585
14. Barlow, R. B., Birdsall, N. J. M. and Hulme, E. C. (1979) Br. J. Pharmacol. *66,* 587–590
15. Wolff, M. E., Baxter, J. D., Kollman, P. A., Lee, D. L., Kuntz, I. D., Bloom, E., Matulich, D. T. and Morris, J. (1978) Biochemistry *17,* 3201–3208
16. Debrunner, P. G. and Frauenfelder, H. (1982) Ann. Rev. Phys. Chem. *33,* 283–299
17. Steiner, R. F., Lambooy, P. K. and Sternberg, H. (1983) Arch. Biochem. Biophys. *222,* 158–169
18. Levinthal, M. (1968) J. Chim. Phys. *65,* 44–45
19. Karplus, M. and Weaver, D. L. (1976) Nature *260,* 404–406
20. Kim, P. S. and Baldwin, R. L. (1982) Ann. Rev. Biochem. *51,* 459–489
21. Stryer, L. (1975) "Biochemistry" W. H. Freeman and Co., San Francisco
22. Koshland, D. E. and Neet, K. E. (1968) Ann. Rev. Biochem. *37,* 359–410
23. Frieden, C. (1979) Ann. Rev. Biochem. *48,* 471–489
24. Fersht, A. R. and Requena, Y. (1971) J. Mol. Biol. *60,* 279–290
25. Kin, Y. D. and Lumry, R. (1971) J. Am. Chem. Soc. *93,* 1003–1013
26. Citri, N. (1973) Adv. Enzymol. *37,* 397–648
27. Morawetz, H. (1972) Adv. Protein Chem. *26,* 243–277
28. Bennett, W. S. and Steitz, T. A. (1978) Proc. Nat. Acad. Sci. *75,* 4848–4852
29. Steitz, T. A., Harrison, R., Weber, I. T. and Leahy, M. (1983) Ciba Foundation Symposium 93 "Mobility and Function in Proteins and Nucleic Acids" 25–46
30. Ohning, G. V. and Neet, K. E. (1983) Biochemistry *22,* 2986–2995
31. Franklin, T. J. (1980) Biochem. Pharmacol. *29,* 853–856
32. Weichman, B. M. and Notides, A. C. (1980) Endocrinology *106,* 434–439
33. Muller, R. E., Traish, A. M. and Wotiz, H. H. (1983) J. Biol. Chem. *258,* 9227–9236
34. Notides, A. C. and Nielsen, S. (1975) J. Steroid Biochemistry *6,* 483–486
35. Jencks, W. P. (1975) Adv. Enzymol. *43,* 219–410
36. Fersht, A. (1977) "Enzyme Structure and Mechnisms", W. H. Freeman, Reading and San Francisco
37. Privalov, P. L. (1979) Adv. Protein Chem. *33,* 167–236
38. Beece, D., Einstein, L. and Frauenfelder, H. (1980) Biochemistry *19,* 5147–5157

Discussion

List of interventions.

1 Vauquelin

1 Intervention of Vauquelin

Receptors are present only in minute concentrations so that they can be directly characterized only by radioligand binding. However, since receptors are proteins, any conformational change might result in differences in accessibility of a certain amount of amino acid residues or peptide bonds towards group-specific reagents. This technique has been successfully applied in combination with radioligand binding for the demonstration of agonist-mediated conformational changes of β-adrenergic and muscarinic acetylcholine receptors. The β-receptors can be inactivated by the alkylating reagent N-ethylmaleimide (NEM) in the presence of agonists but not in the presence of antagonists or in the absence of ligand. These studies also showed that only a certain percentage of the receptors can undergo agonist-mediated conformational changes (i.e. agonist/NEM-sensitive) and hence that the receptors are functionally heterogeneous.

Theoretical Studies of Molecular Recognition and Catalysis by Enzymes*

Harold A. Scheraga

Baker Laboratory of Chemistry, Cornell University, Ithaca, New York 14853-1301, USA

Abstract

The procedures used in conformational energy calculations, and some typical results of such computations on model systems and on polypeptides and proteins, are summarized. It is then shown how application of this methodology to two enzyme-substrate systems (α-chymotrypsin plus oligopeptides and hen egg white lysozyme plus oligosaccharides) leads to computed binding modes and energies that agree with experiment. This agreement attests to the validity of the potential functions and methodology used in the computations, and demonstrates the utility of this methodology in providing an understanding of the interactions that lead to molecular recognition.

I Introduction

Theoretical conformational analysis is being used to provide a molecular basis of recognition, or specificity in biological systems, in particular of: a) the interactions between parts of a polypeptide chain so that they can recognize each other and fold into the biologically active form of the native protein; b) the interactions between two biological molecules, e.g. the recognition of a substrate by an enzyme or of an antigen by an antibody; and c) the interactions that enable biological molecules to recognize cell surfaces. We are approaching this problem by a combined experimental and theoretical approach, but this paper is concerned primarily with our theoretical studies.

We provide here a brief summary of the computational methodology and some examples of calculations on model systems. We also show the results of computations on the conformations of natural polypeptides, fibrous proteins, globular proteins, and binary complexes of enzymes and substrates.

II Methodology

Empirical potential energy functions are used in the theoretical calculations [1–4]. They are represented as interactions between pairs of atoms, and the total energy of a molecular system is taken as a sum over all pair interactions. The potential func-

* This paper has also appeared in Pont. Acad. Sci. Ser. Vat., *55*, 21 (1984)

tions were parameterized by applying them to compute experimental data on gas-phase and crystal structures, and thermodynamic properties, of small oligopeptides and related molecules. While originally derived for polypeptides, they were found to be applicable also to polysaccharides without any change of parameterization [5]. Further, they apply equally well to both intramolecular and intermolecular interactions.

The total potential energy is expressed in terms of contributions from non-bonded, electrostatic, and hydrogen-bonding interactions, and intrinsic torsional potentials, and is embodied in an algorithm referred to as ECEPP (Empirical Conformational Energy Program for Peptides). The details are given by Pottle et al. [3], and the most recent set of parameters is described by Némethy et al. [4]. Procedures are also available for including entropy effects [6–10] and hydration effects [11–13] in the computations. (As shown in Fig. 1 of a paper by Scheraga [14], the entropy arising from flexibility can stabilize a conformation that does not have the lowest energy. While examples of such situations have been observed [7, 10], inversion of stability by the entropy contribution generally occurs only if the minima are quite close in energy.)

In order to find the most stable state of a molecular system, the "conformational energy" (including entropy effects) is minimized. The multiple-minima problem is circumvented by incorporating various constraints into the minimization procedure, e.g. short-, medium- and long-range information, and distance constraints [15–17].

The variables in the minimization procedure are the dihedral angles for rotation about the single bonds of the molecules. When dealing with intermolecular systems such as enzyme-substrate complexes, additional degrees of freedom are involved, viz. overall translation and rotation of the enzyme and substrate as they approach each other in a "docking" process. In such a case, each partner in the complex influences the conformation of the other because of intermolecular interactions.

III Some Calculations with Model Systems

It has been known for some time that the α-helices and β-sheets of proteins have a right-handed twist but the reason for such a twist was unknown. Several years ago, we used conformational energy calculations to predict the handedness of various homopolymer α-helices, and identified the interatomic interactions that determine the right-handed twist of these structures [18, 19]. Recently, we applied the same methodology to the more complicated β-sheets and found that the low-energy structures of such sheets are indeed right-handed, as originally observed for β-sheets in globular proteins by Chothia [20]. Moreover, we identified the interatomic interactions that lead to a right-handed twist [21–25]. In both α-helices and β-sheets, local interactions within these structures (rather than longer-range interactions with the rest of the globular protein) appear to play the dominant role in determining their twist.

We next considered the problem as to how α-helices and β-sheets interact with each other in forming a globular protein. For example, the most favorable arrangement of two α-helices is an anti-parallel one (see Fig. 1) dictated by the most favorable arrangement of the dipoles of the α-helices [26].

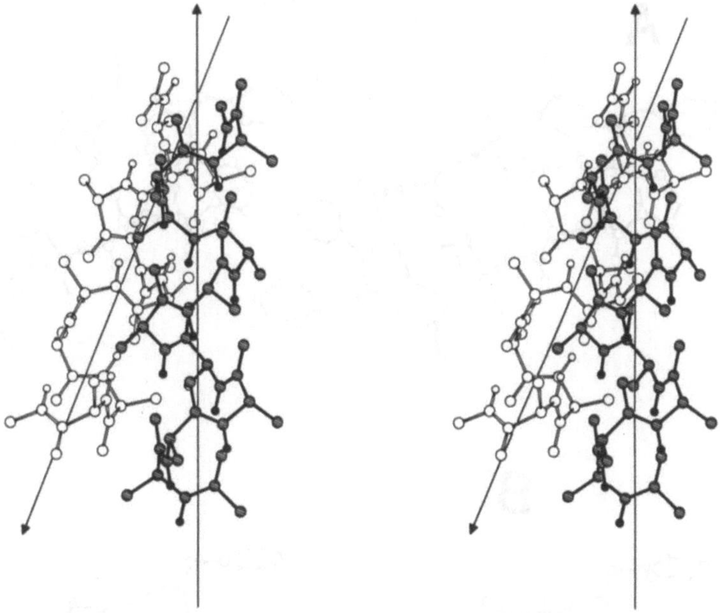

Fig. 1. Stereoscopic picture of two $CH_3CO-(L\text{-}Ala)_{10}-NHCH_3$ α-helices in the lowest-energy packing state (with a torsion angle of − 154°). The helix axes are indicated by arrows, with the head of the arrow pointing in the direction of the C-terminals of each helix [26]

IV Some Calculations on Polypeptides and Proteins

Turning to natural polypeptides and proteins, we may cite our computations on gramicidin S, collagen-like poly(tripeptides), α-lactalbumin, melittin, and bovine pancreatic trypsin inhibitor (BPTI) as examples. In the case of gramicidin S, a cyclic decapeptide, the computed structure [27, 28] was found to agree [29, 30] with a subsequently-determined X-ray structure [31] (see Fig. 2). Likewise, the computed triple-stranded helical structure of poly(Gly-Pro-Pro) [32] (Fig. 3) has cartesian coordinates that agree with those from a single-crystal X-ray diffraction study of $(Pro\text{-}Pro\text{-}Gly)_{10}$ [33] within ± 0.3 Å.

A structure of α-lactalbumin was computed [34] by taking advantage of its sequence homology to lysozyme and using the known structure of the latter protein as a starting point for energy minimization (Fig. 4). While the X-ray structure of α-lactalbumin has not yet been determined, the predicted structure is at least in agreement with various experimental results cited by Warme et al. [34] and some [35] that were obtained subsequently.

By use of a build-up procedure, in which small energy-minimized fragments were combined (with subsequent energy-minimization of the resulting large fragments), it was possible to compute the structure of the 20-residue membrane-bound portion of melittin [36]. The computed structure (Fig. 5) is in satisfactory agreement with experiment [37, 38].

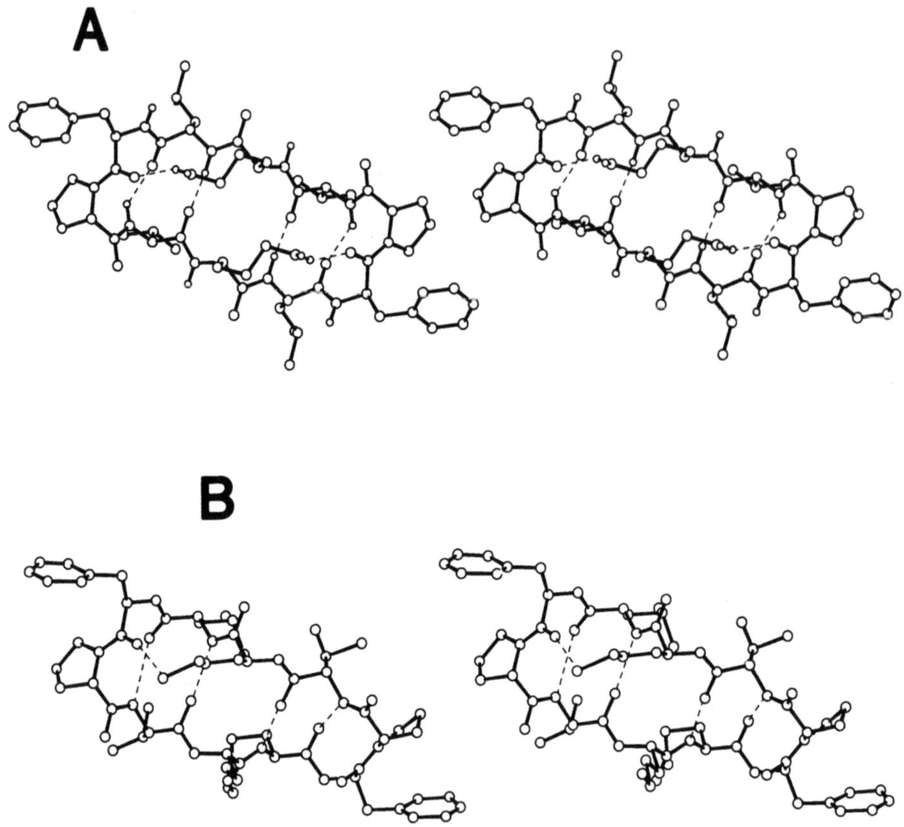

Fig. 2. **A** Computed [27, 28] and **B** X-ray [31] structures of gramicidin S, showing (among other things) a hydrogen bond between the ornithine side chain and the phenylalanine backbone carbonyl group

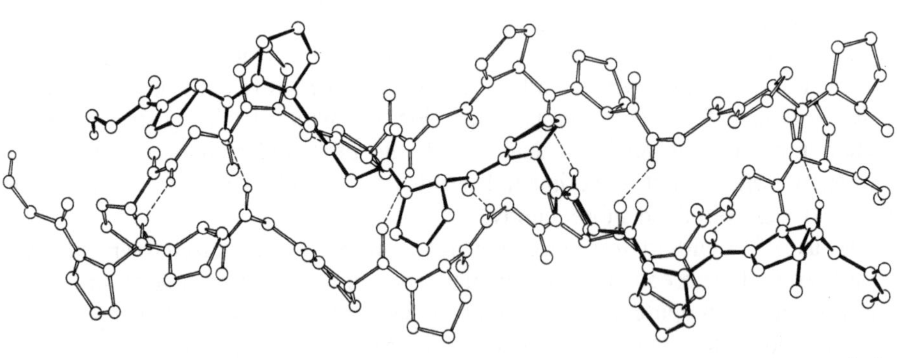

Fig. 3. Computed triple-stranded coiled-coil complex of poly(Gly-Pro-Pro) of lowest energy [32]

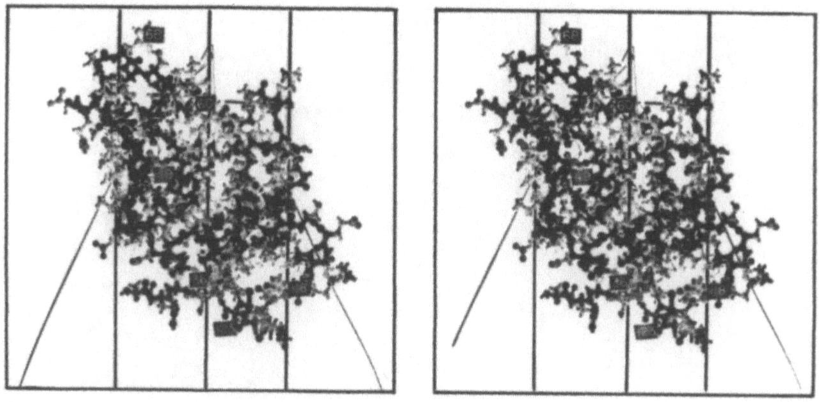

Fig. 4. Stereo view of computed structure of α-lactalbumin [34]

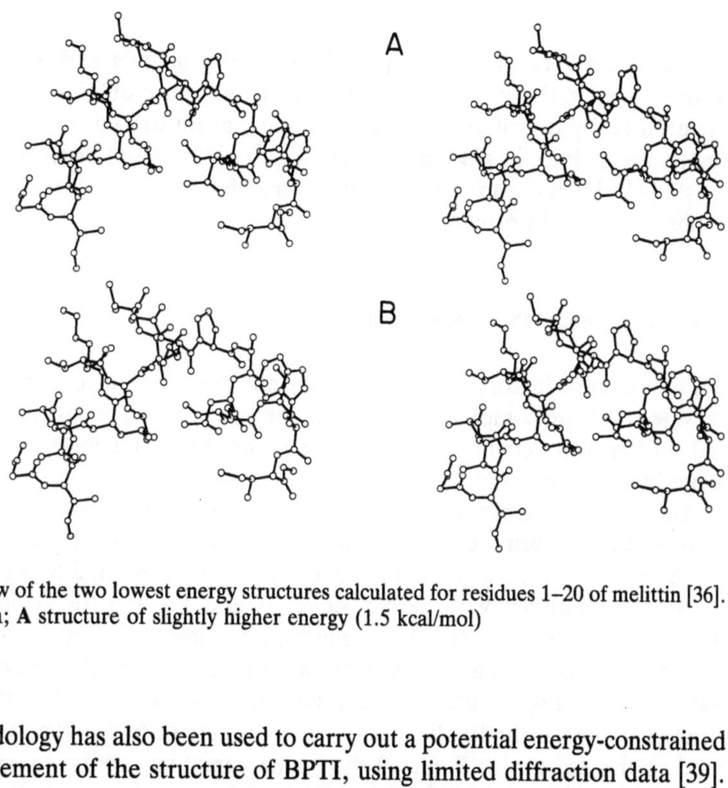

Fig. 5. Stereo view of the two lowest energy structures calculated for residues 1–20 of melittin [36]. **B** global minimum; **A** structure of slightly higher energy (1.5 kcal/mol)

This methodology has also been used to carry out a potential energy-constrained real-space refinement of the structure of BPTI, using limited diffraction data [39]. Figure 6 illustrates how the energy-refined map on the right identifies more of the electron density of Arg-42 than the map on the left.

Finally, by using a series of constraints in an energy minimization, some progress has been made in computing the structure of BPTI [17]. Thus far, a virtual-bond structure of this protein has an r.m.s. deviation of 2.2 Å from the experimental one. Work is in progress to try to improve the computational procedure, and hence the

Fig. 6. Comparison of sections of electron density maps around Arg-42 of BPTI, projected onto the x, y plane [39]. (Left) Map calculated by using the experimental 2.5-Å phases. (Right) Map calculated by using the phases from the structure factor calculation

agreement between the computed and observed structures. Originally, distance constraints (used in these protein-folding algorithms) were obtained by a long series of chemical and physical chemical experiments, as illustrated for three tyrosyl... aspartyl interactions in ribonuclease A (Fig. 7) [40]. More recently, nuclear Overhauser [41, 42] and non-radiative energy transfer [43] measurements are being used to obtain similar information.

V Chymotrypsin Plus Oligopeptides

With these introductory remarks, we may now consider the low-energy structures of enzyme-substrate complexes. In this section, we describe results for α-chymotrypsin and, in the next section, results for lysozyme. Further details may be found in a paper by Scheraga et al. [44].

The action of an enzyme such as chymotrypsin on an oligopeptide substrate may be described in terms of the notation of Schechter and Berger [45] as shown in Fig. 8. Here, P_i represents the residues of the substrate and S_i the subsites of the enzyme that interact with each residue; A and P stand for alanine and phenylalanine, respectively, in the three oligopeptide substrates illustrated. Computations on complexes of the enzyme with these three substrates also provide information about the binding of the pentapeptide AAPAA, wherein the flanking alanine residues influence the binding, and hence the rate of hydrolysis of the PA peptide bond in this substrate. Figure 9 shows the computed lowest-energy structure for PAA in the active site of the enzyme.

Such computations provide details of the interactions between the various parts of the enzyme and substrate and the relative affinities of different substrates. For example, the computed *relative* binding energies of the aromatic amino acids Trp, Tyr and Phe are quite similar to those determined experimentally [46]. In the case of the tryptophan peptide, the calculated coordinates of the indole ring in the active-

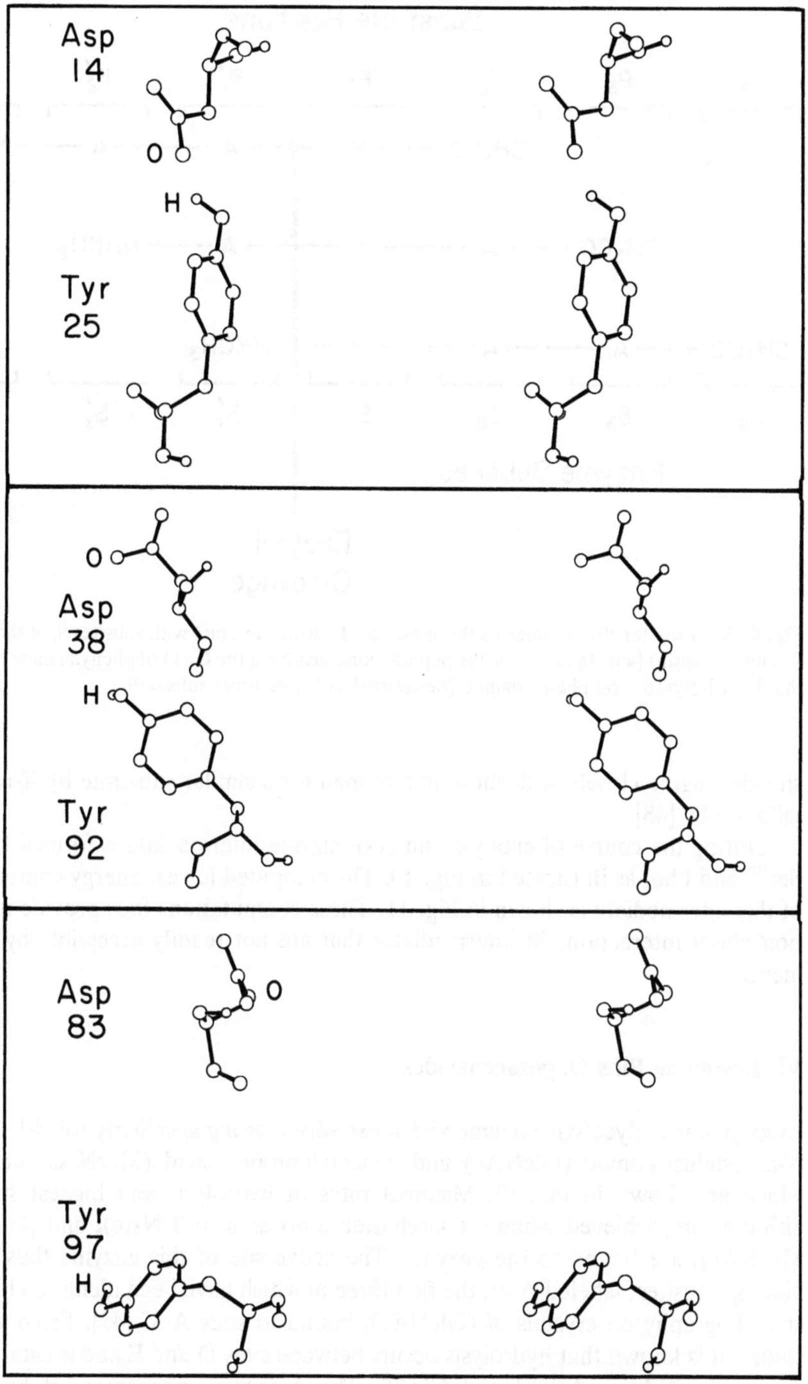

Fig. 7. Illustration of three Tyr . . . Asp interactions in bovine pancreatic ribonuclease A, deduced from various chemical and physical chemical experiments [40] before the X-ray structure was determined. The drawing was prepared from the X-ray coordinates of Wlodawer et al. [47]

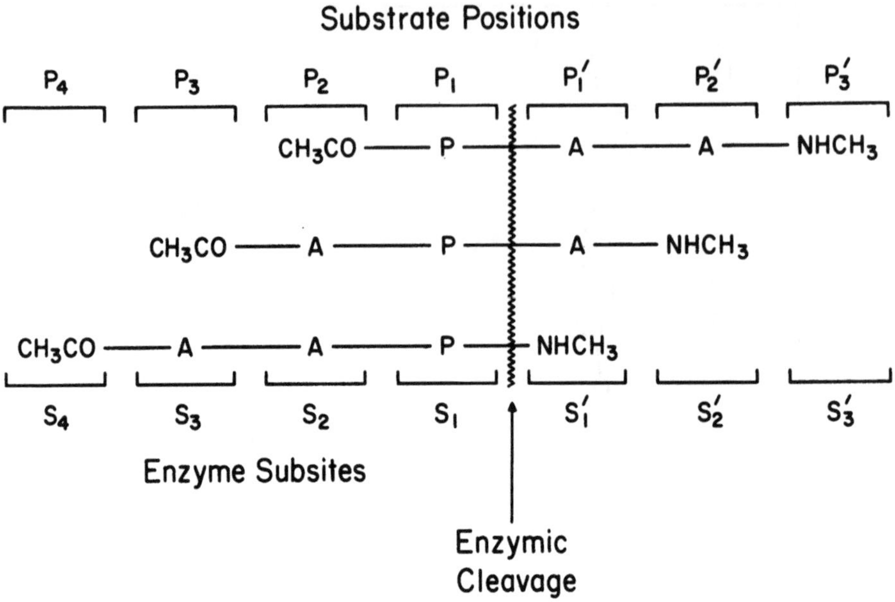

Fig. 8. Notation for the residues of the substrate, P_i, which interact with subsites S_i of the enzyme (α-chymotrypsin) [45]. In all cases, the peptide bond involving the C=O of phenylalanine is the one that is hydrolyzed, and phenylalanine (designated as P_1) occupies subsite S_1

site cleft agree closely with those determined for a similar substrate by X-ray crystallography [48].

During the course of catalysis, an acyl-enzyme intermediate is formed between Ser^{195} and Phe, as illustrated in Fig. 10. The computed lowest-energy conformation of this intermediate is shown in Fig. 11. These computations thus provide information about interactions in intermediates that are not readily accessible by experiment.

VI Lysozyme Plus Oligosaccharides

Lysozyme is a glycolytic enzyme which exhibits a strong specificity for β-1,4-linked N-acetylglucosamine (GlcNAc) and N-acetylmuramic acid (MurNAc) units [49] which are shown in Fig. 12. Maximal rates of hydrolysis and highest substrate affinities are achieved when six saccharide units as in $(GlcNAc)_6$ and $(GlcNAc \cdot MurNAc)_3$ are bound to the enzyme. The active site of this enzyme thus has six binding subsites, labeled A–F, the first three of which have been identified by X-ray crystallography on crystals of $(GlcNAc)_3$ bound to sites A–C [49]. From solution studies it is known that hydrolysis occurs between sites D and E and is catalyzed by two acid residues, Asp 52 and Glu 35. The proposed mechanism of hydrolysis involves the formation of a carbonium ion at C_1 of the D-site residue stabilized by the carboxylate of Asp 52 with Glu 35 serving as a general acid that protonates the oxygen of the departing saccharide [49].

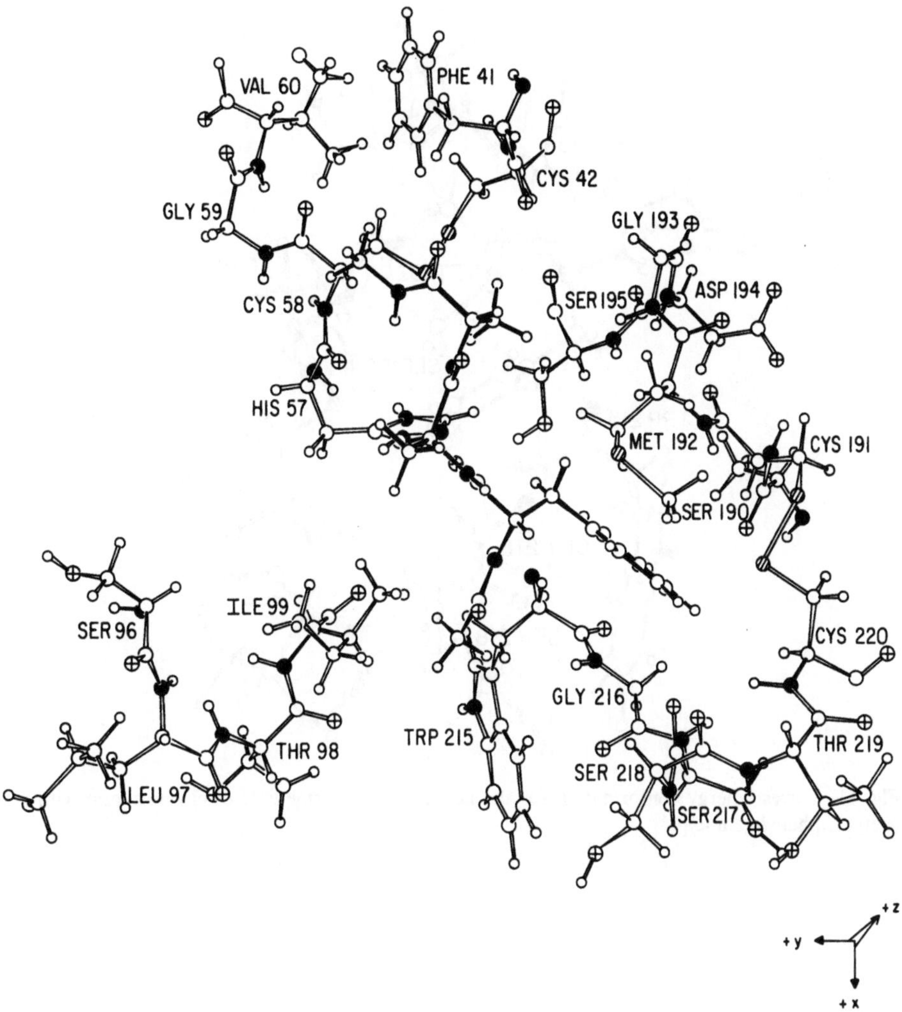

Fig. 9. Lowest-energy conformer for N-acetyl-Phe-Ala-Ala-NH$_2$ in the cleft binding region of α-chymotrypsin [44]

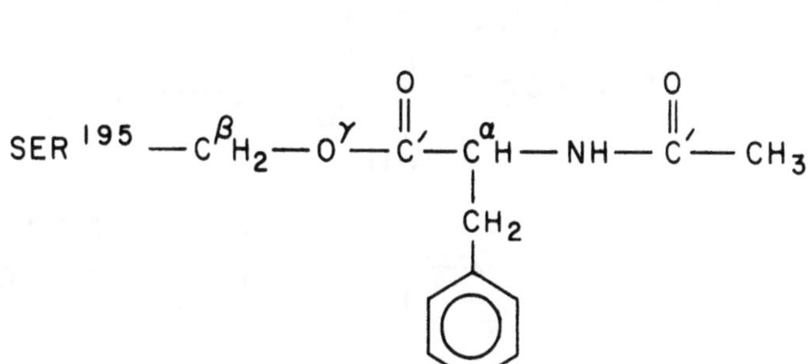

Fig. 10. Acyl-enzyme intermediate for N-acetyl-L-phenylalanine

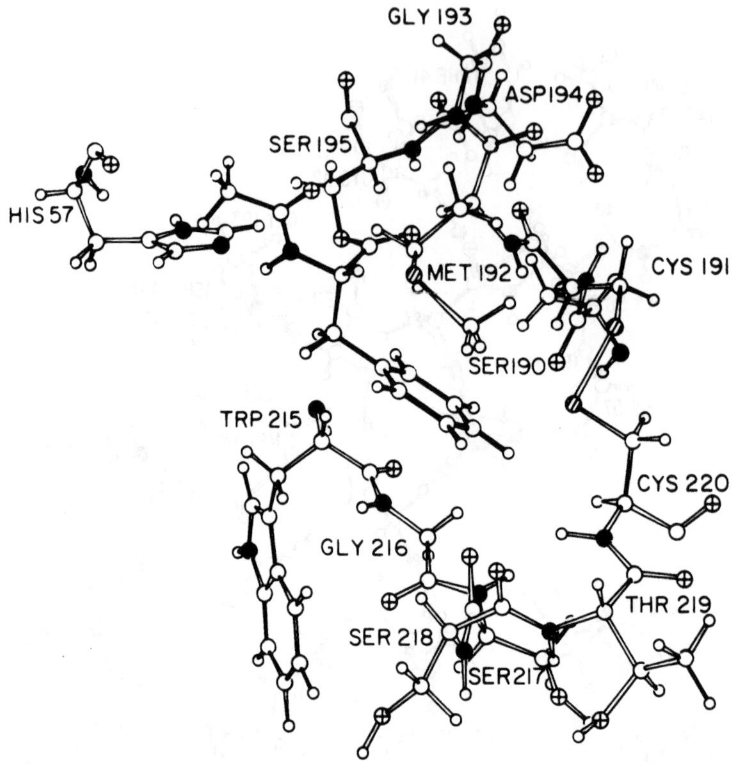

Fig. 11. Lowest-energy conformation for the acyl-enzyme intermediate of chymotrypsin with N-acetyl-L-phenylalanine [44]

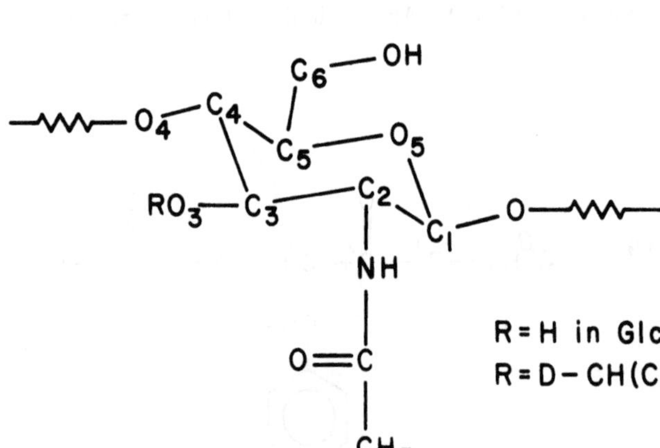

Fig. 12. Structures of GlcNAc and MurNAc

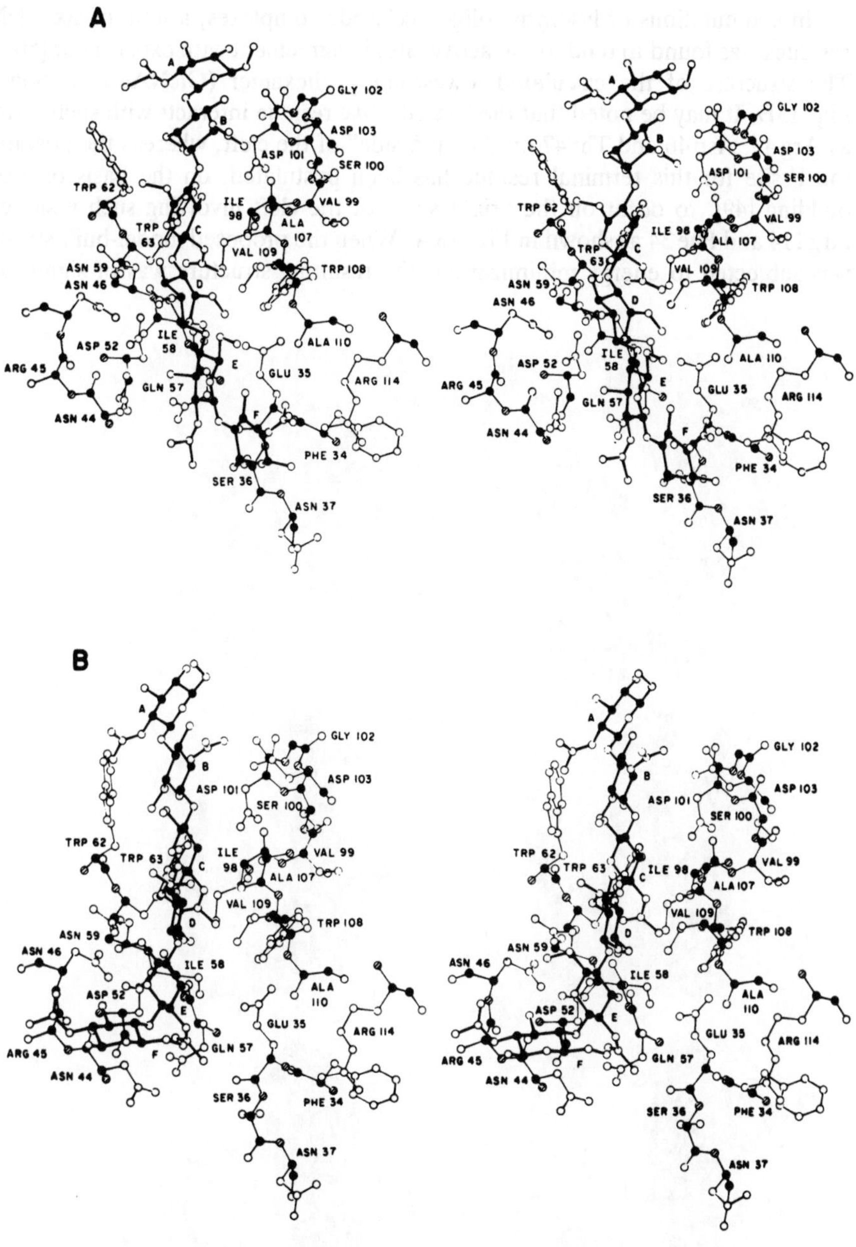

Fig. 13. Stereo views of: **A** Geometry- and energy-refined model-built structure binding to the right side of the active-site cleft. **B** Calculated lowest-energy structure of $(GlcNAc)_6$ bound to the left side of the active-site cleft [50, 51]

In computations of lysozyme-oligosaccharide complexes, a total of six GlcNAc residues was found to bind to the active site, in agreement with experiment [50, 51]. The structure of the calculated lowest-energy hexamer $(GlcNAc)_6$ is shown in Fig. 13B. It may be noted that the last GlcNAc residue interacts with such residues as Arg 45, Asn 46 and Thr 47 on the "left side" of the cleft, whereas the position of the F site for this terminal residue has been postulated, on the basis of model-building [49], to occur on the "right side" of the cleft involving such residues as Arg 114 and Phe 34 as shown in Fig. 13A. When the proposed model-built structure was subjected to energy minimization, the resulting structure was of significantly

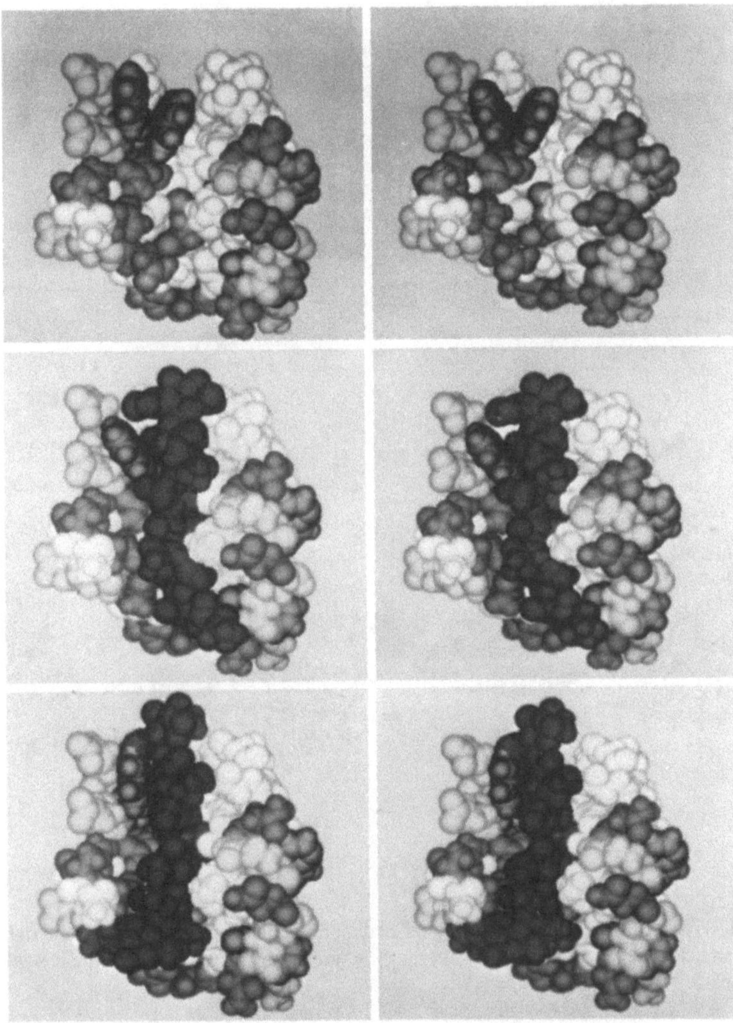

Fig. 14. Stereo views of space-filling models of: *(top)* the active site of native hen egg white lysozyme; *(middle)* energy-minimized model-built hexamer $[(GlcNAc)_6]$ bound to the active site *(right-sided mode);* and *(bottom)* lowest-energy hexamer bound to the active site *(left-sided mode)* [51]

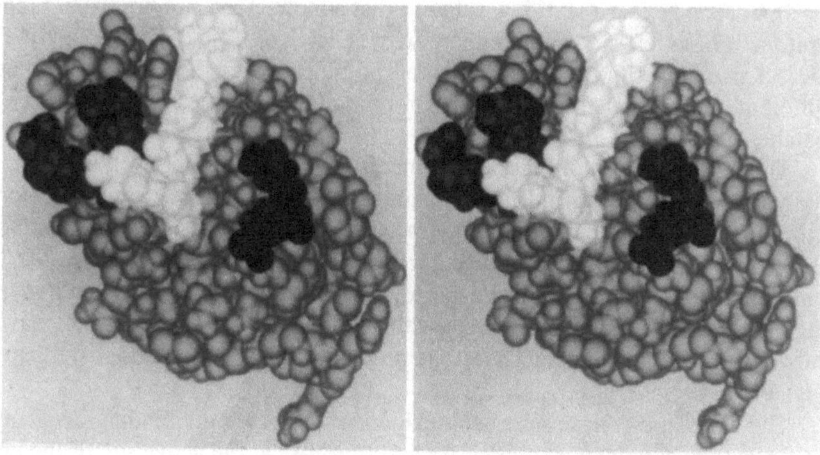

Fig. 15. Stereo view of the computed structure of the low-energy complex between $(GlcNAc)_6$ and the hen egg white lysozyme (same complex as in the bottom of Fig. 14). The dark shaded regions contain the residues involved in left- and right-sided binding, respectively [52]

higher energy [51]. The two lowest-energy left- and right-sided $(GlcNAc)_6$ binding conformations are shown with space-filling models in Fig. 14.

In order to provide an experimental test of these calculations, i.e. to decide whether $(GlcNAc)_6$ binds preferentially to the left or right side, the following experiments were carried out [52]. These take cognizance of the fact that residues on the left and right side differ in different species of lysozymes, as indicated in Table 1 and illustrated in Fig. 15.

In ringed neck pheasant lysozyme (RNPL), Lys and His are substituted for Asn and Arg at positions 113 and 114, respectively. Therefore, the affinity of $(GlcNAc)_6$ for RNPL should differ from that for hen egg white lysozyme (HEWL) if binding occurs on the right side, but should not differ if binding occurs on the left side. From experiments on the kinetics of hydrolysis of this substrate by both enzymes, it appears that the values of the Michaelis-Menten parameter K_M are similar for both enzymes [52]. This suggests that the left-sided binding mode is preferred.

Table 1. Some residues involved in binding sites in lysozymes

Left Side	Right Side
Arg 45	Phe 34
Asn 46	Asn 113 (Lys)[b]
Thr 47	Arg 114 (His)[b]
Asp 48[a]	
Arg 68[a]	

[a] Additional residues included in epitope
[b] Substitutions in RNPL. All other residues are those in HEWL

In a second experiment, it was found that a monoclonal antibody that binds to an epitope on the left side can compete with $(GlcNAc)_6$ for the active site of HEWL [52]. This competition is not due to steric blockage of the left side because two different monoclonal antibodies can bind simultaneously to HEWL, at the two dark patches indicated in Fig. 15.

The results of both of these experiments suggest that $(GlcNAc)_6$ binds preferentially to the left side, as indicated by the computations. These results demonstrate the utility of conformational energy calculations in providing an understanding of the interactions involved in molecular recognition. They also indicate the importance of a *combined* theoretical/experimental approach to gain such an understanding. For example, the experiments with HEWL and RNPL would not have been carried out without the need to test the predictions of the calculations summarized in Fig. 14.

The calculations have also been extended to include the binding conformations of alternating copolymers of GlcNAc and MurNAc at the active site of the enzyme. The calculated lowest-energy structure for MurNAc · GlcNAc · MurNAc bound to sites B, C, D is in excellent agreement with the X-ray crystal structure of this complex [53]. If either the calculated or the X-ray structure is extended into the lower active site with GlcNAc residues it is found that it binds with its F site residue on the left side of the cleft. The alternating hexamer copolymer $(GlcNAc \cdot MurNAc)_3$, however, binds in a preferred right-sided mode (similar to the homopolymer in Fig. 13A) because of the absence of a favorable hydrogen bond in site F between the 3-OH of the substrate (the –OH of GlcNAc is now replaced with a lactic acid side chain in MurNAc) and the backbone >C=O of Arg 45; also, the lactic acid side chain makes unfavorable steric contacts with Arg 45. In addition, from an investigation of the allowed binding sites for MurNAc residues, it appears that this residue can be accommodated in sites B, D and F only, in good agreement with experiment [53].

Approximately one-third of the binding energy for hexasaccharide complexes is attributable to the interactions of the N-acetyl groups of the substrate with the enzyme (mainly with the three active site tryptophans 62, 63 and 108 in sites B, C, D, and through hydrogen bonding with Asn 59 and Ala 107 in site C and with Glu 35 in site E), in agreement with experimental results. Polymers of saccharides without the N-acetyl group (e.g. glucose) are thus predicted to bind with low affinities, again in agreement with experiment [49]. The calculated hydrogen bonding scheme (hydrogen bonding accounts for about one-fourth of the total energy of stabilization) is identical in site C to that inferred from X-ray crystallography.

VII Conclusions

It may be seen from the results described for α-chymotrypsin and lysozyme that conformational energy calculations can predict the three-dimensional structures of enzyme-substrate complexes. Information is obtained about the various pairwise interactions between enzyme and substrate that contribute to the ground-state binding energy. In fact, the binding energy results from a sum over *many* small pairwise interaction energies rather than a few large specific contributions; thus, the assumption of discrete sites (Fig. 8) is an oversimplification. In addition, the calculations can predict the relative affinities of different substrates for a particular enzyme.

All of these calculations are based on the use of ECEPP potentials [1, 4], which have been obtained from crystal and gas-phase data. These potentials are ideally suited for computation of the structures of enzyme-substrate complexes because the interactions between enzymes and substrates are the same as those between the molecules of a crystal. They may therefore be employed (as we have done in the case of chymotrypsin and lysozyme) to identify the crucial interactions that lead to recognition. Once these interactions are known they may be used to construct, from theoretical considerations alone, substrates and inhibitors that can bind with the highest affinities to the active site of the enzyme.

A related important question, discussed elsewhere [54], concerns the nature of the catalytic step, subsequent to the binding step. This question has thus far been approached by quantum mechanical calculations on model systems that simulate the binary enzyme-substrate complexes discussed here.

Figure 16 is a schematic representaion of the kinds of problems that can be treated by the methodology described in Section II. Direct application to X-ray data leads to low-energy structures [39, 55, 56]. Such refined structures provide starting points for computing the conformations of homologous proteins [34, 56], which in turn can be used (in a docking process) to treat the binding of enzymes to substrates (the process characterized by K_M), the binding of drugs and hormones to receptors, and the binding of proteins to nucleic acids. With sufficient experimental information about structure-activity relationships, this technique can be applied to map the active site of an unknown receptor. The same energy minimization algorithms can be used (together with constraints) in protein folding studies. Two potential extensions of this methodology are to intermolecular interactions in multi-subunit proteins such as hemoglobin, and to the structures and energetics of activation complexes (the process characterized by k_{cat}). In the latter case, the computations must allow for bond stretching and bond angle bending, in addition to the variation of dihedral angles that is allowed for in all of the other computations.

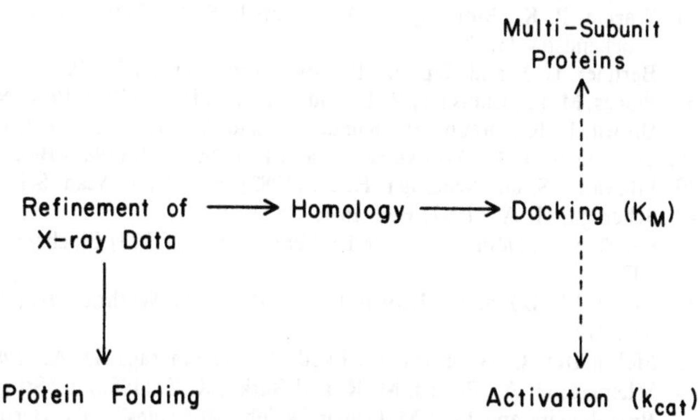

Fig. 16. Various problems treated by energy minimization. The solid arrows pertain to computations already carried out, while the dashed arrows pertain to potential new applications

References

1. Momany, F. A., McGuire, R. F., Burgess, A. W. and Scheraga, H. A. (1975) J. Phys. Chem. *79,* 2361
2. Dunfield, L. G., Burgess, A. W. and Scheraga, H. A. (1978) J. Phys. Chem. *82,* 2609
3. Pottle, C., Pottle, M. S., Tuttle, R. W., Kinch, R. J. and Scheraga, H. A. (1980) J. Computational Chem. *1,* 46
4. Némethy, G., Pottle, M. S. and Scheraga, H. A. (1983) J. Phys. Chem. *87,* 1883
5. Pincus, M. R., Burgess, A. W. and Scheraga, H. A. (1976) Biopolymers *15,* 2485
6. Gibson, K. D. and Scheraga, H. A. (1969) Physiol. Chem. and Phys. *1,* 109
7. Gibson, K. D. and Scheraga, H. A. (1969) Proc. Natl. Acad. Sci. U.S. *63,* 242
8. Gō, N. and Scheraga, H. A. (1969) J. Chem. Phys. *51,* 4751
9. Gō, N. and Scheraga, H. A. (1976) Macromolecules *9,* 535
10. Gō, N., Lewis, P. N. and Scheraga, H. A. (1970) Macromolecules *3,* 628
11. Hodes, Z. I., Némethy, G. and Scheraga, H. A. (1979) Biopolymers *18,* 1565
12. Meirovitch, H. and Scheraga, H. A. (1980) Macromolecules *13,* 1406
13. Meirovitch, H. and Scheraga, H. A. (1981) Macromolecules *14,* 340
14. Scheraga, H. A. (1981) in "Structural Aspects of Recognition and Assembly in Biological Macromolecules", ed. Balaban, M., Elsevier, p. 87
15. Meirovitch, H. and Scheraga, H. A. (1981) Macromolecules *14,* 1250
16. Meirovitch, H. and Scheraga, H. A. (1981) Proc. Natl. Acad. Sci. U.S. *78,* 6584
17. Wako, H. and Scheraga, H. A. (1982) J. Protein Chem. *1,* 5, 85
18. Ooi, T., Scott, R. A., Vanderkooi, G. and Scheraga, H. A. (1967) J. Chem. Phys. *46,* 4410
19. Yan, J. F., Momany, F. A. and Scheraga, H. A. (1970) J. Am. Chem. Soc. *92,* 1109
20. Chothia, C. (1973) J. Mol. Biol. *75,* 295
21. Chou, K. C., Pottle, M., Némethy, G., Ueda, Y. and Scheraga, H. A. (1982) J. Mol. Biol. *162,* 89
22. Chou, K. C. and Scheraga, H. A. (1982) Proc. Natl. Acad. Sci., U.S. *79,* 7047
23. Scheraga, H. A., Chou, K. C. and Némethy, G. (1983) in "Conformation in Biology", ed. Srinivasan, R. and Sarma, R. H., Adenine Press, p. 1
24. Chou, K. C., Némethy, G. and Scheraga, H. A. (1983) J. Mol. Biol. *168,* 389
25. Chou, K. C., Némethy, G. and Scheraga, H. A. (1983) Biochemistry *22,* 6213
26. Chou, K. C., Némethy, G. and Scheraga, H. A. (1983) J. Phys. Chem. *87,* 2869
27. Dygert, M., Gō, N. and Scheraga, H. A. (1975) Macromolecules *8,* 750
28. Némethy, G. and Scheraga, H. A. (1984) Biochem. Biophys. Res. Commun. *118,* 643
29. Rackovsky, S. and Scheraga, H. A. (1980) Proc. Natl. Acad. Sci. U.S. *77,* 6965
30. Scheraga, H. A. (1985) Annals. N.Y. Acad. Sci. *439,* 170
31. Hull, S. E., Karlsson, R., Main, P., Woolfson, M. M. and Dodson, E. J. (1978) Nature *275,* 206
32. Miller, M. H. and Scheraga, H. A. (1976) J. Polymer Sci., Polymer Symp. *54,* 171
33. Okuyama, K., Tanaka, N., Ashida, T. and Kakudo, M. (1976) Bull. Chem. Soc. Japan *49,* 1805
34. Warme, P. K., Momany, F. A., Rumball, S. V., Tuttle, R. W. and Scheraga, H. A. (1974) Biochemistry *13,* 768
35. Berliner, L. J. and Kaptein, R. (1981) Biochemistry *20,* 799
36. Pincus, M. R., Klausner, R. D. and Scheraga, H. A. (1982) Proc. Natl. Acad. Sci. U.S. *79,* 5107
37. Brown, L. R., Braun, W., Kumar, A. and Wuthrich, K. (1982) Biophys. J. *37,* 319
38. Terwilliger, T. C., Weissman, L. and Eisenberg, D. (1982) Biophys. J. *37,* 353
39. Fitzwater, S. and Scheraga, H. A. (1982) Proc. Natl. Acad. Sci. U.S. *79,* 2133
40. Scheraga, H. A. (1967) Fed. Proc. *26,* 1380
41. Leach, S. J., Némethy, G. and Scheraga, H. A. (1977) Biochem. Biophys. Res. Commun. *75,* 207
42. Braun, W., Bösch, C., Brown, L. R., Gō, N. and Wuthrich, K. (1981) Biochim. Biophys. Acta *667,* 377
43. McWherter, C. A., Haas, E., Leed, A. and Scheraga, H. A. (1986) Biochemistry, in press
44. Scheraga, H. A., Pincus, M. R. and Burke, K. E. (1982) in "Structure of Complexes Between Biopolymers and Low Molecular Weight Molecules", eds. Bartmann, W. and Snatzke, G., John Wiley, Chichester, p. 53
45. Schechter, I. and Berger, A. (1967) Biochem. Biophys. Res. Commun. *27,* 157

46. Platzer, K. E. B., Momany, F. A. and Scheraga, H. A. (1972) Int. J. Peptide and Protein Res. *4,* 201
47. Wlodawer, A., Bott, R. and Sjolin, L. (1982) J. Biol. Chem. *257,* 1325
48. Steitz, T. A., Henderson, R. and Blow, D. M. (1969) J. Mol. Biol. *46,* 337
49. Imoto, T., Johnson, L. N., North, A. C. T., Phillips, D. C. and Rupley, J. A. (1972) in "The Enzymes", Academic Press, New York and London, Vol. VII, p. 665
50. Pincus, M. R., Zimmermann, S. S. and Scheraga, H. A. (1977) Proc. Natl. Acad. Sci. U.S. *74,* 2629
51. Pincus, M. R. and Scheraga, H. A. (1979) Macromolecules *12,* 633
52. Smith-Gill, S. J., Rupley, J. A., Pincus, M. R., Carty, R. P. and Scheraga, H. A. (1984) Biochemistry *23,* 993
53. Pincus, M. R. and Scheraga, H. A. (1981) Biochemistry *20,* 3960
54. Pincus, M. R. and Scheraga, H. A. (1981) Accts. Chem. Res. *14,* 299
55. Warme, P. K. and Scheraga, H. A. (1974) Biochemistry *13,* 757
56. Swenson, M. K., Burgess, A. W. and Scheraga, H. A. (1978) in "Frontiers in Physicochemical Biology", ed. Pullman, B., Academic Press, p. 115

Discussion

List of interventions

1 Intervention of Breslow

How does your calculation square with the literature reports that there is a positive energy destabilizing interaction of the ring D side chains with lysozyme?

1.1 Scheraga

Your question implies that there is distortion of the D ring, burying it deep in the active-site cleft. However, Nathan Sharon and coworkers have demonstrated that the D ring is not buried deep in the cleft – and our computed structures (*without* distortion of the D ring) are in agreement with those of Sharon et al. Distortion of the D ring would raise the energy and make such a binding mode less favorable.

2 Intervention of Luisi

What is the relation between the energy calculated by you (or Dr. Karplus) and the experimental energies of the chemists, like free energy, or enthalpy? This question also seems relevant in view of the fact that you compare your calculated conformations with those existing in solution or in the crystal, where the folding is determined by free energy minimization.

2.1 Scheraga

Only relative, not absolute, energies have a meaning in our calculations aimed at determining the structure of a protein molecule. Therefore, the values of the energy cannot be compared with the experimental energies or free energies. However, calculated and experimental *structures* of individual molecules *can* be compared. The reason that the calculations do not give *absolute* energies is that some parts of the intramolecular energies are *justifiably* omitted from the computations. For example, the interaction between C_1 and C_3 in C_1–C_2–C_3 is independent of rotation about the C_1–C_2 bond. Therefore, we don't compute it. On the other hand, intermolecular (in contrast to intramolecular energies) can be compared. For such comparison to be meaningful, however, all components of the energy must be included. In our calculations of enzyme-substrate complexes, we did not include the hydration free energy, and therefore we cannot compare absolute values of free energies. However, the relative values do have meaning, and the calculated relative binding energies of aromatic peptides to chymotrypsin agree with relative values of K_M's. Also, as I indicated in my talk, calculated structures of enzyme-substrate complexes agree with experimental results.

3 Intervention of Lehn

The mellitin tetramer forms a voltage-dependent chloride channel in membranes. Your calculations might allow you to study the gating mechanism which controls chloride passage.

3.1 Scheraga

This is certainly a very interesting problem, and it is feasible to examine such a tetramer by our computational methods. However, simply because one does not have enough time to work on all interesting problems, we have decided to concentrate on a related problem – and that is the structure of bacterial rhodopsin. Since this protein presumably consists of seven packed α-helices, we hope to be able to apply our methodology for treating assemblies of α-helices to the rhodopsin-structure problem.

4 Intervention of Huber

a) Is the docking problem solved as one might guess from the lysozyme-hexa NAG docking calculations?
b) Prediction of homologous proteins from the structure of parent compounds would be extremely valuable. I would like to have a testable example.

4.1 Scheraga

I don't know what you mean by "solved". I have presented a method, and results, for treatment of the intermolecular (in addition to the intramolecular) interaction energies between two molecules. Each partner in the pair can influence the conformation of the other. The resulting computed most favorable (left-side) binding mode of the lysozyme-hexa NAG complex has received some experimental confirmations, as I indicated in my talk. In this sense, the docking problem is solved. However, every docking calculation itself involves a non-trivial calculation, which is more complex the larger the partners. We have docked a small-molecule substrate into an enzyme. If you want to extend this to dock BPTI on trypsin, the calculations would be more involved, but the important point is that our methodology, or some variant thereof, can be applied even to this problem. When the crystallographers succeed in solving the structure of α-lactalbumin, we will have a test of our computed structure of this protein. We are currently trying to compute the structure of thrombin from those of several serine proteases. During the course of the computations, we have made elementary comparisons of the structures of trypsin, chymotrypsin and elastase, and could treat one of these as an unknown structure in order to test the methodology.

5 Intervention of Wodak

The problem of simulating the "docking" between two proteins, is by no means trivial, as you claim. Proteins do not behave like rigid bodies at all and their interaction involves, at the very least, rearrangement of amino acid side chains at the interaction interface. It remains true however, that if one assumes that conformational changes are limited to such rearrangements (that may be justified in some cases, such as in Trypsin-BPTI for example) the main contribution to the association free energy between two proteins may be correctly evaluated using a set of simplifying assumptions. The "docking" of a protein and a flexible ligand is a more difficult problem to treat in computer simulations than the one described above, since the essential energetic contributions to the intermolecular interaction cannot be evaluated without a computationally prohibitive survey of the conformational space of the flexible ligand as well as of side chains on the protein surface. You claim to have overcome these problems in the docking of lysozyme with oligosaccharides, but you give us no details on your computational approach. The secret of your success lies in biasing the computational procedure towards the correct answer. This may have little to do with solving the docking problem but may help to make predictions.

5.1 Scheraga

I gave no details of the computational approach because my talk was not concerned with describing computational techniques but with discussing some results and some problems that can be solved by computational methodology. However, the computational approach is explained in great detail in paper that we have published with Matthew Pincus (see M. R. Pincus and H. A. Scheraga, Accts. Chem. Res., *14,* 299 (1981) for references). In these computations, we definitely did not regard the partners in the complex as rigid bodies but, instead, explored the conformational space of the flexible ligand and of the flexible side chains on the enzyme. The important point is that the intermolecular interactions modify the low-energy conformations of the separate partners. In other words, the conformations of both the ligand and the enzyme are allowed to change, and do change, in response to intermolecular interactions, when the separated partners are brought together to form the complex. There was no bias in seeking the low-energy conformation of the complex. It was the intermolecular interaction energies that determined the low-energy structure.

6 Intervention of Schulten

A new algorithm which enables one to search for the global minima of Gibbs functions with many independent variables, the method of Monte Carlo annealing, has been introduced recently by Kirkpatrick et al. This Monte Carlo algorithm involves an effective temperature which determines the acceptance of random searching events. The new aspect of this method is a statistical mechanical analysis of the searching events which allows an optimal control of the effective temperature and provides information on the degree of degeneracy of the most stable states. The method has been successfully applied to systems involving a very large number of degrees of freedom, and therefore deserves consideration for the problem of protein folding. [S. Kirkpatrick, C. D. Gelatt, Jr. and M. P. Vecchi, "Optimization by Simulated Annealing", IBM Thomas J. Watson Research Center, Yorktown Heights, New York, 1982.]

6.1 Scheraga

The algorithm to which you referred (S. Kirkpatrick, C. D. Gelatt, Jr. and M. P. Vecchi, Science, *220,* 671 (1983)) involves a Metropolis Monte Carlo procedure. The primary problem with a Metropolis type search of the conformational space of a protein is that the random search constitutes a Markov process, i.e. the search has no memory. Hence, the algorithm does not retain past information that can be used for an improvement of the subsequent search. Recently, D. Vanderbilt and S. G. Louis, J. Computational Physics, *56,* 259 (1984), have introduced an algorithm that automatically adapts the step size of the Metropolis search to the *local* topography of the conformational space. This appears to be an improvement over the standard Metropolis algorithm. However, knowledge of only the *local* topography may be inadequate. We have developed an algorithm [G. H. Paine and H. A. Scheraga, Biopolymers *24,* 1391 (1985)] that searches the conformational space *globally*, and then adjusts its Monte Carlo search to emphasize the regions of conformational space of high probability. One difficulty that we see with the "annealing" process of Kirkpatrick et al. is that the Monte Carlo search may slowly become frozen in a local minimum. Furthermore, there needs to be a criterion to know when one has reached the global minimum. Vanderbilt and Louis have addressed this problem briefly. Their criterion, however, does not assure that the algorithm has reached the global minimum. We do agree that an annealing process deserves consideration, and incorporation of the method into our adaptive Monte Carlo algorithm may be possible.

7 Intervention of Williams

Which lysozyme structure did you start from? Which lysozyme structure did you compare in bound systems? It is necessary to use the highest resolution structures since the side chains "move" during refinement.

A second point of interest is that the study of the cytochrome c surface using probes reveals multiple sites and possible movements on the surface. These sites and the surface diffusion make the discussion of docking likely to be difficult.

7.1 Scheraga

We used the X-ray coordinates of hen egg-white lysozyme reported by David Phillips for the free enzyme. However, in computations on the enzyme-substrate complexes, the side chains were allowed to move in response to interaction energies between the enzyme and substrate.

Your cytochrome c results are reminiscent of our computations on chymotrypsin-peptide complexes. We actually found four, not one, binding sites of low energy in the neighborhood of the active-site cleft in our docking calculations. One of these four is the actual active site, and the other three are nearby. We have speculated that all four sites can serve to "lure" the substrate, which would then move to the nearby actual active site by surface diffusion.

Communications

Why Water?

A. R. Ubbelohde

University of London, Imperial College of Science and Technology, Department of Chemical Engineering and Chemical Technology, London SW7 2AZ, United Kingdom

Why Water?

At this Solvay 1983 Conference we are being invited to consider some very intricate peculiarities of the structure of distinctive proteins which are found to be functionally effective in specific molecular operations. Rapid advances in molecular biology challenge our understanding of how distinctive structuring of atoms in a protein can so narrowly guide and control certain chemical reactions of biological importance. Modern instrumental observations of molecules down to nanometer precision should become of enormous importance in elucidating such precise catalytic effects. Our constant aim must be to identify distinctive functions of key structures in molecules of biological importance.

My own contribution as a chemical physicist to this aim may appear somewhat of an intrusion in a discussion dedicated to biological problems. My main point is that most of the specific molecular functions of proteins would become badly smudged, or would even be obliterated, in the absence of water which is normally present and which acts as supporting fluid for the various more distinctive assemblies of atoms in biological catalysts[1]. I can only make quite brief remarks on my theme "Why Water?". Each of these remarks summarizes an invitation to a research programme in biological chemical physics.

Many experts in molecular biology do, of course, acknowledge the complexity of the role of water, but claim that they have themselves quite enough to answer for concerning their own fascinating studies of protein structures. They tend to label the background role of water as the supporting fluid merely as "very complicated". This will not do. Probing a little deeper into the role of fluids could give the molecular biologist some valuable research tools. Note that *quasi-gaseous* models of fluids (as first constructed by Van der Waals to account for the structure of any liquid near its critical temperature) are fairly useless for describing fine details of molecular packing which can be so important in relation to the movement of neighboring molecules or parts of a macromolecule in a protein. A much more informative modern

[1] An interesting general consideration is whether mere removal of the supporting fluid e.g. by "cold-drying" in ultra-high vacuum, might simply desiccate any biological molecule *reversibly*, or whether such desiccation would necessarily denature certain specialized functions more or less permanently, in the same way as would raising the temperature

approach is to probe the quasi-crystalline or anti-crystalline packing of molecules in the Molten State of Matter, i.e. quite near to the freezing point (T_f) of the liquid when pure. This modern approach has the enormous advantage, in relation to the problems under discussion at this conference, that the crystal structure into which the pure liquid passes on freezing is often known quite accurately from various kinds of diffraction measurements. Starting from this known crystal structure, the sequence of molecular operations which transform it and lead to the molten state for any particular substance can be described with some confidence. Generally, the changes of enthalpy (H_f) and entropy (S_f) in these operations can be considered additively [1]. Enthalpy changes can often be calculated or estimated from our knowledge of intermolecular forces. Different classes of melt arise according to whether these forces are only short range (< 4 nm) or whether they also include long range and cooperative interactions between the molecules. According to their dominant character, modern classifications group all known fluids into four or five principal types [1]. However, melts with short range intermolecular forces only, which comprise about 90% of all known fluids, include all those of biological importance.

We may consider a sequence of disordering operations on the parent crystal to bring it into the molten state. We introduce point defects statistically into an ideal lattice by successive disordering

i) of positional location of the molecules with respect to their parent crystal lattice;
ii) of orientational aspect with respect to neighboring molecules in the condensed state of matter;
iii) of configurational isomerization (when the molecules are flexible).

The outcome, though more or less strongly disordered can still be referred to the parent crystal lattice, i.e. the liquid is *quasi crystalline.* However, if the molecules can form short range hydrogen bonds, based on atoms of low atomic number such as N, O, F, S the quasi crystalline fluid generally rearranges its molecules so as to decrease the enthalpy by

iv) forming *anti-crystalline clusters* of varying size with the molecules collapsed together so as to satisfy the hydrogen bonding more completely.

Figure 1 illustrates how H_2O molecules arrange their hydrogen bonds so as to pack into anticrystalline clusters. Single H_2O molecules, or anticrystalline clusters of only a few molecules, can force their way into crevices of appropriate size in a folded protein molecule, of which many examples have been shown to us at this Solvay Conference. By doing so, *hybrid hydrogen bond structures* are formed at the cost of moderate local changes of entropy and enthalpy and moderate changes of intermolecular spacings. Consequent chemical reactions controlled by the protein catalyst can be sensitively affected, and can be given quite unusual enthalpy and entropy factors.

This emphasis on the role of water – both as single molecules and in anticrystalline clusters – as a *spacer* bears a suggestive relation to sensitive ways of modifying the chemical reactions controlled by any specific organic catalyst.

i) The balance between single molecules and anti-crystalline clusters is known to be sensitively controlled by temperature. This may explain why biologically controlled reactions tend to be generally smothered at temperatures higher than

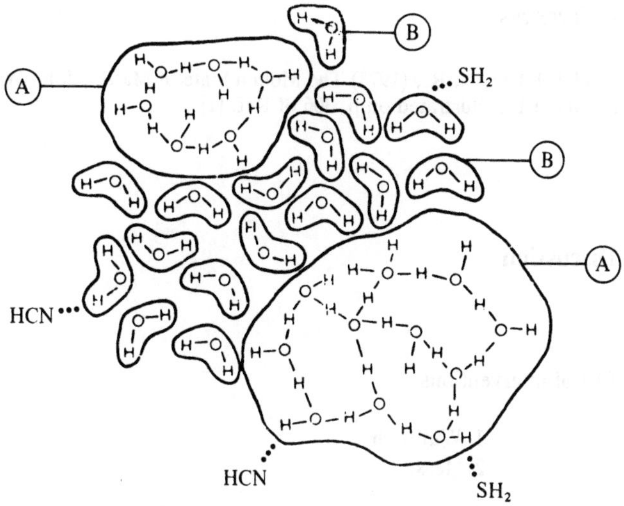

Fig. 1. *Anti-crystalline* clusters of H_2O molecules Ⓐ together with isolated H_2O molecules Ⓑ. In the absence of protein the distribution of Ⓐ and Ⓑ is sensitively responsive to the temperature [°C] and the applied pressure [kilobar].

When a flexible protein is introduced, its tangled network with amino acids in specific sequence may fold itself round and encapsulate regions of water. If any of the polar groups in the protein can get near enough to Ⓐ or Ⓑ of the water molecules, hybrid hydrogen bonds... may form, with a loosening of the structures such as Ⓐ but a tightening of packing near Ⓑ. In this way water molecules can support the enzyme network sensitively.

Additive polar molecules if sufficiently small, such as HCN, H_2O_2, H_2S, NH_3 may replace Ⓑ H_2O molecules in very dilute concentrations. Consequent modification of hybrid hydrogen bonds and electrostatic forces can affect enzyme catalysis sensitively. Only H_2S and HCN are illustrated

ambient. Close examination of such reactions which still operate at, say, 100 °C (if any) might throw rewarding light on hybrid hydrogen bond structures still effective at such temperatures.

ii) Similar remarks refer to effects of applying several thousand atmospheres to the same biological systems.

iii) Additives (say less than 10^{-3}) mole fraction, which have sensitive effects on anti-crystalline properties of pure water at ambient temperatures, might also show sensitive interference with protein catalyzed reactions. Obvious examples are H_2O_2, H_2S, NH_3, $(NH_2)_2$, HCN, CH_3NO, CH_3OH, $(CH_2OH)_2$ or glycerol. Inorganic trace control additives to examine are Li OH, CO_2, N_2O.

Finally, it may be noted that water contained in fine capillaries with walls of organic materials is well known to have its thermal properties profoundly modified, usually by steep lowering of the freezing point. Water within cell walls likewise tends to have its association into clusters strongly modified. Various physical examples have been studied (see [1]). The possibilities now indicated refer to the course of chemical reactions as controlled by proteins, themselves supported by water either as single H_2O molecules, or as anti-crystalline clusters.

References

1. Ubbelohde, A. R., (1978) The Molten State of Matter, John Wiley & Sons
2. Figure 1 is illustrated on p. 306 of Ref. [1]

Discussion

List of interventions

1 Wüthrich
2 Hess

1 Intervention of Wüthrich

Professor Ubbelohde's appeal that more consideration should be given to the water which surrounds biopolymer chains might be generalized to say that the influence of the physiological environment should be more extensively investigated. I refer to the brief description of glucagon conformations presented yesterday: in three different environments, this molecule adopts three markedly different conformations. This behavior of glucagon is indicative of what can be expected quite generally for non-globular polypeptides and also for the surface of globular proteins, which is the part of the molecular structure that is most directly involved in intermolecular interactions. What we really need are experimental data on molecular conformations in a wider variety of environments than can be investigated in single crystals. Such experiments can also be expected to provide improved calibration of the parameters for theoretical studies of biopolymers in such environments.

2 Intervention of Hess

The interaction of water with proteins is indeed of decisive importance for our understanding of enzyme catalysis and protein conformation, in the case of soluble as well as membrane-bound proteins. We distinguish the influence of the bulk water from locally bound water taking part in acid-base catalysis of active centres of proteins. In a series of studies Rafi Korenstein and I analyzed the reactivity of bacteriorhodopsin toward the water structure and with respect to its state of hydration. This membrane-bound retinal protein of salt bacteria *(H. halobium)* undergoes a photochemical reaction cycle upon illumination, and vectorially transfers protons through the bacterial membrane in a coupled reaction. Green light excitation leads to bleaching of the purple pigment, reflecting an overall series of five thermal intermediates which can be recorded by suitable spectroscopy [for reference, see Summary in Bacteriorhodopsin, B. Hess, D. Kuschmitz and M. Engelhard, In: Membranes and Transport 2, Ed.: A. N. Martonosi, Plenum Publishing Corporation, New York, pp. 309–318 (1982)]. The purple membrane can be isolated in open membrane sheets and analyzed by following the rate of formation and decay of a relatively stable intermediate (M-412), which is coupled to the transfer of protons through the membrane.

In the course of our studies we found that the kinetics of formation and decay of M-412 are independent of the bulk water structure, be it ice or supercooled water in H_2O or D_2O. As

Table 1. Relaxation times of the decay and formation kinetics of the M-412 intermediate of bacteriorhodopsin excited by laser flash at 585 nm in a pulse of 1 μs length. The M-412 intermediate was analyzed spectroscopically at 420 nm

The decay and formation kinetics of M_{412} transient in water								
	H_2O				D_2O			
	ice $\tau_{1/2}$ (ms)	water $\tau_{1/2}$ (ms)	$\tau_{1/2}^{water}/\tau_{1/2}^{ice}$	t°C	ice $\tau_{1/2}$ (ms)	water $\tau_{1/2}$ (ms)	$\tau_{1/2}^{water}/\tau_{1/2}^{ice}$	t°C
formation of M_{412} transient	0.48 ± 0.03	0.52 ± 0.03	1.08 ± 0.11	-2°	1.67 ± 0.09	1.83 ± 0.06	1.10 ± 0.04	+2°
	0.48 ± 0.07	0.57 ± 0.04	1.21 ± 0.11	-3°	1.74 ± 0.12	1.99 ± 0.14	1.15 ± 0.02	+1°
decay of M_{412} transient	71.3 ± 9.6	86.3 ± 12.8	1.21 ± 0.05	-2°	80.2 ± 8.0	95.1 ± 11.2	1.19 ± 0.06	+2°
	68.4 ± 3.5	103.6 ± 15.9	1.51 ± 0.16	-3°	76.7 ± 6.8	112.7 ± 16.4	1.49 ± 0.26	+1°

can be seen from the table, the differences in the half-times of the decay and formation kinetics are negligible. From this analysis it follows that the phase structure of bulk water is not of essential importance for the photocycle kinetics in this membrane-bound protein. Indeed, the analysis of bacteriorhodopsin in solid ice proved to be useful in studying quite a number of other problems.

On the other hand, the influence of the state of hydration on the photocycle of bacteriorhodopsin is remarkable. In order to study its effect, thin layers of purple membrane were prepared by drying concentrated suspensions of purple membranes in water on a glass slide at room temperature and atmospheric pressure. Variable hydration of the thin layers was obtained by equilibrating the preparations with various relative humidities produced by saturated salt solutions. The average thickness of the preparation was 1–3 m as determined by scanning electron microscopy. The glass slide containing the preparation was inserted into a cuvette suitable for spectroscopy and equilibrated with the required specific humidity before analysis.

Flash photolysis spectroscopy showed that bacteriorhodopsin undergoes a complete photocycle in the absence of any bulk solvent when equilibrated with relative humidities of the air. Under these conditions the same photochemical intermediates were observed as those recorded if purple membrane fragments are suspended in water or salt solutions. Preparations equilibrated with 94% relative humidity showed the same kinetics and included the same transients as in purple membrane fragments suspended in water. The relaxation time for the M-412 decay is slowed down by lowering the hydration state of the preparation. At 90% relative air humidity the first relaxation time of the decay process was in the order of 3.5 μs; at relative humidity of 0% (= 10^{-3} torr) the half-time was 2.3 msec. The results show that the amount of absorbed water on the purple membrane determined the relaxation time of the M-412 decay. Changes of hydration cause a change of relaxation times over four orders of magnitude. The precursor intermediate of the M-412 component (the μs L-component) is only sensitive to a minor degree and the transition of the nanosecond intermediate (K-component) did not react to the changes in hydration to any extent.

From our experiments it can be concluded that the state of hydration sets a conformational state in the protein which controls the decay kinetics of the relatively stable M-412 intermediate in the protein, whereas the earlier shortlived intermediates (the K- and L-components) are not affected to any critical degree. Thus, the early events are obviously occurring within a microenvironment of the protein which is isolated from interaction with the surface of the membrane protein structure.

This interpretation is further supported by the observation of a decoupling of the cis-trans-isomerization process in bacteriorhodopsin occurring in the dark as a function of hydration as well as by the fact that the photochemical bleaching of the M-412 component is observed at the lowest humidity state.

The results indicate that the photochemical reaction in bacteriorhodopsin is independent of the state of the bulk solution and is influenced only by the microenvironment of the reaction system which operates in a state of quasi-isolation (ref. Rafi Korenstein and B. Hess, FEBS Lett. *82*, 7–11 (1977); Rafi Korenstein and B. Hess, Nature *270*, 5633, 184–186 (1977)).

Experimental Studies of Protein-Dynamics Through Ligand Binding

H. Frauenfelder

Department of Physics, University of Illinois at Urbana-Champaign, 1110 Westgreen Street, Urbana, Illinois 61801 (USA)

For many years we have been studying the binding of dioxygen and carbon monoxide to various heme proteins such as myoglobin and separated hemoglobin chains. The main emphasis of this work has been on finding and understanding general concepts and on constructing a simple model that explains all observed facts. Here I will, without going into any details, describe some of the main results.

1 Experimental Aspects

We study ligand binding and protein dynamics using four different techniques, flash photolysis [1–5], Fourier-transform infrared spectroscopy [6, 7], X-ray diffraction [8, 9], and SQUID susceptometry [10]. With all techniques, we attempt to observe the heme proteins and their reactions over as wide a range of parameters as possible. In flash photolysis, for example, we follow the binding process from 25 ps to about 1 ks (over 13 orders of magnitude in time), at temperatures from 2 to 330 K, at pressures up to 2 kbar, and in various solvents. We demand from our model that it describe all data, not just a conveniently chosen subset.

2 Ligand Binding – Experimental Results

The time and temperature dependence of ligand binding is studied through flash photolysis. The sample, for instance MbCO, is placed in a cryostat. A laser pulse breaks the bond between the heme iron and the ligand, the ligand dissociates and later rebinds. Photodissociation and rebinding are monitored with a light beam at a wavelength where MbCO and Mb absorb differently (e.g. 436 nm).

The binding of CO to Mb exhibits characteristic features that are found, with changed details, in the binding of CO and O_2 to all heme proteins. The general characteristics can be summarized as follows:

i) Below about 180 K, a single binding process is observed. We denote this internal (geminate) process by I. As shown for one example in Fig. 1, process I is nonexponential in time and is approximately given by

$$N(t) \approx (1 + t/t_0)^{-n} . \qquad (1)$$

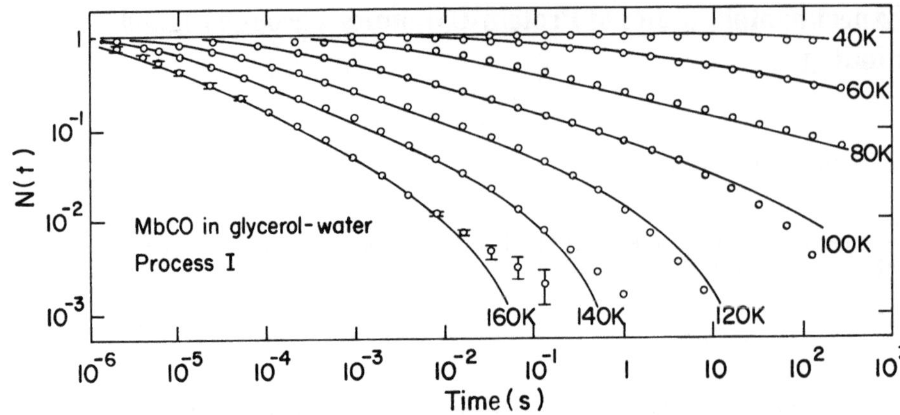

Fig. 1. Binding of CO to Mb at low temperatures. (After ref. [1].) N(t) denotes the fraction of Mb molecules that have not rebound a CO molecule at the time t after photodissociation

Here N(t) is the fraction of heme protein molecules that have not rebound a ligand at time t after photodissociation, and n and t_0 are temperature-dependent parameters [1]. Process I is independent of ligand concentration in the solvent.

ii) Above about 180 K, binding becomes biphasic. The second rebinding process, denoted by M, is also independent of ligand concentration in the solvent and is approximately proportional to $t^{-1/2}$.

iii) Above about 220 K, a third rebinding process can be seen. This process is approximately exponential in time and its rate is proportional to the ligand concentration in the solvent. We denote this process by S, for "solvent process".

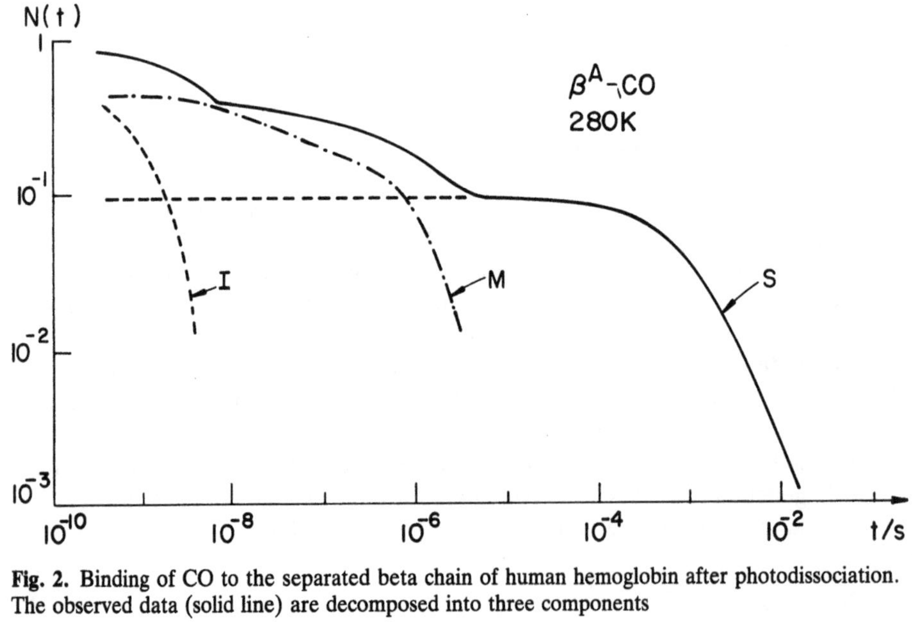

Fig. 2. Binding of CO to the separated beta chain of human hemoglobin after photodissociation. The observed data (solid line) are decomposed into three components

In Fig. 2 N(t) is shown for the binding of CO to the beta chain (β^A) of normal human hemoglobin. The separation of N(t) into the three components I, M, and S is indicated. With increasing temperature the solvent process becomes more dominant. In MbCO at 300 K, the amplitude of I has become unobservable and M can only be detected with very careful measurements [11].

3 The Sequential Model

All data of the type shown in Fig. 1 and 2 can be explained with a simple model in which it is assumed that a ligand, coming from the solvent, enters the protein matrix, diffuses or percolates through the matrix, enters the prominent cavity at the heme (heme pocket) and then forms the covalent bond at the heme iron [1]. In this model the three processes, I, M, and S observed after photodissociation can be described as follows. At low temperatures, the photodissociated ligand cannot leave the heme pocket and binds directly from there. The resulting internal process I is first-order and does not depend on ligand concentration. At higher temperatures some of the photodissociated ligands move from the pocket into the protein matrix and return from there, giving rise to the matrix process M. At still higher temperatures some ligands diffuse through the pocket and the hydration shell into the solvent; all ligands in the solvent then compete for the vacant binding site, resulting in the solvent process S.

4 Concepts and Phenomena

The detailed studies and quantitative fitting of the binding data yield some general concepts that probably are valid not just for the simple binding process, but may apply to most or all biomolecules.

i) Distributed Barriers. Why is binding at low temperatures nonexponential in time as shown in Fig. 1 and expressed through Eq. 1? The simplest and only explanation that we have been able to find involves a distribution of barrier heights. The final binding step, the formation of the covalent bond between the heme iron and the ligand in the heme pocket, is activated. If the barrier between the unbound and the bound state does not have a unique height, but if the activation enthalpy H is different in different proteins, N(t) can be written as

$$N(t) = \int dH\, g(H) \exp(-k(H, T)t)\ . \tag{2}$$

Here g(H)dH is the probability of finding a protein with barrier height between H and H + dH, and k(H, T) is the rate coefficient corresponding to a barrier of height H. With Eq. 2 the observed data can be fitted over a wide temperature range, and the activation energy spectra, g(H), are characteristic for a particular heme protein [1, 2, 4].

ii) Conformational Substates. The explanation of the fractal time dependence (Eq. 1 through Eq. 2) raises a new question: Why do not all proteins have the same barrier height? We believe that the answer is connected in a profound way with the construction and function of biomolecules [1, 12] and that a protein cannot be in a unique state of lowest energy, but must have a highly degenerate ground state. To justify this assertion, consider a protein in a given conformation, an arrangement of all atoms that permits the protein to execute a particular function. Some of the hydrogen bonds that stabilize the protein can be shifted without changing the overall binding energy appreciably. Many of the side groups can be rotated without affecting the overall energy. The protein can thus assume a large number of conformational substates, all with essentially the same energy and all performing the same biological function. The existence of such substates can explain the observed distribution in activation enthalpy. Assume that each substate possesses a particular barrier height. At low temperatures, transitions among substates will be very slow or absent. Each protein molecule is frozen into a particular substate with corresponding barrier height, and the observed binding rate for process I will be given by Eq. 2. At high temperatures, however, transitions among the substates can be very fast and the formation of the covalent bond at the heme iron occurs with an average rate

$$\langle k \rangle = \int dH\, g(H)\, k(H) \,. \tag{3}$$

The existence of conformational substates can also be seen in many other experiments, particularly through the determination of the Debye-Waller factor in X-ray diffraction [8, 9, 13, 14] and through the Mössbauer-Lamb factor in the Mössbauer effect [15–17].

iii) Quantum-Mechanical Tunneling. The rate coefficient k(H, T) in Eq. 2 describes bond formation within the heme pocket. Above about 40 K, k(H, T) can be adequately described by an Arrhenius relation,

$$k(H, T) = A \exp(-H/RT) \,. \tag{4}$$

Below 40 K, however, binding is much faster than predicted by Eq. 4 [2, 18]. Such a deviation is expected; at very low temperatures the system forms the covalent bond by tunneling through the enthalpy barrier rather than by overcoming it classically. At first, tunneling of a molecule as heavy as CO is surprising. The barrier between pocket and heme iron is so small, however, that tunneling must be observable. Indeed, comparison of the tunneling rates for ^{12}CO and ^{13}CO in Mb reveals the isotope effect characteristic for the tunnel effect [6].

iv) Validity of the Sequential Model. In the detailed discussion of the sequential model sketched in Sect. 3, we assume that the innermost barrier that dominates ligand binding at low temperature is still present and active at room temperature and that the room temperature rate coefficient for the innermost binding step can be calculated from Eqs. 1, 2. It is difficult to test this assumption in the binding of CO or O_2 to Mb because the innermost process I can only be observed to about 160 K. Above this temperature the matrix and the solvent process mask the direct rebinding. Fortunately, tests are possible with the separated beta chains of normal Hb (β^A) and the mutant Hb Zürich (β^{ZH}). The peak enthalpies of the innermost barrier for binding of CO are considerably smaller in β^A and β^{ZH} than in Mb (Mb: 10 kJ/mol,

β^A: 3.2 kJ/mol, β^{ZH}: 1.4 kJ/mol). Process I consequently is much faster in β^{ZH} and β^A than in Mb and competes better with the matrix process. Flash photolysis experiments with β^{ZH} [5] and with β^A (to be published) prove that process I is still present at 300 K. The extrapolation with Eq. 2 is valid for β^A; for β^{ZH}, the observed fast binding component I at 300 K is about a factor 3 slower than the extrapolated rate. These observations imply that process I, the formation of the Fe-ligand bond in the heme pocket, is similar at 300 K and at low temperatures. The deviation in β^{ZH} is very likely due to its structure; in contrast to Mb and β^A, the pocket in β^{ZH} is open to the solvent and the reorganization after photodissociation is large [19, 20]. In Mb, where process I cannot be seen above about 160 K, the presence of the innermost barrier can be ascertained by a different approach: the pH and the pressure dependence of the association coefficient at 300 K both agree with the corresponding pH and pressure dependence of process I, extrapolated from low to high temperatures. All observations thus can be fitted within the sequential model.

v) Control. The sequential model implies that the binding (association) rate is controlled at the heme [4]. The protein influences the bond formation at the heme iron indirectly, most likely through the structure on the proximal side and through the distal histidine.

vi) The Matrix Process. In the first version of the sequential barrier model [1] we assumed that the motion through the protein matrix, from the solvent to the pocket, is governed by two barriers. Data on many heme proteins show that the matrix process always displays a time dependence that is close to $t^{-1/2}$. It is therefore not clear what the best description is. We can treat the motion through the matrix as diffusion, as a random walk with a small number of steps, or as a random walk with a preferred pathway. The existing data do not yet permit a decision among these possibilities, and more and better data will be required for a solution of the problem. Moreover, the role of the hydration shell remains to be elucidated.

vii) Dynamics. So far, no connection to protein dynamics has been made. It is clear, however, that protein motion plays an essential role in the entire binding process and that studies of the binding yield information about protein dynamics. The point was made long ago by Perutz [21] that a rigid protein would not permit ligand association and dissociation: the average X-ray structure shows no hole or channel through which the ligand can move. The protein must relax in order to permit ligand binding. Calculations by Case and Karplus lead to the same conclusion [22]. The role of protein motion is seen most clearly in binding studies if the solvent viscosity is varied [3]: the observed binding kinetics depends strongly on solvent viscosity. With increasing solvent viscosity it becomes more difficult for a photodissociated ligand to move from the heme pocket to the solvent. While the importance of protein dynamics thus is evident even for such a simple process as ligand binding, the general fȩatures remain to be elucidated.

viii) Reaction Theory. In our earlier work [1] we naively used Eyring's form of transition state theory (TST) to evaluate our data. TST assumes equilibrium and unit transmission coefficient and the friction in the medium is not taken into account. The results sketched in paragraph (vii) show, however, that ligand binding is strongly affected by solvent viscosity, and Gavish and Werber [23] have demonstrated that enzyme activity depends on viscosity. Fortunately, there exists a better theory. In 1940, Kramers treated chemical reactions in the presence of friction [24] and his results have been verified and extended by an army of theoretical physicists and chemists (e.g. [25, 26]). There is no question that Kramers's approach is superior to TST for reactions in condensed phases. The application of Kramers's theory to ligand binding (and presumably to other biomolecular reactions also) leads to smaller values of the activation enthalpy and more reasonable values of the activation entropy than TST [3]. Moreover, these considerations demonstrate that biomolecules can adjust their reaction rates through changes in viscosity – a possibility that may be particularly important in membranes.

References

1. Austin, R. H., Beeson, K. W., Eisenstein, L., Frauenfelder, H. and Gunsalus, I. C. (1975) Biochemistry *14,* 5355–5373
2. Alberding, N., Chan, S. S., Eisenstein, L., Frauenfelder, H., Good, D., Gunsalus, I. C., Nordlund, T. M., Perutz, M. F., Reynolds, A. H. and Sorensen, L. B. (1978) Biochemistry *17,* 43–51
3. Beece, D., Eisenstein, L., Frauenfelder, H., Good, D., Marden, M. C., Reinisch, L., Reynolds, A. H., Sorensen, L. B. and Yue, K. T. (1980) Biochemistry *19,* 5147–5157
4. Doster, W., Beece, D., Bowne, S. F., DiIorio, E. E., Eisenstein, L., Frauenfelder, H., Reinisch, L., Shyamsunder, E., Winterhalter, K. H. and Yue, K. T. (1982) Biochemistry *21,* 4831–4839
5. Dlott, D. D., Frauenfelder, H., Langer, P., Roder, H. and DiIorio, E. (1983) Proc. Natl. Acad. Sci. USA *80,* 6239–6243
6. Alben, J. O., Beece, D., Bowne, S. F., Eisenstein, L., Frauenfelder, H., Good, D., Marden, M. C., Moh, P. P., Reinisch, L., Reynolds, A. H. and Yue, K. T. (1980) Phys. Rev. Lett. *44,* 1157–1160
7. Alben, J. O., Beece, D., Bowne, S. F., Doster, W., Eisenstein, L., Frauenfelder, H., Good, D., McDonald, J. D., Marden, M. C., Moh, P. P., Reinisch, L., Reynolds, A. H., Shyamsunder, E. and Yue, K. T. (1982) Proc. Natl. Acad. Sci. USA *79,* 3744–3748
8. Frauenfelder, H., Petsko, G. A. and Tsernoglou, D. (1979) Nature (London) *280,* 558–563
9. Hartmann, H., Parak, F., Steigemann, W., Petsko, G. A., Ponzi, D. R. and Frauenfelder, H. (1982) Proc. Natl. Acad. Sci. USA *79,* 4967–4971
10. Roder, H., Berendzen, J., Bowne, S. F., Frauenfelder, H., Sauke, T., Shyamsunder, E. and Weissman, M. B. (1984) Proc. Natl. Acad. Sci. USA *81,* 2359–2363
11. Henry, E. R., Sommer, J. H., Hofrichter, J. and Eaton, W. A. (1983) J. Mol. Biol. *166,* 443–451
12. Austin, R. H., Beeson, K., Eisenstein, L., Frauenfelder, H., Gunsalus, I. C. and Marshall, V. P. (1974) Phys. Rev. Lett. *32,* 403–405
13. Artymiuk, P. J., Blake, C. C. F., Grace, D. E. P., Oatley, S. J., Phillips, D. C. and Sternberg, M. J. E. (1979) Nature *280,* 563–568
14. Petsko, G. A. and Ringe, D. (1984) Ann. Rev. Biophysics *13,* 331–371
15. Keller, H. and Debrunner, P. G. (1980) Phys. Rev. Lett. *45,* 68
16. Parak, F., Frolov, E. N., Mössbauer, R. L. and Goldanskii, V. I. (1981) J. Mol. Biol. *145,* 824
17. Cohen, S. G., Bauminger, E. R., Nowik, I. and Ofer, S. (1981) Phys. Rev. Lett. *46,* 1244
18. Alberding, N., Austin, R. H., Beeson, K. W., Chan, S. S., Eisenstein, L., Frauenfelder, H. and Nordlund, T. M. (1976) Science *192,* 1002–1004
19. Tucker, P. W., Phillips, S. E. V., Perutz, M. F., Houtchens, R. and Caughey, W. S. (1978) Proc. Natl. Acad. Sci. USA *75,* 1076–1080
20. Phillips, S. E. V., Halle, D. and Perutz, M. F. (1981) J. Mol. Biol. *150,* 137–141
21. Perutz, M. F. and Matthews, F. S. (1966) J. Mol. Biol. *21,* 199–202
22. Case, D. A. and Karplus, M. (1979) J. Mol. Biol. *132,* 343–368

23. Gavish, B. and Werber, M. M. (1979) Biochemistry *18,* 1269–1275
24. Kramers, H. A. (1940) Physica *7,* 284–304
25. Skinner, J. L. and Wolynes, P. G. (1978) J. Chem. Phys. *69,* 2143–2150
26. Hanggi, P. (1983) J. Stat. Phys. *30,* 401–412

Topic 3

Synthetic Models of "Hosts"

Designed Host-Guest Relationships

D. J. Cram

Department of Chemistry and Biochemistry, University of California at Los Angeles, Los Angeles, California 90024, USA

1 Introduction

Complexing partners conveniently divide into two important classes. Hosts contain convergently-arranged binding sites, and are synthetic counterparts of the receptor sites of enzymes, nucleic acids, proteins of the immune system, or ionophores. Guests possess divergently-arranged binding sites, and are the synthetic counterparts of substrates, inhibitors, or drugs. Metal cations also serve as guests. Complexes are composed of hosts and guests held together in solution in a definite structural relationship. Solvation is usually non-structured complexation which competes with structured complexation between hosts and guests. Forces available for complexation are: hydrogen-bonding; ion-pairing; pi-acid to pi-base attractions; metal ion-to-ligand attractions; van der Waals attractions; and the entropic component of desolvation.

Multiple binding sites are needed for the structuring of complexes since the binding energy and orienting power at a single contact site is low compared to the energy of a covalent bond. The design of complexes is aided by two guiding principles. The *principle of complementarity* states that "to complex, hosts must have binding sites which cooperatively contact and attract the binding sites of guests without generating strong nonbonded repulsions." The *principle of preorganization* states that "the smaller the changes in organization of host, guest and solvent required for complexation, the stronger the binding."

Selectivity in complexation depends on mutual structural recognition of a particular host and a particular guest. Guests are usually simple and available, whereas hosts must be designed and prepared. Structural parameters used to control the properties of hosts include: general sizes, shapes and degrees of rigidity; locations and kinds of binding sites, steric or chiral barriers, spacing groups, catalytic groups, solubility- and solvation-controlling groups, and conformation-controlling groups.

We here discuss structural recognition as applied to three kinds of complexation:

1) complexation of alkylammonium and of the alkali metal cations, the latter being important in organic synthesis when ion-paired to carbanions in nonpolar media;
2) chiral recognition in complexed transition states for organometallics undergoing addition and substitution reactions;
3) rate accelerations due to complexation between reactants.

Corey-Pauling-Koltun (CPK) molecular models guided our design of complexes.

Crystal structures of over 30 complexes provided the most refined criteria for conformity between concept and result. The more important tests were whether or not the systems performed the chemical tasks set for them. The focus of this paper is on the latter. Although most of our time is spent synthesizing hosts, no attention is given here to that challenging task.

2 Complexation of Alkylammonium and Alkali Metal Ions

The free energies of complexation ($-\Delta G^0$ values) of a variety of lipophilic hosts have been measured by the picrate salt extraction technique [1, 2]. The distribution of picrate salts between D_2O and $CDCl_3$ is measured in the absence and presence of host in the organic layer. The $-\Delta G^0$ values for associations of hosts and salts in $CDCl_3$ are calculated from the differences in the amounts of salts drawn into the $CDCl_3$ layer at 25 °C. The scale ranges from 5 or 6 to about 16 kcal mol^{-1}, but has been extended by the distribution of guest salt between a host of known and one of unknown but higher binding power. The $-\Delta G^0$ value of the better binding host is calculated from the position of equilibrium [2]. The scale has been extended to ~ 23 kcal mol^{-1}, at which point decomplexation rates became too slow to provide equilibration in the systems studied. With these methods, $-\Delta G^0$ values have been determined for hosts binding the picrate salts of Li^+, Na^+, K^+, Rb^+, Cs^+, NH_4^+, $CH_3NH_3^+$ and t-$BuNH_3^+$.

Charts I, II, III and IV illustrate the kinds of hosts that have been designed and synthesized and the relationships between their structures and abilities to bind lithium, sodium, potassium and cesium picrates, respectively. The differences in structural recognition by a particular guest in complexing a variety of possible hosts are measured by the differences in the $-\Delta G^0$ values listed below each structure. For lithium picrate, the values range from >23 to <6 kcal mol^{-1}; for sodium picrate, from 19.2 to <6 kcal mol^{-1}; for potassium picrate, from 17.1 to <6 kcal mol^{-1}; for cesium picrate, from 13.1 to <6 kcal mol^{-1}. The principles of complementarity and preorganization are illustrated repeatedly in the structure-binding relationships found in these charts [3]. Only a few relationships will be discussed.

Spherand (1) has a rigidly preorganized cavity visible in its crystal structure (2), which is complementary to Li^+ and Na^+ and which it binds with >23 and 19.2 kcal mol^{-1} [4]. This host completely rejects all other ions. Its generally spherical cavity is defined by six octahedrally arranged oxygens whose unshared electrons occupy roughly converging orbitals. The six oxygens are completely shielded from solvation by the six methyl and six aryl groups, so the dielectric properties are between those of a vacuum and those of a hydrocarbon. Compound (3) differs from (1) in composition by only two hydrogen atoms which terminate a chain of six anisyl-type units. Unlike (1), which has only a single conformation, (3) has over 1000 possible conformations. Unlike the ligating sites of (1) those of (3) are subject to solvation. Compound (3) binds Li^+ and Na^+ with <6 kcal mol^{-1} [2]. The crystal structure of (4) shows that it possesses an organization and cavity very similar to that of (1) [5]. However, the electronic character of its six fluorines resemble those of neon, making their unshared electron pairs noncomplementary to the ligating propensities of Li^+ and Na^+. Although *superbly organized* for binding, (4) shows no binding whatsoever because of its lack of electronic complementarity.

Chart I. $-\Delta G^0$ (kcal/mol), $CDCl_3$, 25 °C, for complexing lithium picrate

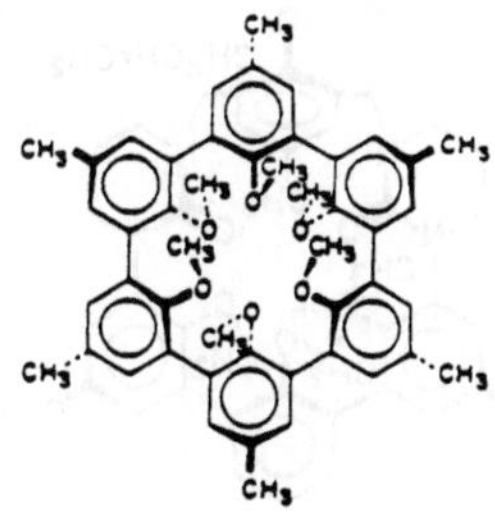

>23

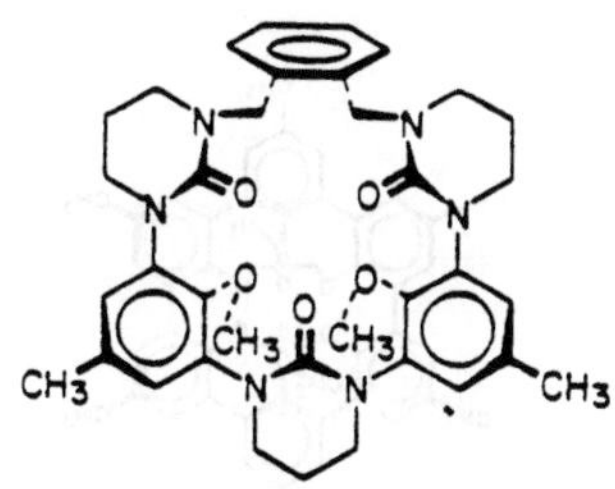

18.3

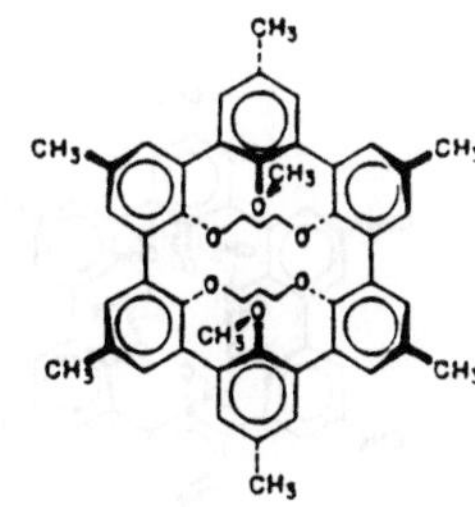

16.8

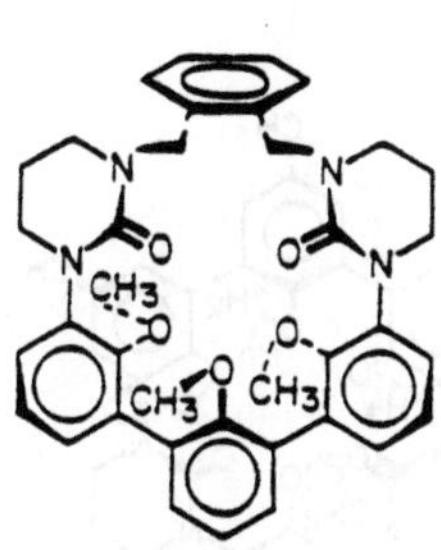

16.6

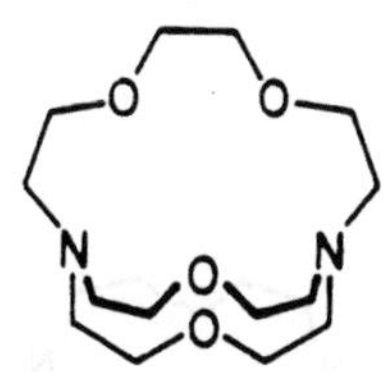

16.6

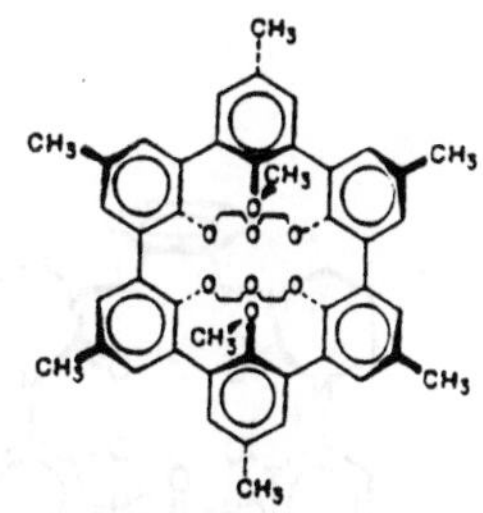

15.9

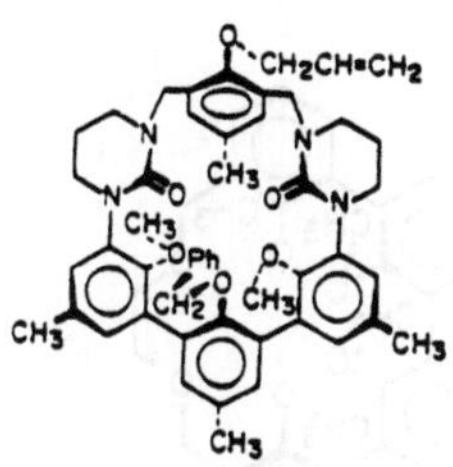

12.6

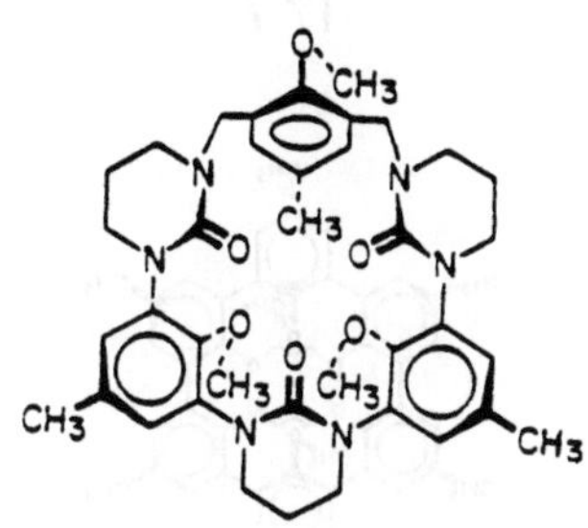

12.1

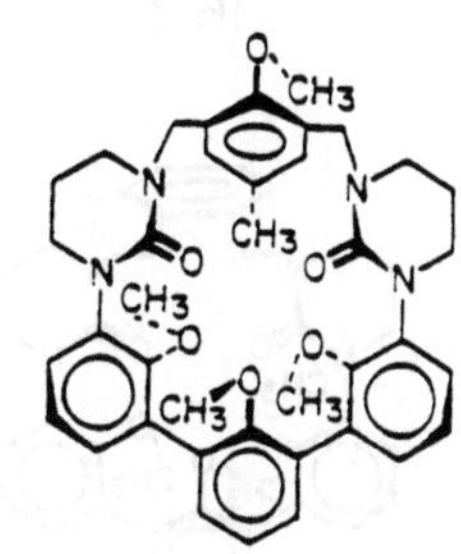

12.0

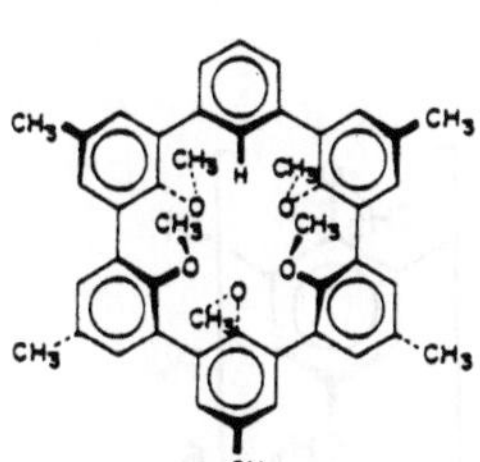

10.4

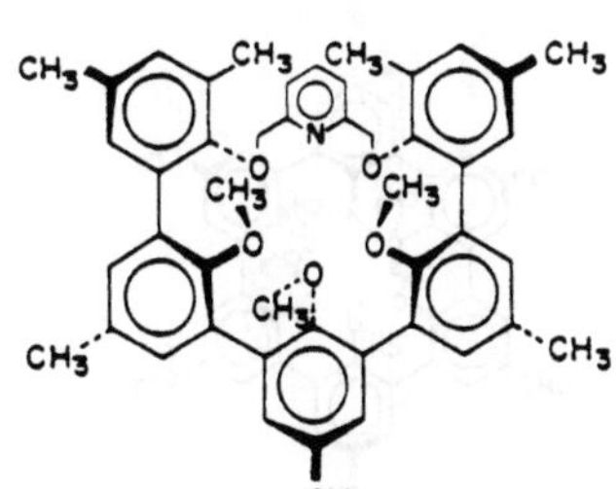

8.6

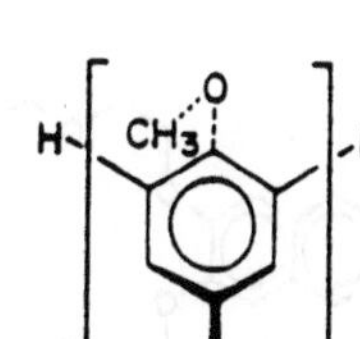

<6

Chart II. $-\Delta G^0$ (kcal/mol), $CDCl_3$, 25 °C, for complexing sodium picrate

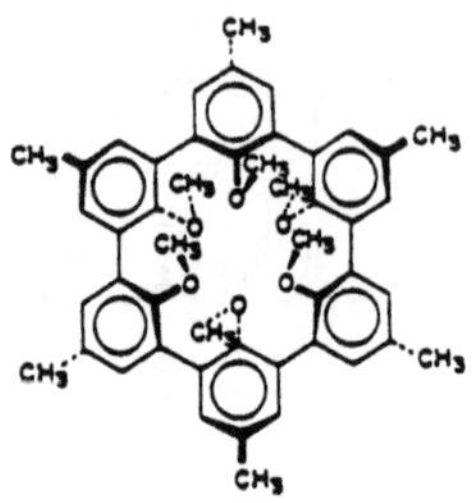

19.2

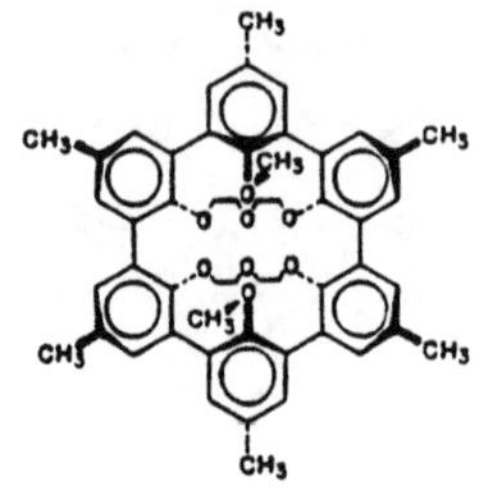

18.7

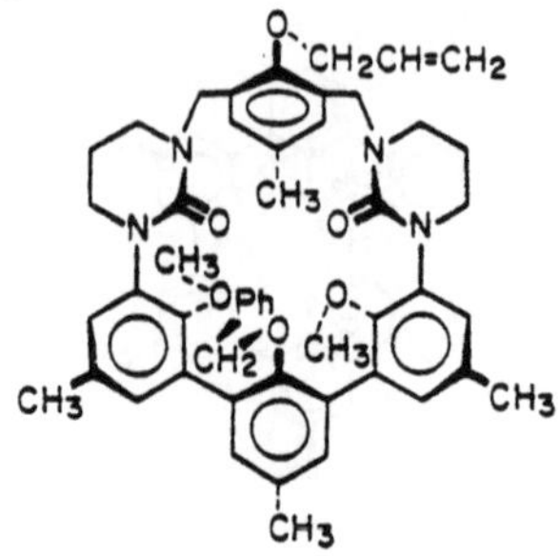

16.5

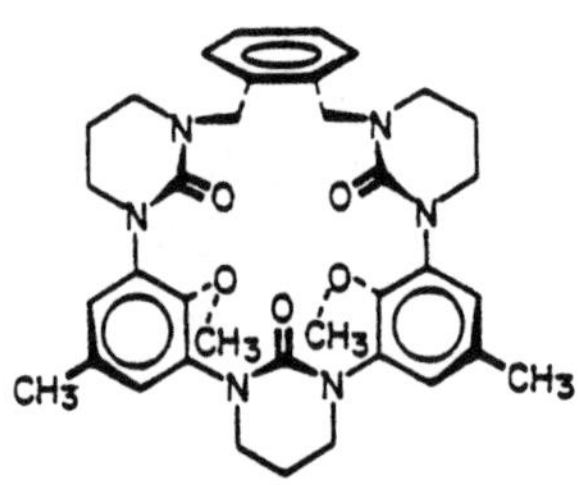

16.3

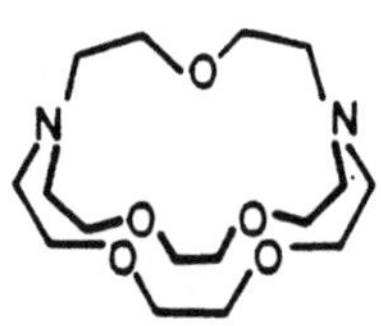

17.7

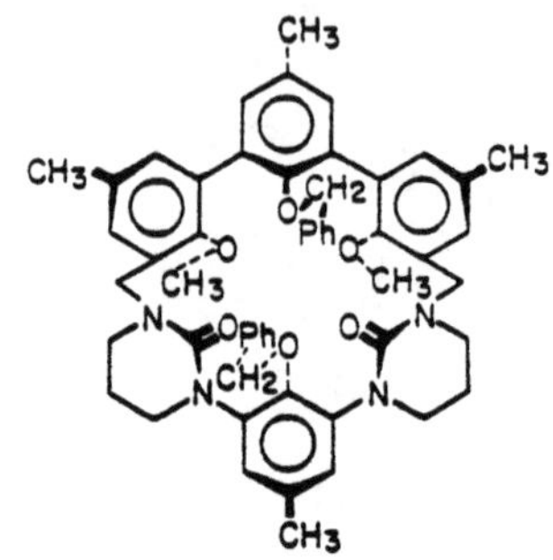

15.9

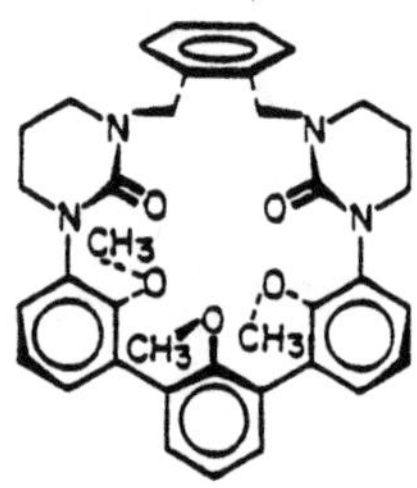

15.4

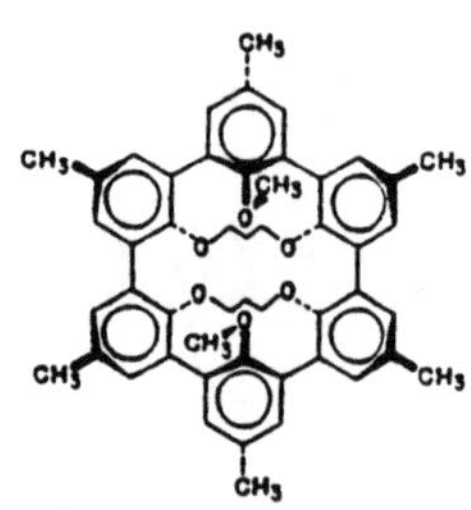

13.6

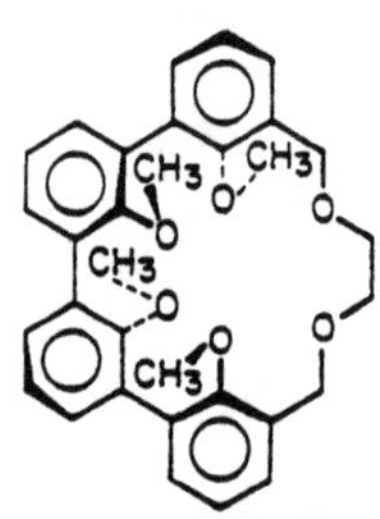

13.4

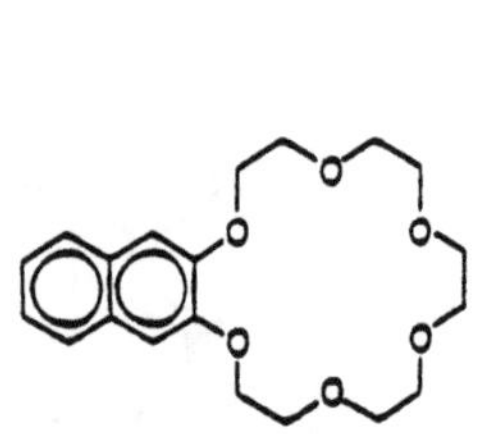

8.3

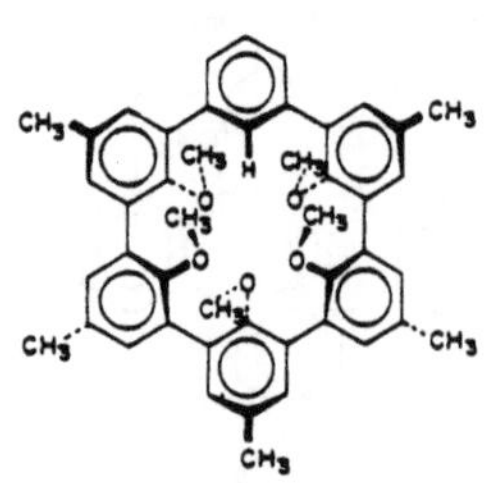

6.6

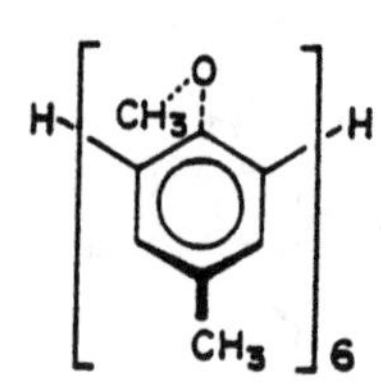

<6

Chart III. $-\Delta G^0$ (kcal/mol), $CDCl_3$, 25 °C, for complexing potassium picrate

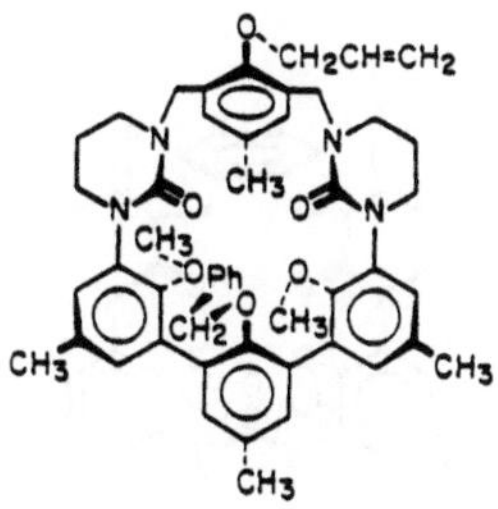

17.1

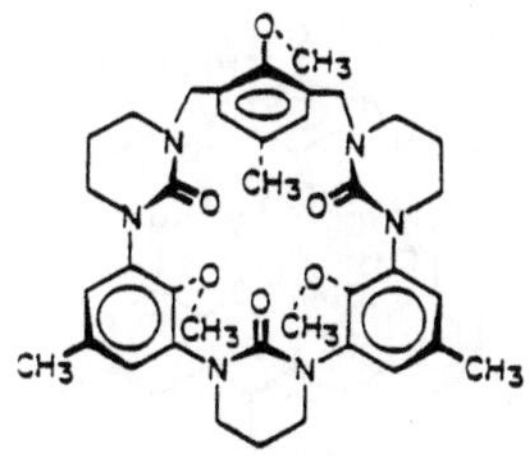

15.6

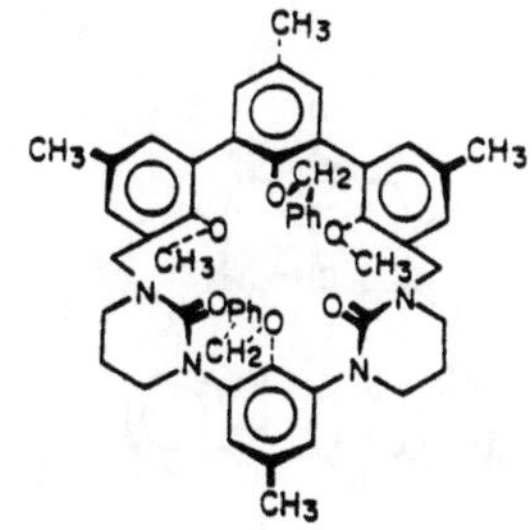

13.1

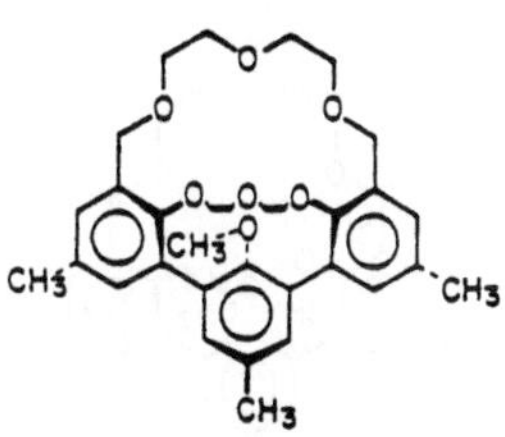

13.7

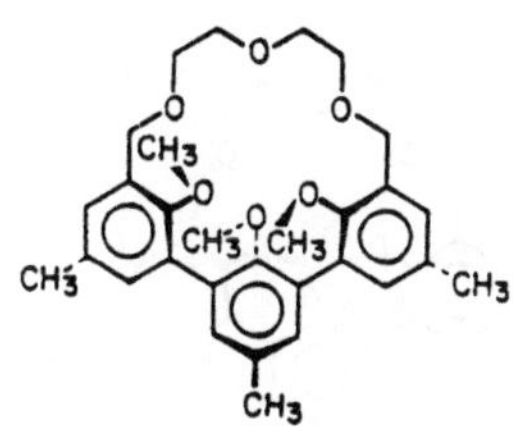

11.8

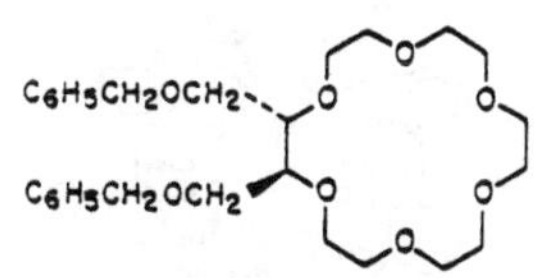

11.4

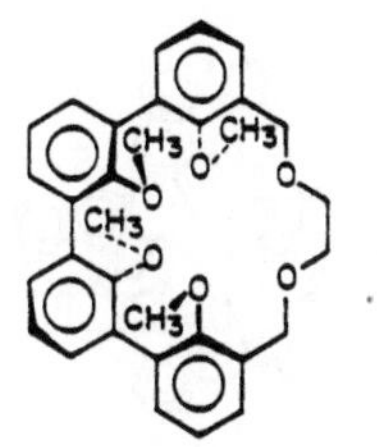

10.9

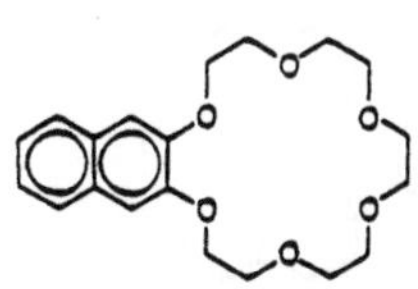

10.8

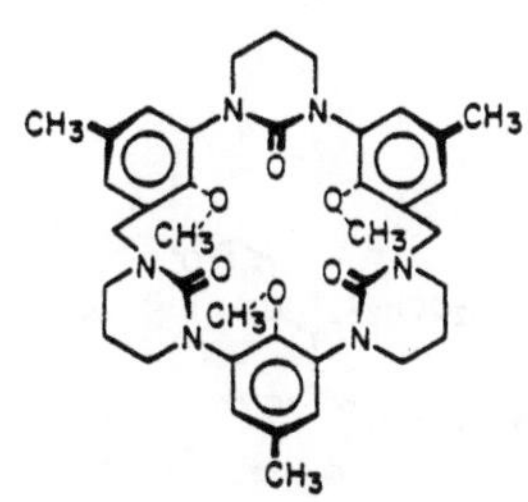

10.4

<6

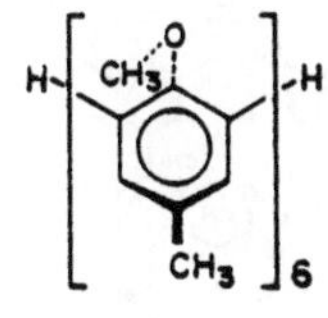

<6

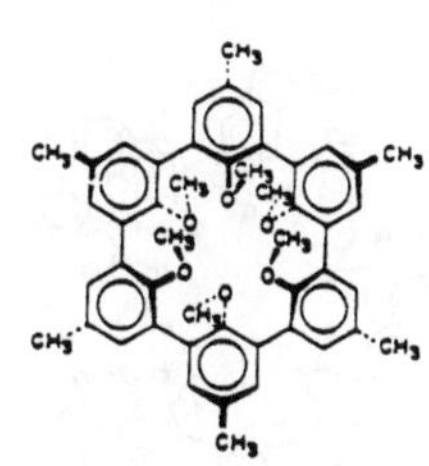

<6

Chart IV. $-\Delta G^0$ (kcal/mol), $CDCl_3$, 25 °C, for complexing cesium picrate

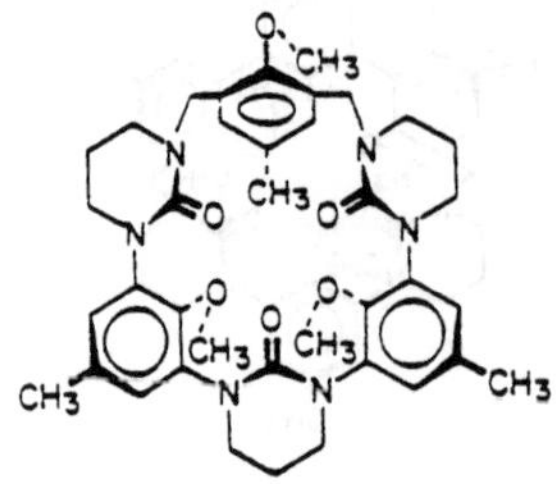

13.1

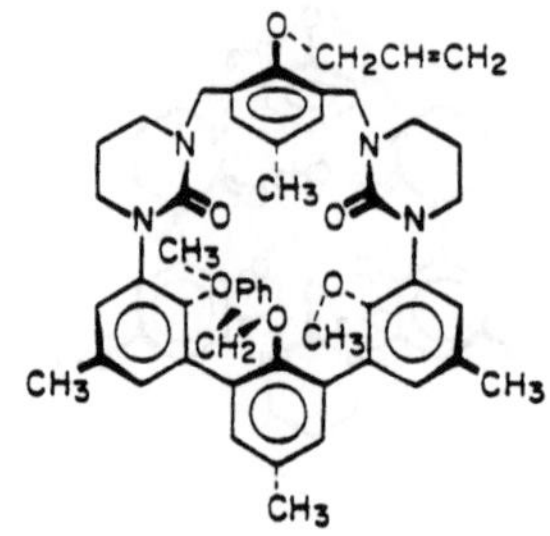

12.4

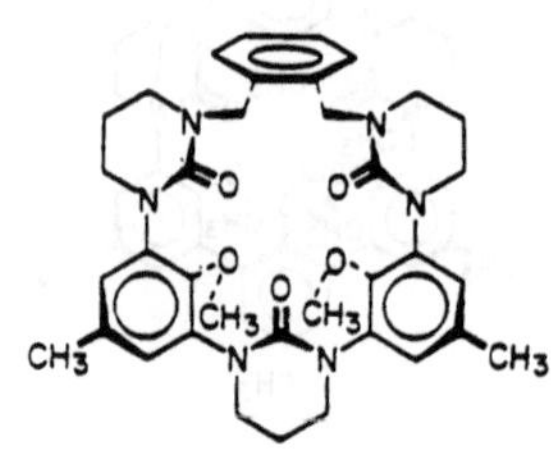

11.8

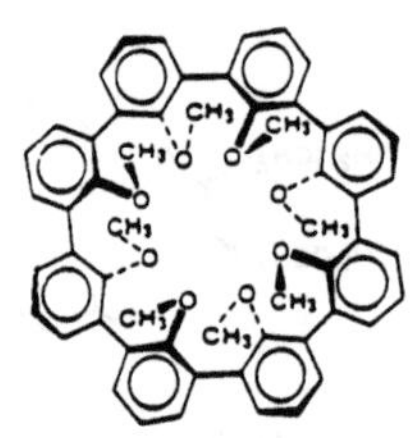

11.6

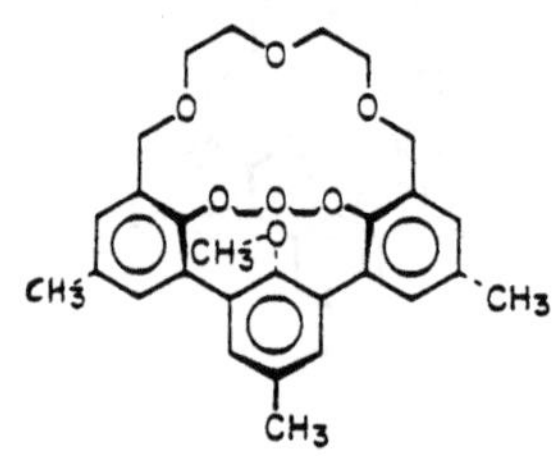

9.8

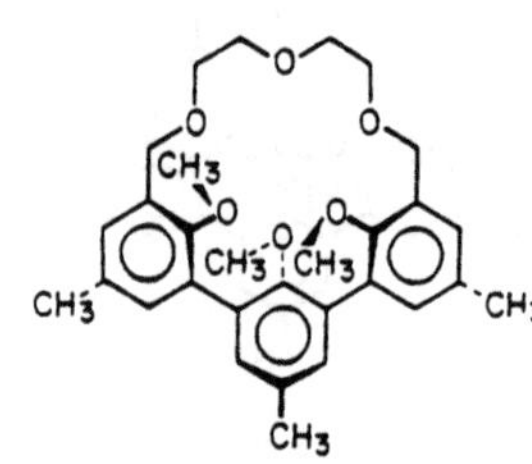

9.0

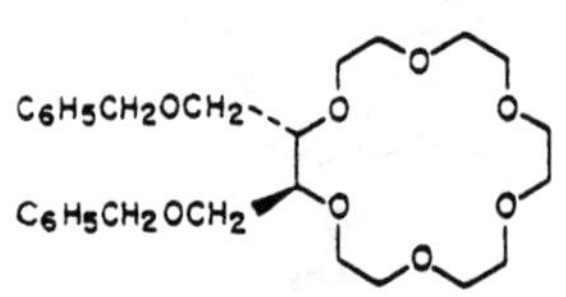

8.5

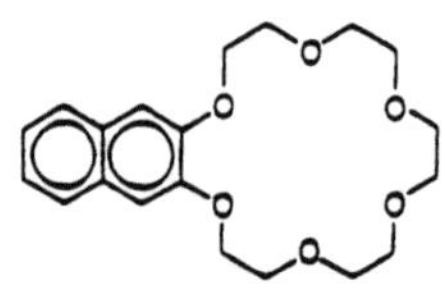

8.3

8.2

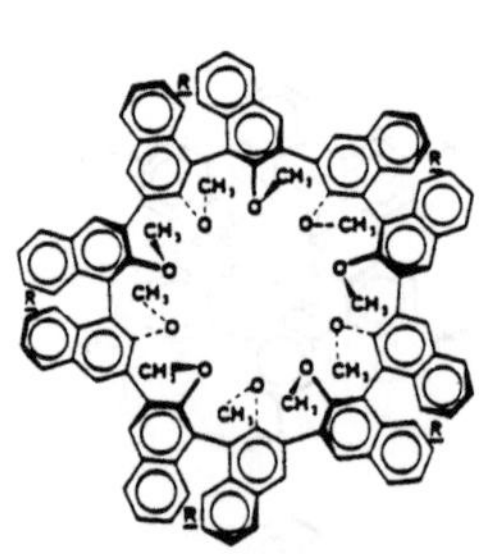

<6

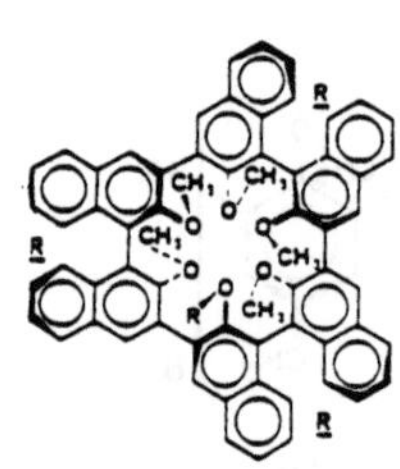

<6

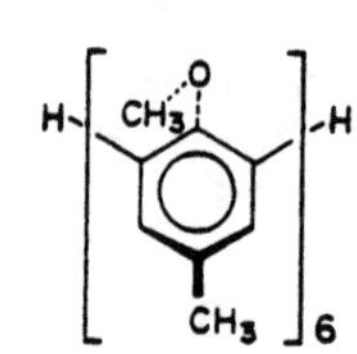

<6

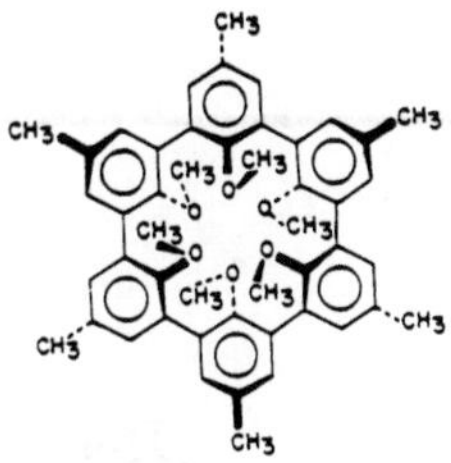

(1)

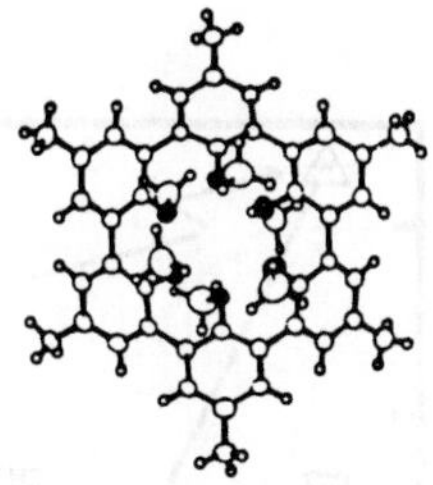

(2) (snowflake)

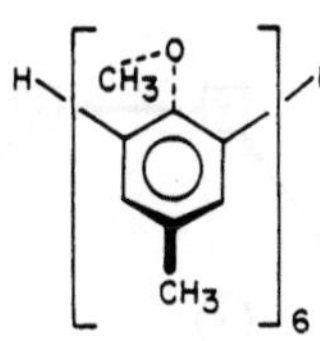

(3)

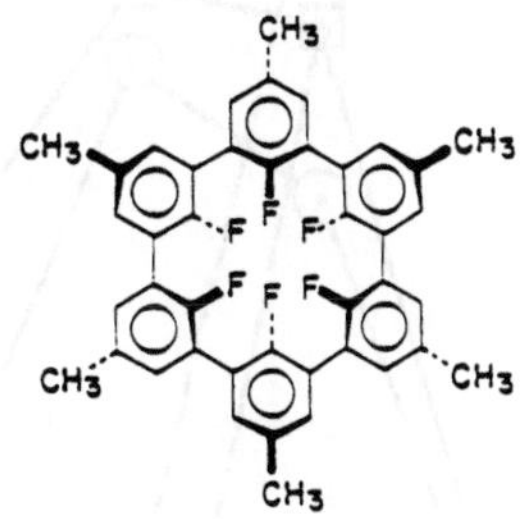

(4)

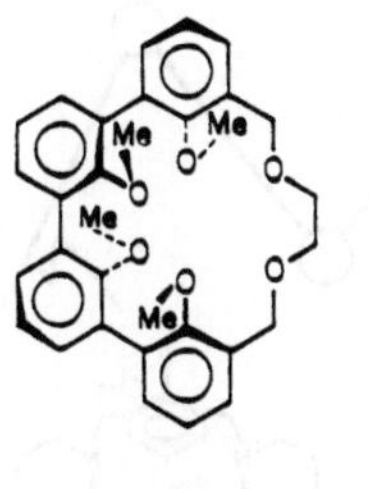

(5)

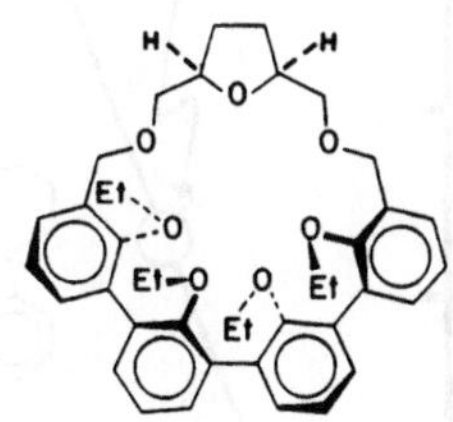

(6)

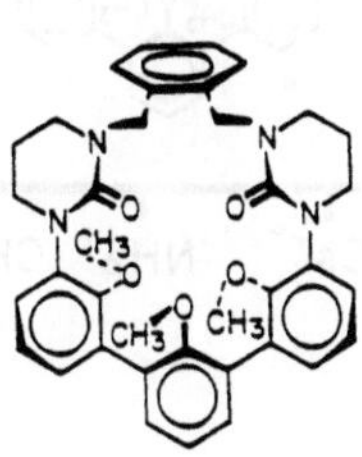

(7)

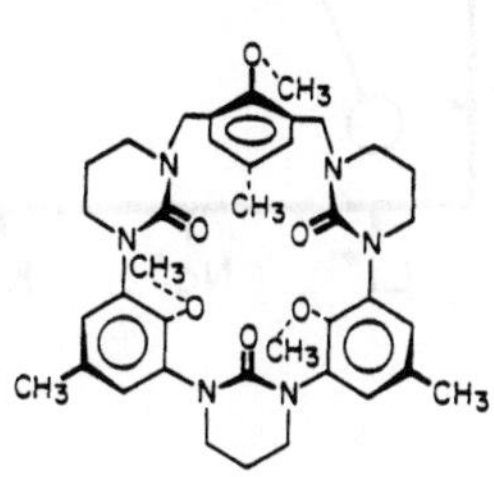

(8)

Chart V.

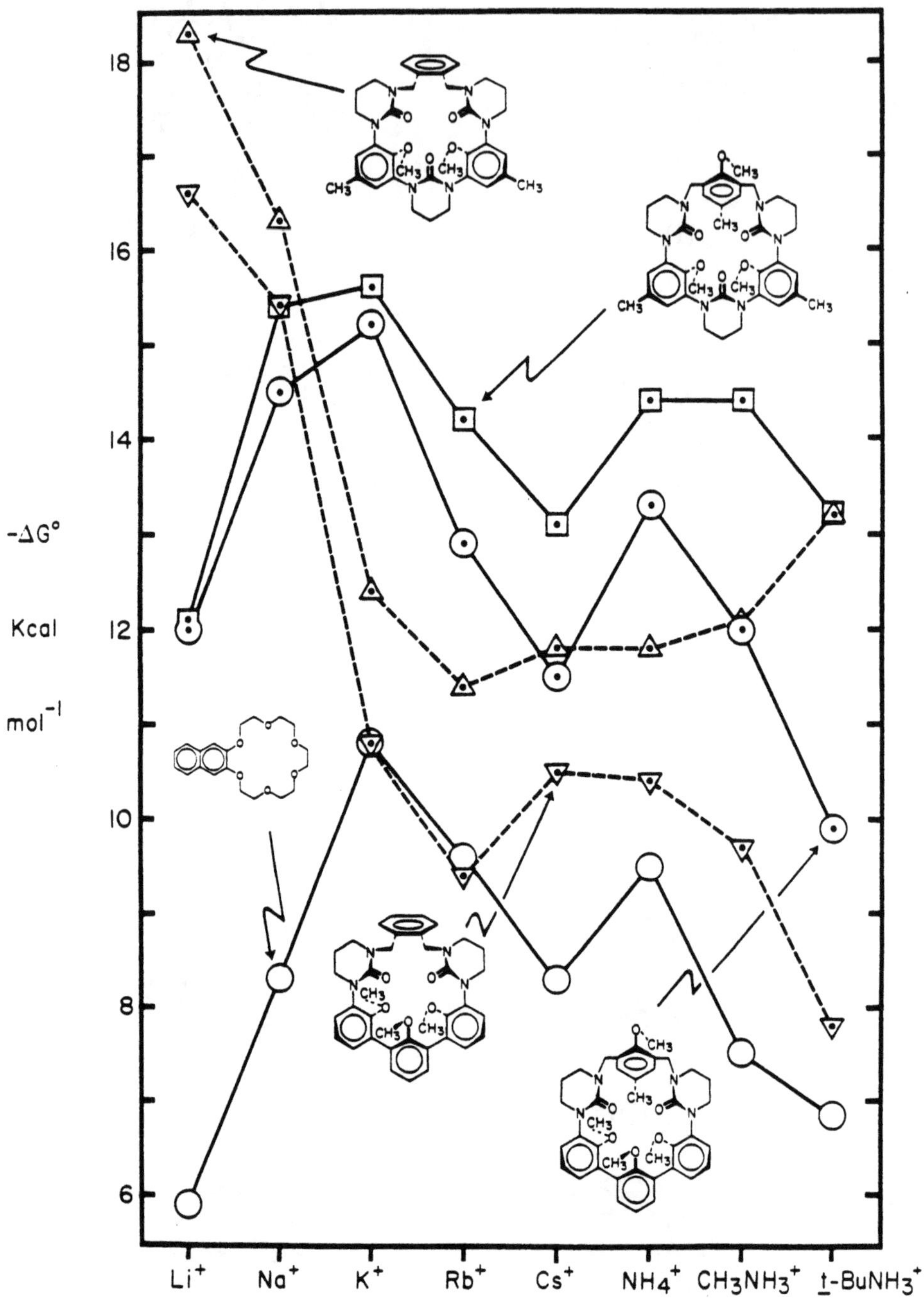

By blending different degrees of preorganization and complementarity, high structural recognition can be achieved. For example, hemispherand (5) binds Na^+ 6.3 kcal mol^{-1} better than Li^+ and 2.8 kcal mol^{-1} better than K^+. Enlargement of the ring and introduction of the rigidifying tetrahydrofuran and ethyl groups, as in (6), provides a selectivity of K^+ over Na^+ by 4.5 kcal mol^{-1} and of $CH_3NH_3^+$ over t-$BuNH_3^+$ by $\gg$4.3 kcal mol^{-1} [6]. Spherand (7) contains two cyclic urea units highly organized by the constraints of the 19-membered ring, the rigid *o*-xylyl group, and the spacial requirements of three anisyl-like groups. It binds Na^+ better than K^+ by 4.6 kcal mol^{-1}. The tris-urea host (8) has a 20-membered macroring and contains six binding sites well spaced for strong binding. It contains just enough conformational mobility to adapt to the eight ions studied. The $-\Delta G^0$ values range only from a low of 12.1 to a high of 15.6 kcal mol^{-1}. Chart V illustrates another way of correlating structure-binding relationships [7].

The *rates of extraction* of guest picrate salts from water into chloroform containing dissolved hosts is very fast for the chorands (crowns), the urea spherands, and the hemispherands, but is slow for the anisyl spherands [3]. The values of rate constants for picrate salts complexing the anisyl spherands in $CDCl_3$ (homogeneous) at 25 °C is 10^5 to 10^6 s^{-1} mol^{-1} [2], and for complexing the urea spherands is 10^{10} to 10^{11} s^{-1} mol^{-1} [8]. The decomplexation rate constants in $CDCl_3$ at 25 °C for the anisyl spherands varies from 10^{-4} to $< 10^{-12}$ s^{-1} [2], and for the urea spherands from 10^2 to 10^4 s^{-1} [8]. Thus, rates of extraction, of complexation, and of decomplexation are all subject to manipulation by structural design and synthesis of hosts just as are the related complexation free energies.

3 Chiral Catalysis

Most of the carbon-carbon bond-making reactions useful in organic synthesis involve carbanion addition or substitution reactions. Usually these reactions are conducted in nonpolar media in which the carbanions are ion-paired or partially covalently bonded to Li^+, Na^+, K^+, Cs^+ or Mg^{2+}. Often prochiral carbanions ion-paired with metal cations are converted to products containing chiral centers in these reactions. Thus the possibility exists for ligating organometallics with enantiomerically pure hosts that provide asymmetric environments for carbon-carbon bond-making reactions. The object is to make optically pure products from achiral starting materials by stoichiometric or turn-over catalysis. Progress on this approach is reported here.

Chorand (9) is a weak but highly asymmetric binder of K^+ ions which contains a C_2 axis. The structure of a complex between (9) and a potassium carbanionide ion pair is independent of which side of the macroring is occupied by the carbanion moiety. When racemic (10) is treated in toluene at −80 °C with methyl vinyl ketone, no reaction occurs, even when KOBu-*t* is present. The catalyst remains undissolved. When (9) is present, the complexed base goes into solution, and by proton transfer generates a planar carbanion ion-paired to host-bound potassium. The planar carbanion is "sided" because only one of its two faces is open to attack by nucleophiles, the other side being blocked by a naphthalene ring. As a result, product (11) is essentially optically pure, and its configuration is that predicted by molecular model examination [9]. The turn-over number is 10, so true catalysis is observed. Formula (12) is a conceptualization of the type of transition state involved, which rationalizes the results.

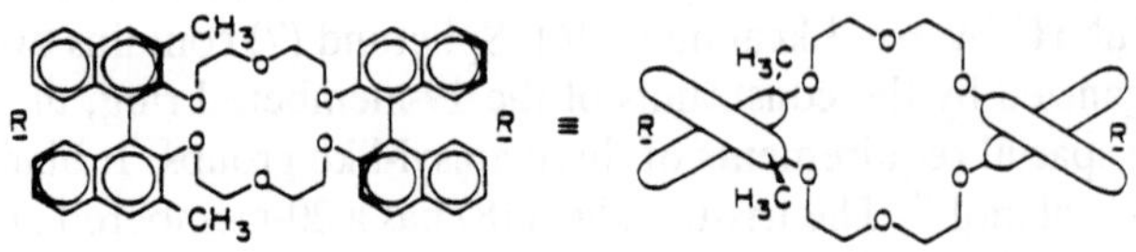

(9)

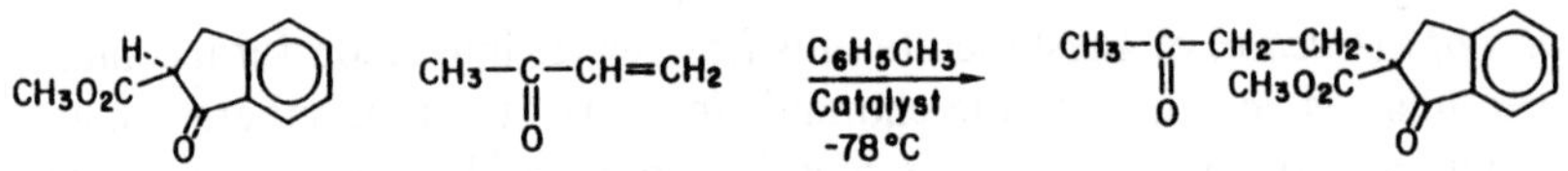

1 mol 3 mol

(10) (11)

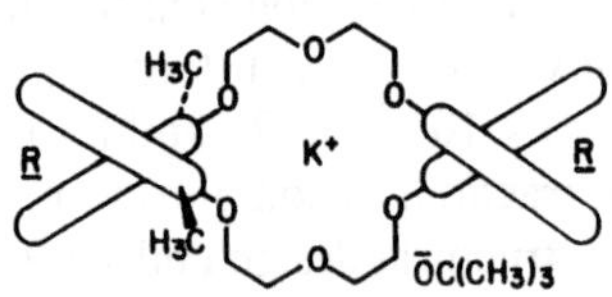

Catalyst

Turnover	Yield	ee	R/S	Δ(ΔG‡), kcal mol⁻¹
10	48%	~99%	~100	~2.0

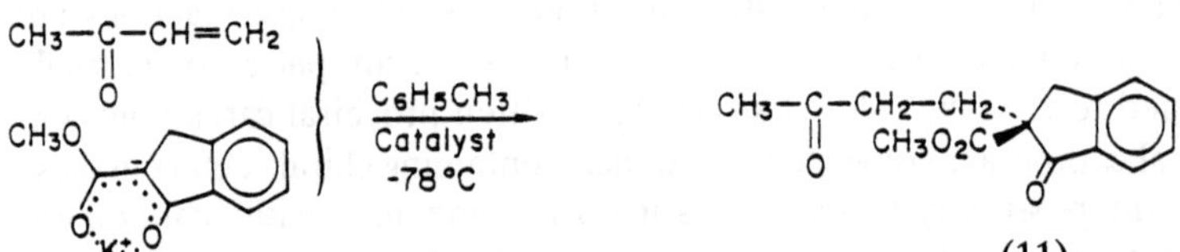

(11)

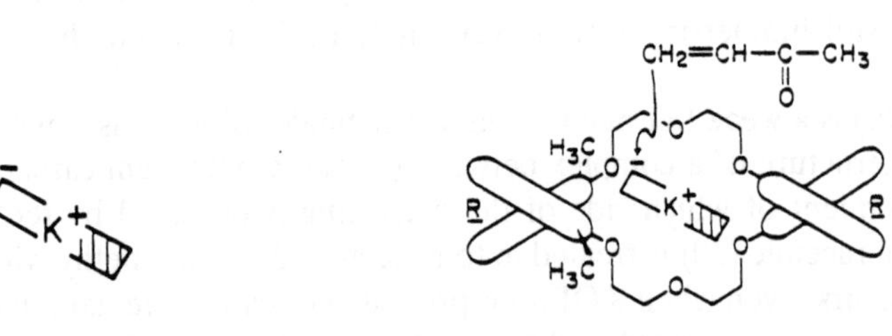

(12)

Other more strongly binding chiral catalysts are under examination, for which (13–15) are prototypes [10, 6].

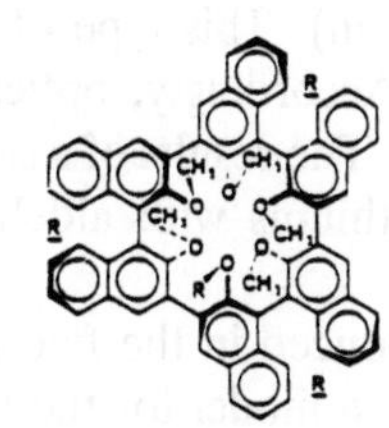

(13)

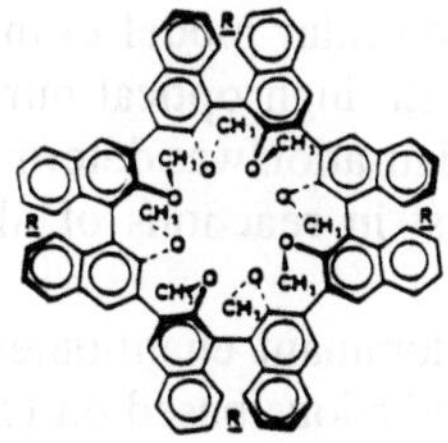

(14)

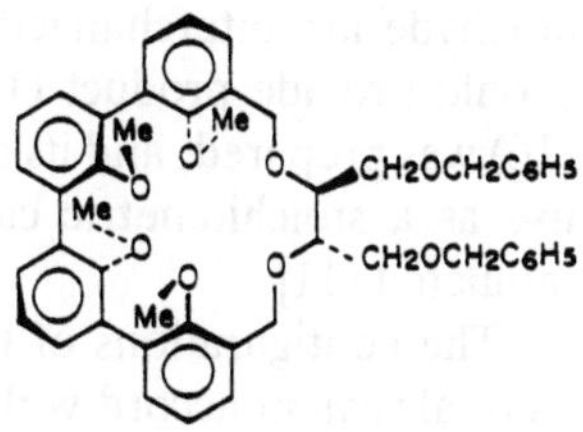

(15)

Tetramethylethylenediamine is known to activate alkyllithium reagents. Compound (16) is a highly asymmetric analog which contains three C_2 axes. Thus, alkyllithiums complexed to (16) and to aldehyde reactants should have either structure (17) or a much less stable diastereoisomeric structure in which H and R′ of the

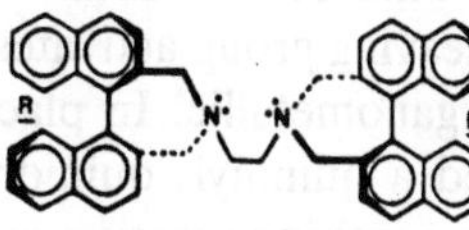

(16) (host)

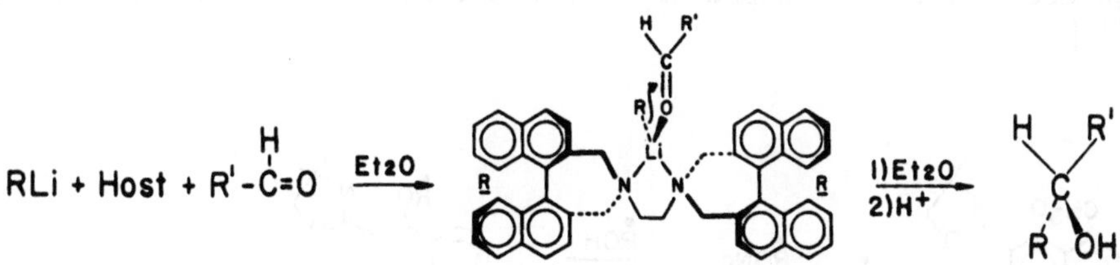

(17) (18)

$$R'\text{-}CH{=}O \;+\; R\text{-}Li \bullet \text{Host} \xrightarrow[\text{2) }H^+]{\text{1) }Et_2O,\ -120°C} R'\text{-}CH(OH)\text{-}R \;+\; \text{Host}$$

R'	R	Yield	ee	(R)/(S)
C_6H_5	$CH_3CH_2CH_2CH_2$	73%	92%	24
$CH_3CH_2CH_2CH_2$	C_6H_5	47%	49%	0.33
C_6H_5	$CH_3CH_2CH_2$	63%	61%	4
C_6H_5	CH_3CH_2	75%	67%	5
C_6H_5	CH_3	35%	36%	2.1

aldehyde are interchanged (molecular model examination). This type of complex should provide product (18) with high optical purity. Accordingly, optically pure (16) was prepared, and its configuration was determined. The results of a study of its use as a stoichiometric catalyst in reactions of alkyllithiums with aldehydes are tabulated [11].

The configurations of the dominant enantiomer produced in the five reactions were always in accord with predictions based on (17) as a model for the transition states in question. However, the model predicts a much higher degree of asymmetric induction than was observed. Unfortunately, the rates of the reactions catalyzed by (16) are not faster than the uncatalyzed reactions by large enough factors to channel the reactions completely through (17). As a result, both catalyzed and uncatalyzed reactions occur, and their relative rates depend on the character of RLi [11].

Enantiomerically pure binaphthyls or other biaryls provide good asymmetric barriers for incorporation into hosts. We devised an approach to their synthesis through use of naturally occurring chiral auxiliaries. Use was made of the Meyers reaction, in which a hindered oxazoline attached to an aromatic ring ortho to a methoxy or halogen leaving group activates the system toward aromatic nucleophilic substitution by an organometallic. In place of the normally used methoxy leaving group, we substituted a quininyl, quinidinyl, menthyl, or α-fenchyl group, as in compounds (19). The resulting reaction intermediates and transition states became diastereomeric, and the enantiomeric products were therefore formed at different rates. In other words, the asymmetric leaving group induced asymmetry in the product. Substituted binaphthyls of 94–48% enantiomeric purity were observed [12].

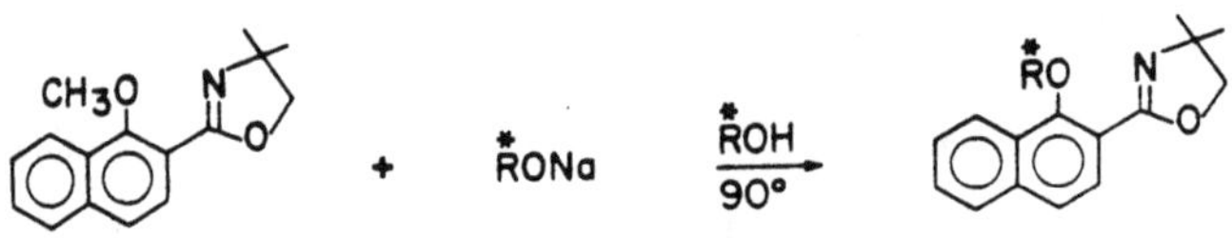

(19)

*R = Menthyl

*R = α-Fenchyl

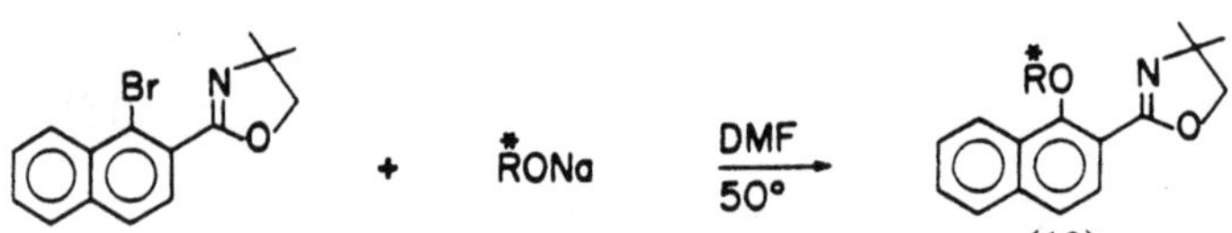

(19)

*R = Quininyl

*R = Quinidinyl

*R = Menthyl

*R = α-Fenchyl

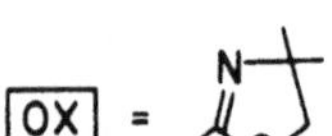

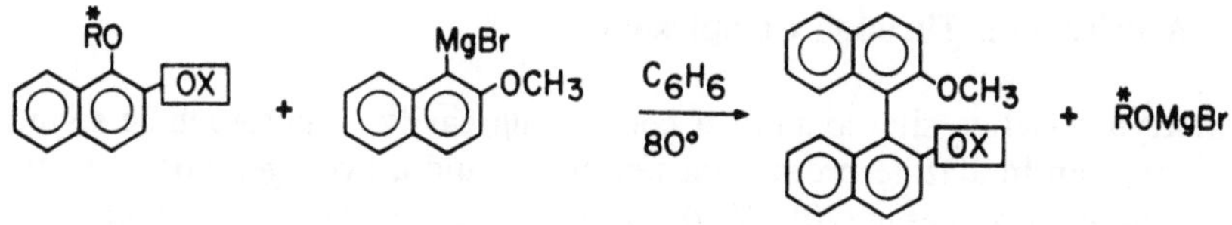

$\overset{*}{R}$	% Yield	Optical purity, % ee	Δ(ΔG‡) (kcal mol⁻¹)
Quininyl	7	94 (S)	2.6
Quinidinyl	27	84 (R)	1.7
Menthyl	53	78 (R)	1.5
α-Fenchyl	65	48 (R)	0.8

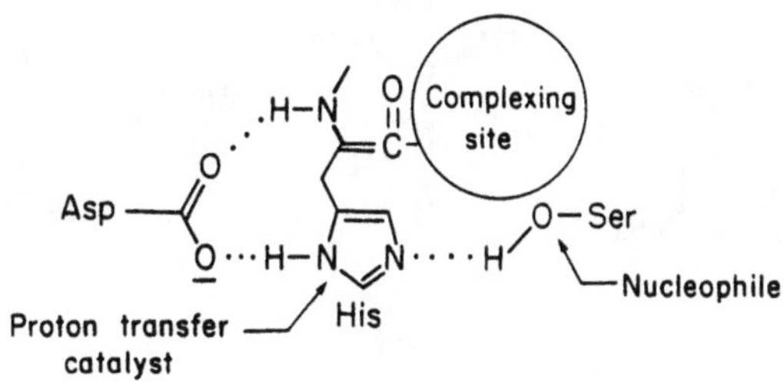

(20)

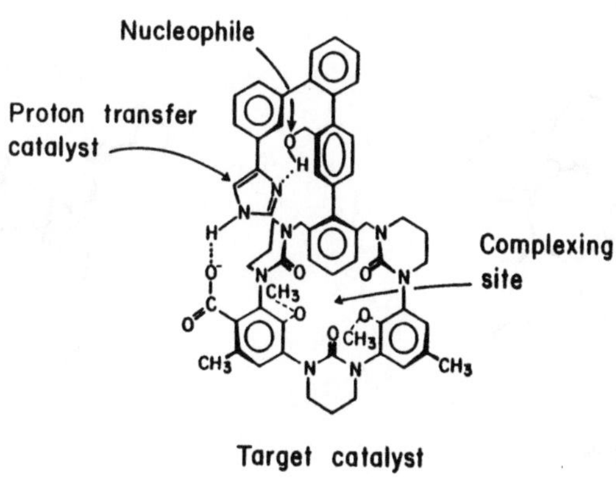

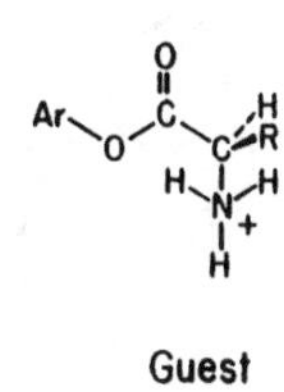

(21) (22)

4 Kinetic Acceleration Through Complexation

The chymotrypsin-active site contains a complexing cavity, a nucleophilic serine hydroxyl group, an imidazole proton transfer agent, and a hydrogen-bonding carboxylate group organized as in (20). We designed a series of transacylase mimics, of which (21) is representative. In models, (21) is complementary to guests such as (22). Possible syntheses of (21) are over 30 steps, so we developed an incremental approach to the problem [13].

Host (8) was first prepared, and was found to strongly bind t-BuNH$_3^+$ picrate in $CDCl_3$ with a $-\Delta G^0$ value of 13.2 kcal mol^{-1}. The complex with t-BuNH$_3$ClO$_4$ gave the expected crystal structure drawn in projection formula (23) and portrayed in machine drawings (24) and (25). The complex shows the expected tripod hydrogen bonding between the NH_3^+ and the three C=O groups of the host [7]. Kinetic studies demonstrated that the rates of decomplexation were very fast on the human time scale [8].

Host (26) was then prepared, and was also found to be a powerful binder of $CH_3NH_3^+$ and Na^+ ions. When mixed with (26), the perchlorate salt of alanyl p-nitrophenoxide (27) complexed and acylated (26) to give (28) which was fully characterized. The kinetics of the reaction were studied in $CHCl_3$ at 25 °C and the

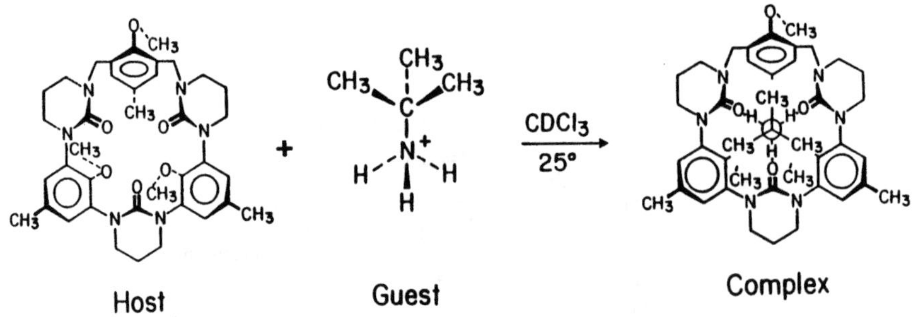

(8) (23)

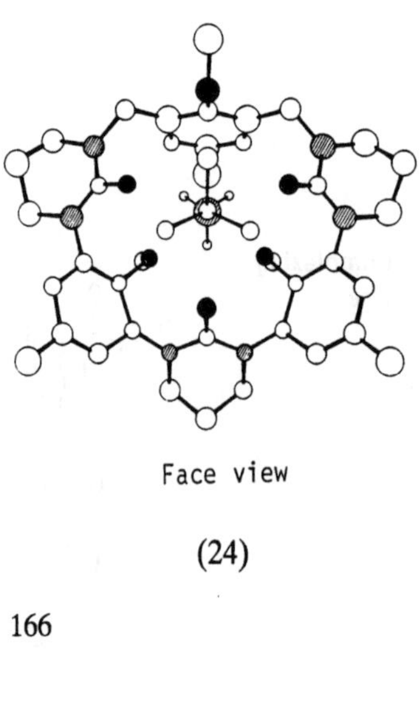

(24)

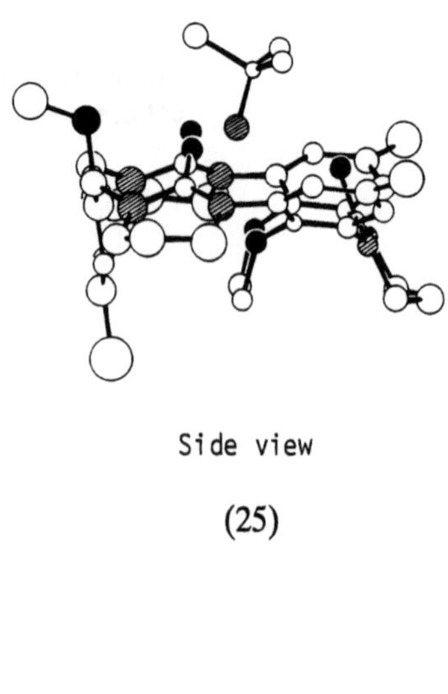

(25)

$-\Delta G°$, Kcal mol^{-1}, in $CDCl_3$ at 25°C

Picrate salts

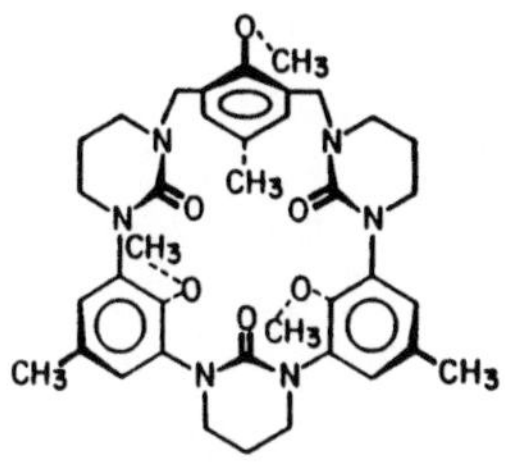

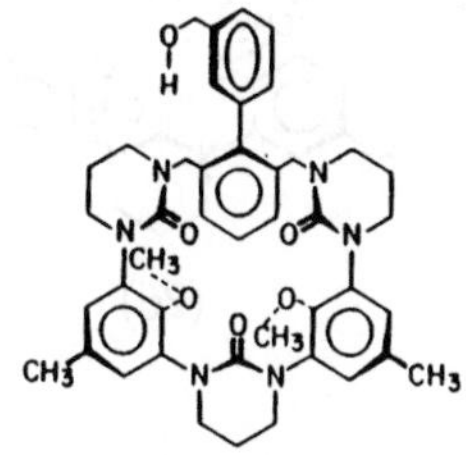

	(8)	(26)
$CH_3NH_3^+$	14.4	12.7
Na^+	15.3	13.6

reaction was found to be first order in R_3N/R_3NHClO_4 buffer ratio (R_3N is diisopropylethylamine). Thus $ArCH_2O^-$ is the reactive nucleophile. The reaction showed saturation kinetics characteristic of enzyme systems. Under the same conditions the noncomplexing model compound, 3-phenylbenzyl alcohol, was acylated at a rate too slow to measure but one on which higher limits could be put. Comparisons of the second order rate constants for acylation of (26) and its noncomplexing model showed that (26), by collecting and orienting the reactants through complexation, accelerated the rate by $\sim 10^{11}$. The *reference states* in this comparison are *uncomplexed* (26) and its model on the one hand and the rate-limiting transition states for transacylation on the other. The latter is complexed in the case of (26) and uncomplexed in the case of the model. When (26) was acylated in the presence of one mole of $NaClO_4$, the Na^+ acted as a competitive inhibitor and reduced the rate by 10^2 [13].

The high magnitude of this 10^{11} value points to the importance of collection and orientation by complexation to rate acceleration in a medium such as $CHCl_3$. If the crystal structure of complex (23) is used to extrapolate by means of molecular models to the structure of the complex between (26) and (27), then the CH_2OH of the former is beautifully proximate to the carbonyl group of the latter. We visualize the transition state as a complex held together and stabilized by $NH_3^+ \cdots (O{=}C)_3$ and by $(CH_2O \cdots C \overset{\cdot\cdot\cdot}{-} O)^-$ as the binding sites. The transition state for the reaction of (27) with 3-phenylbenzyl alcohol is held together only by $(CH_2O \cdots C \overset{\cdot\cdot}{-} O)^-$ as a binding site.

The interesting question arises as to the relative contributions of the effects of collection and orientation to the 10^{11} rate factor. An estimated constant of $K_a \sim 10^9$ for the association of (26) with (27) is used in the calculation of this factor. This value provides no measure of the contribution that collection makes to the rate factor of 10^{11}. The K_a value is high because of strong enthalpic driving forces that stabilize both host and guest by complexation and which change their electronic characters accordingly. For example, the acidity of the NH_3^+ group of the guest decreases by several powers of ten during complexation, which by a reduced inductive effect must reduce the electrophilicity of the carbonyl group of the ester. The

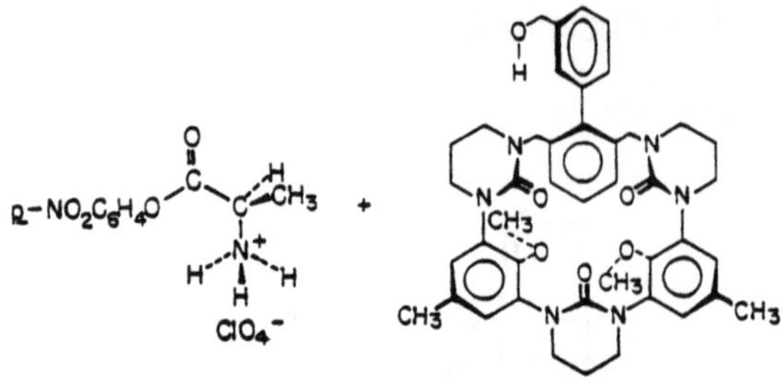

(27) (26)

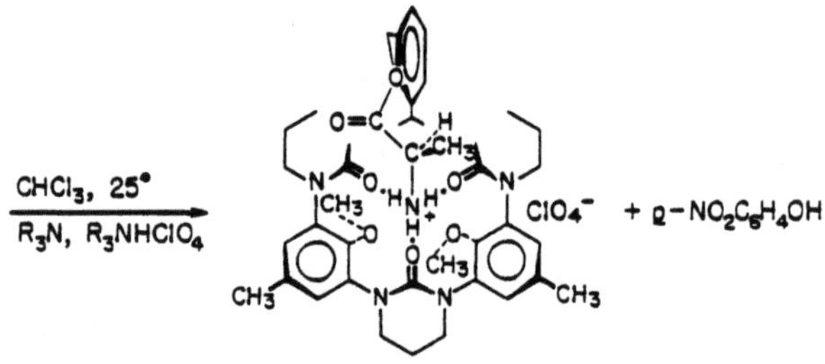

(28)

Relative Rates of Acylation

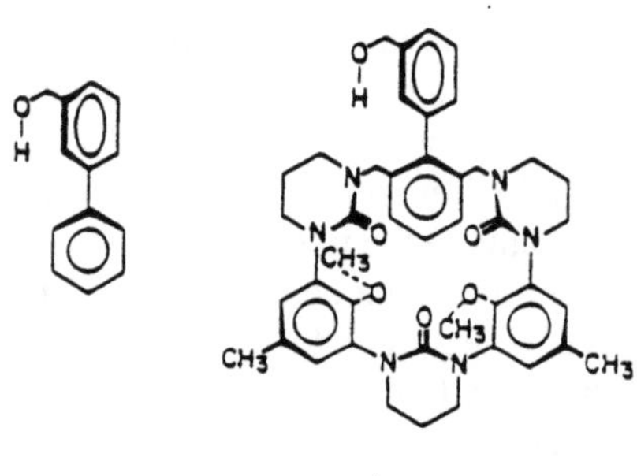

1 ~10^{11} (no $NaClO_4$)

~10^{9} (1 mole $NaClO_4$)

pKa of the CH_2OH group of the host is undoubtedly somewhat modified by complex formation, but it is impossible to guess the outcome of opposing electronic and steric effects. The solvation free energies of uncomplexed host and guest on the one hand and of complex on the other are entirely different, as are the steric effects involved with complexed vs. noncomplexed transacylation. All of these effects make 3-phenyl-benzyl alcohol ester (27) and the attendant transition state for acylation poor models for complex (26) · (27) and its attendant transition state for transacylation. We conclude that K_a integrates collecting and orienting effects, and that it is impossible to dissect the 10^{11} rate acceleration into contributions from each of these two effects.

The presence of an equal molar amount of $NaClO_4$ in the reaction of (26) with (27) reduced the acylation rate by $\sim 10^2$. The Na^+ ion, by complexing (26) in competition with (27), in effect reduced the concentration of (26) · (27) complex, and reduced the observed reaction rate accordingly [13].

5 Acknowledgments

We warmly thank the Division of Basic Sciences, Department of Energy (Contract DOE EY 76-S-03-0034, P.A. 218) for support of the research reported in Sect. 2 that involved binding of metal ions by the hosts free of cyclic urea units. We warmly thank the National Institutes of Health (GM 12640) and the National Science Foundation (CHE 81-09532) for support of the remaining research. The results reported here reflect the thoughtful and hard work of the coworkers mentioned in the next section.

References

1. R. C. Helgeson, G. R. Weisman, J. L. Toner, T. L. Tarnowski, Y. Chao, J. M. Mayer, D. J. Cram, J. Am. Chem. Soc. *101* (1979) 4928–4941
2. G. M. Lein, D. J. Cram, J. Chem. Soc. Chem. Commun. (1982) 301–304
3. D. J. Cram, K. N. Trueblood, Topics in Current Chemistry, Host Guest Complexation Chemistry I, Springer-Verlag, Berlin 1981, pp. 43–106
4. D. J. Cram, G. M. Lein, T. Kaneda, R. C. Helgeson, C. B. Knobler, E. Maverick, K. N. Trueblood, J. Am. Chem. Soc. *103* (1981) 6228–6232
5. D. J. Cram, S. B. Brown, T. Taguchi, M. Feigel, E. Maverick, K. N. Trueblood, J. Am. Chem. Soc. *106* (1984) 695–701
6. S. P. Artz, D. J. Cram, J. Am. Chem. Soc. *106* (1984) 2160–2171
7. D. J. Cram, I. B. Dicker, C. B. Knobler, K. N. Trueblood, J. Am. Chem. Soc. *104* (1982) 6828–6830
8. T. Anthonsen, D. J. Cram, J. Chem. Soc. Chem. Commun. (1983) 1414–1416
9. D. J. Cram, G. D. Y. Sogah, J. Chem. Soc. Chem. Commun. (1981) 625–628
10. R. C. Helgeson, J.-P. Mazaleyrat, D. J. Cram, J. Am. Chem. Soc. *103* (1981) 3929–3931
11. J.-P. Mazaleyrat, D. J. Cram, J. Am. Chem. Soc. *103* (1981) 4585–4596
12. J. M. Wilson, D. J. Cram, J. Am. Chem. Soc. *104* (1982) 881–884
13. D. J. Cram, H. E. Katz, I. B. Dicker, J. Am. Chem. Soc. *106* (1984) 4987–5000

Discussion

List of interventions

1 Intervention of Jencks

Can you say why the rate acceleration of 10^{11} is so large? What is the ratio of the first-order rate constant for the catalyzed reaction to the second-order rate constant for the control reaction, which gives the "effective molarity" the entropic advantage of the catalyzed reaction?

1.2 Cram

The ratio is greater than 125, and is only a minimum value since the second order rate constant for the model reaction was too slow to measure. We could only place a higher limit on the second order rate constant for the model reaction. I believe the comparison of two second order rate constants to be more sound in our system than the "effective molarity" treatment implicit in your question. Our host-catalyst and noncomplexing model compound resemble one another stereoelectronically, and they have a common guest reactant (the uncomplexed guest-ester salt). The electronic characters of both host and guest change markedly upon complexation. For example, the pKa of the RNH_3^+ group of the guest increases by at least four units, the pKa of the $ArCH_2OH$ group of the host also probably changes, and the nucleophilicity of the carbonyl group of the complexed guest is undoubtedly less than that of the uncomplexed guests. In the face of such large electronic changes accompanying complexation, I do not believe that the "entropic advantage of the catalyzed reaction" can be extracted from our data. The enthalpic changes that accompany complexation are just too large in our system.

2 Intervention of Breslow

Part of the 10^{11} can be accounted for by the fact that the reference substrate is deprotonated by the buffer, so that the reference reaction is artificially slowed. Also, the reaction itself has a half-time of 40 minutes. In water with the same buffer, the reaction is 1500 times faster without any catalyst. Thus the complexing reagent shows high selectivity, but the rate achieved is rather slow.

3 Intervention of Williams

Have you thought about comparing cavities and their constructions in:

a) Clathrates – are they as selective as your molecular systems?
b) Silicates – they are very selective for organic oxidations, e. g. zeolites, and can be chosen for their cavity sizes.
c) Polymolybdates – these select ions such as phosphate.
d) Synthesized capped porphyrins.
   In these systems very different atoms are used – not necessarily carbon atoms. Would not such systems have an advantage as a basis for catalysis?

3.1 Cram

My thoughts about cavities and their constructions in clathrates, silicates, polymolybdates and synthesized capped porphyrins are as follows: each kind of system is bound to have its advantages and disadvantages. Our kinds of systems differ from those you mention in the following respects:

1) We have designed and synthezised (constructed) our systems essentially from scratch, whereas those you mention are either naturally-occurring or are modifications thereof. There is an infinite number of designable organic systems. Thus in principle they can serve an infinite number of purposes.
2) Our systems are discrete molecular entities, whereas all but the last of those you mentioned are polymers, or have cavities defined by adjacent molecules in a crystal lattice.
3) The solubility characteristics of our systems are designable, whereas all but the last of the systems you mentioned are not. The clathrates, for example, disintegrate upon dissolution, and the zeolytes are insoluble in all solvents.
4) Our systems are inspired by the evolutionary model, which plays the dual role of guide and goal. The evolutionary model has myriads of variations, and thus provides the synthetic-organic designer with an open-ended challenge. The systems you mentioned are subject to very limited modifications. The main advantage of the inorganic cavities for use in catalysis is their stability.

4 Intervention of Hruby

Perhaps you can approach the problem of turnover in your kinds of systems by combining the findings from your compounds having metal binding sites with the findings from your enzyme mimics. For example, in your ion binding work you have some more flexible cases where you have sacrificed some specifications but still have high binding energies. So perhaps you could design a more flexible binding site (which would be useful for turnover) with a more rigid, directed catalytic site for maximum catalysis. Is this kind of approach being examined?

4.1 Cram

Your question addresses the advantage of rigidity versus flexibility of hosts, Preorganization, which is associated with rigidity, certainly gives higher binding and specificity, provided the conditions of complementarity between host and guest are fulfilled. With catalytic hosts the guests change their structures, which requires the hosts to be flexible enough to accommodate such changes. Ordinarily the changes do not involve the binding sites of either host or guest. In such cases, we believe that the binding sites should be rigid and the catalytic groups flexible enough to be complementary to the transition states of the reaction coordinate. In developing strategies for host design, we think of a rigid support structure containing the binding site(s) with more flexible catalytic sites appended. Our partial transacylase mimic, if anything, may be too rigid for turnover, and we may have to provide more flexibility to the nucleophilic hydroxyl and the proton-transferring imidazole and carbonyl groups. The beauty of the total synthetic approach is that the structural features of catalysts are subject to structural design and manipulation.

5 Intervention of Kellogg

One of your lovely compounds had a cavity appropriate for ferrocene. Does ferrocene indeed complex and if so is it held rigidly or does it spin in the cavity as happens with the cyclodextrins?

5.1 Cram

No, the experiment has not yet been carried out. This will be done soon.

Molecular Recognition: Design of Abiotic Receptor Molecules*

J. M. Lehn

Institut Le Bel, Université Louis Pasteur – Strasbourg, Collège de France – Paris, France

Molecular recognition may be considered to form, together with *molecular catalysis* and *transport,* the first level of functional properties displayed by the supramolecular species resulting from the interaction of substrates with their receptor molecules [1–3]. The investigation of molecular recognition processes and the design of artificial molecular receptors clearly represent an interface between chemistry and biology. The goal of the chemist is to devise systems which present such functions with highest efficiency and selectivity, finding inspiration in the ingenuity of the biological processes and encouragement in their demonstration that very high efficiencies, selectivities and rates can indeed be attained.

Biomimetic chemistry may simplify a complicated biological process, separate it into its components, reproduce a given feature in a *model* system in order to manipulate it and to study its response to perturbations. Chemistry thus helps to *understand* the biological processes. Although these are no doubt of much higher complexity than any chemical one achieved to this day, the field of *abiotic chemistry* is even broader than that of the chemical processes actually realized in biological systems, since it is not limited by the constraints of life, death and evolution. Whereas nature obviously had no reason to find answers to problems which were not raised in the course of evolution, chemistry is free to ask all questions, to *invent* new systems and processes and to continuously create its own object, as already noted by Marcelin Berthelot more than a century ago.

Our own work has dealt with the properties of the supermolecules formed by *macropolycyclic systems* which, owing to their architectural and functional plasticity, are especially attractive for designing both biomimetic and abiotic organic receptors for inorganic as well as organic substrates. They delineate molecular cavities into which substrates may penetrate, forming supramolecular inclusion complexes, *cryptates,* whose geometry and properties may be regulated via receptor design. Receptor design requires that the double complementarity of geometry and interactions with the substrate be satisfied, and that a desired flexibility/rigidity balance be planned.

We shall briefly describe various aspects of our own work on molecular recognition and on the design of abiotic molecular receptors covering inorganic or organic substrates of various geometries and cations as well as anions. Only a selection of results and references will be given here. The concepts underlying these studies and more detailed descriptions of the specific systems investigated may be found in other reports ([1–4] and references therein). The presentation will go from *monotopic*

* Dedicated to Professor Elkan R. Blout on the occasion of his 65th birthday

receptors, which contain a single site for substrate complexation, towards *polytopic receptors,* possessing several binding subunits.

Spherical Recognition

The simplest recognition process is that of spherical substrates which may be cationic (alkali and alkaline-earth ions) or anionic (halide ions).

In the inclusion complexes, the cryptates, formed by the macrobicyclic ligands (1–3) with numerous metal cations, the substrate is contained in the cavity of the complexed molecule in the *in-in* conformation (4) [1]. The stabilities of the optimal alkali and alkaline earth cryptates are several orders of magnitude higher than those of natural or synthetic macrocyclic ligands [5]. These cryptates present size selectivity, the preferred cation being (operationally) that whose size best fits the size of the cavity. Thus, as the bridges of the macrobicycle are lengthened on going from

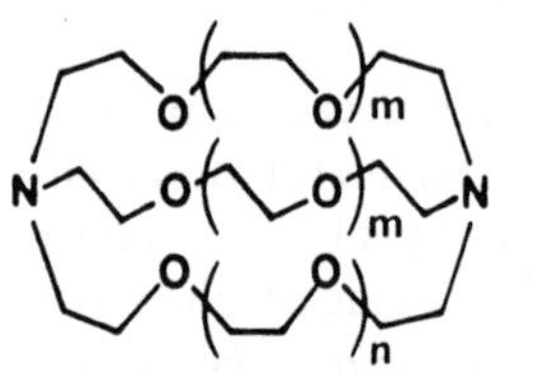

(1) [2.1.1] m = 0, n = 1
(2) [2.2.1] m = 1, n = 0
(3) [2.2.2] m = n = 1

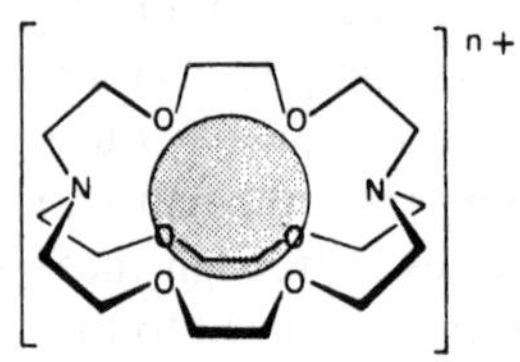

(4)

cryptand (1) to (2) and to (3), the most strongly bound ion becomes respectively Li^+, Na^+ and K^+. Also, the very high preference of (3) (2.2.2) for Sr^{2+} and Ba^{2+} with respect to Ca^{2+} is unique. Suitable structural modifications have allowed one to control the relative complexation of alkaline-earth and alkali cations achieving M^{2+}/M^+ selectivity appreciably higher than that displayed by natural macrocycles [6]. The spherical macrotricyclic ligand (5) binds strongly and selectively the larger spherical cations, forming the strongest known Cs^+ complex, (6) [7].

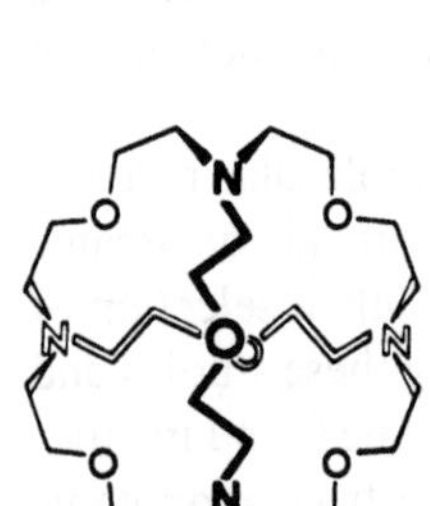

(5)

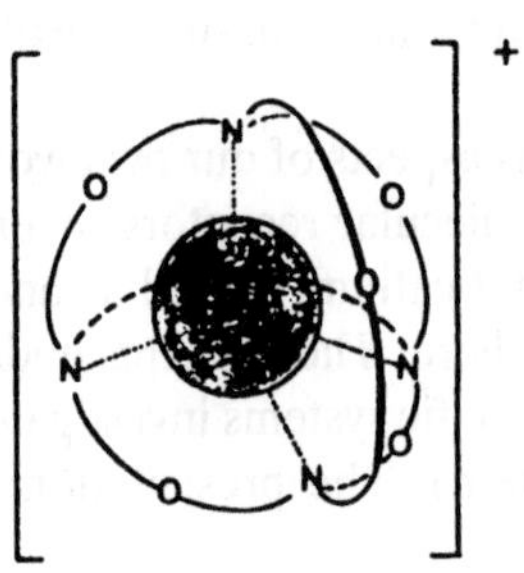

(6)

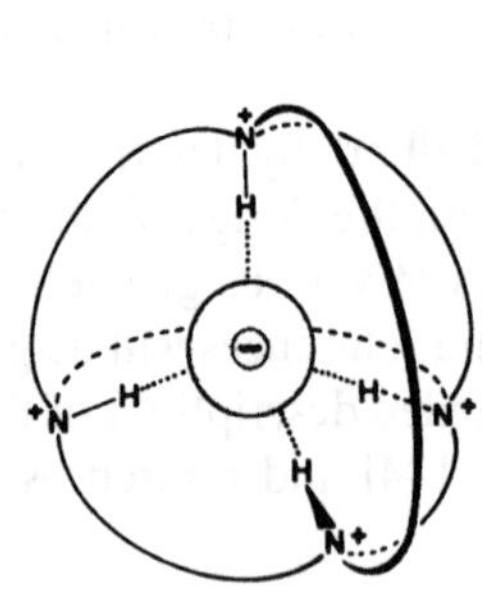

(7)

Anion cryptates are formed by macrotricycles like (5) in their tetraprotonated state with the spherical halide anions [8]. (5)-4H^+ binds the chloride ion very strongly and very selectively, giving the [$Cl^- \subset$ (5)-4H^+] cryptate (7), but does not complex other types of anions. These properties are unique at present with respect to both synthetic and natural halide binding sites, very little being known about the latter. Non-complementarity between an ellipsoidal cryptand and the spherical halides results in appreciable ligand distortions in the cryptates formed and in lower binding constants [9, 10] (see also below).

Thus, cryptands (1–3) and (5) display *spherical recognition* towards spherical cations and anions. The special complexation properties encountered clearly result from the macropolycyclic nature of these ligands. They define a *cryptate effect* characterized by high stability and selectivity of complexation, slow exchange rates and efficient shielding of the bound substrate from the environment.

Tetrahedral Recognition

Selective binding of a tetrahedral substrate requires the construction of a receptor molecule possessing a tetrahedral recognition site. This may be achieved by positioning four suitable binding sites at the corners of a tetrahedron and linking them via six bridges. Such a structure is formally of the macrotricyclic type (8). It is

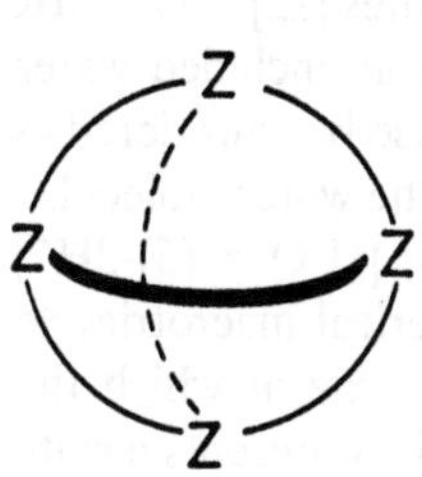

(8)

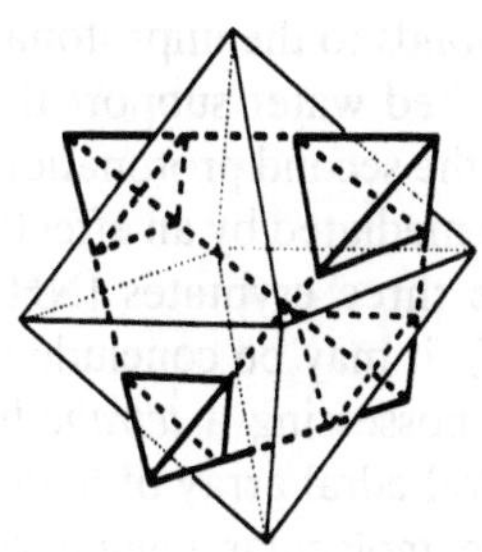

(9)

realized in the spherical macrotricyclic cryptand (5) which contains four nitrogen sites and six oxygen sites located respectively at the corners of a tetrahedron and of an octahedron (9) [7]. Ligand (5) forms a very strong and very selective complex with the tetrahedral NH_4^+ cation [11]. The stability ($K_s \sim 10^6$ in aqueous solution) is higher than that of the complexes of the natural and synthetic macrocycles by a factor of about 10^5. Furthermore, the selectivity with respect to binding of K^+ is also very large (factor of about 250), much larger than that of any other ligand reported to date. These unique features may be attributed to the formation of the *ammonium cryptate* [$NH_4^+ \subset$ (5)] (10) which presents a high degree of structural and binding complementarity between the NH_4^+ substrate and the receptor (5). NH_4^+ fits into the cavity of (5) and is held by a tetrahedral array of $^+$N–H···N hydrogen bonds with the four nitrogen sites and by electrostatic interactions with the six oxygen sites. Again these properties are characteristic of the cryptate structure and are much less pronounced with less highly connected ligands. The strong binding of NH_4^+ by (5) also confers on NH_4^+ in the cryptate an effective pK_a which is about six units higher than that of free NH_4^+; this effect has relevance for pK_a changes in enzyme active sites and biological sites and illustrates how large may be the changes in substrate properties brought about by binding.

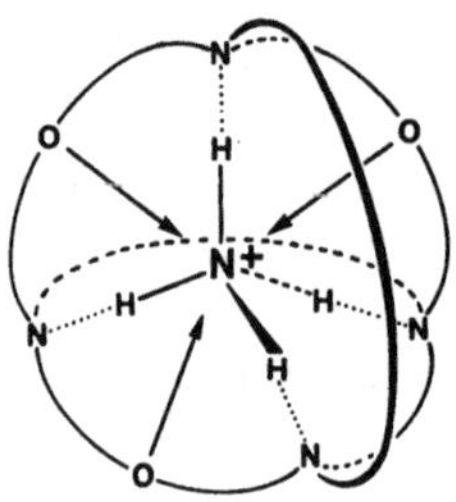

(10)

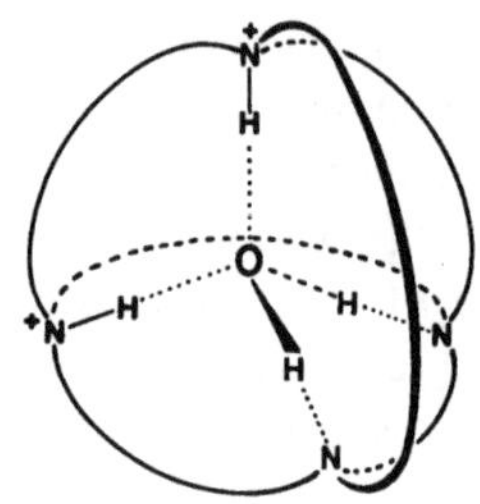

(11)

A study of the protonation features of (5) indicated that the second protonation was easier than the first one and that proton exchange was very slow. The results agree with the formulation of the diprotonated species as a *water cryptate* [H_2O $\subset$ (5)-2H^+] (11), in which the water molecule is held in a tetrahedral array of hydrogen bonds, accepting two $^+$N–H···O bonds from the protonated nitrogen sites and donating two O–H···N bonds to the unprotonated nitrogen sites [12]. ^{17}O-NMR measurements with ^{17}O-enriched water support the presence of an included water molecule. The facilitation of the second protonation of (5) may also be considered as a *positive cooperativity* effect mediated by an effector molecule, the water molecule.

Considering together the three cryptates [NH_4^+ $\subset$ (5)] (10), [H_2O $\subset$ (5)-2H^+] (11) and [Cl^- $\subset$ (5)-4H^+] (7), it may be concluded that the spherical macrotricycle (5) is a molecular receptor possessing a *tetrahedral recognition site* in which the substrates are bound in a tetrahedral array of hydrogen bonds. It represents a state of the art illustration of the molecular engineering involved in abiotic receptor chemistry.

Central and Lateral Discrimination

Extending from complexation of NH_4^+ to the binding of substituted ammonium cations, macrocyclic polyether receptors of type (12), bearing a variety of lateral substituents attached to the central core, present a number of interesting features [13]. The tetracarboxylate (12 a) forms by far the strongest complexes of any known macrocyclic polyether. Very marked *central discrimination* is observed between primary ammonium ions and more highly substituted ones, since only the former possess a $-NH_3^+$ group which may penetrate into the molecular cavity and form hydrogen bonds to the oxygen sites; this selectivity is of special interest in view of the selective binding of biologically active primary ammonium ions like noradrenaline and norephedrine with respect to their N-methylated derivatives adrenaline and ephedrine.

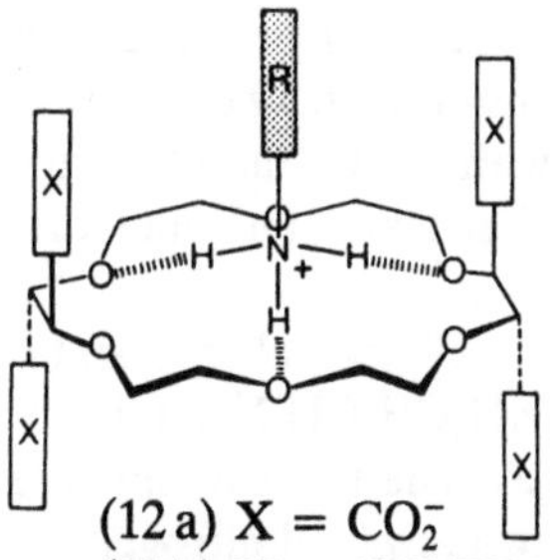

(12a) X = CO_2^-
(12b) X = $CONYY^-$

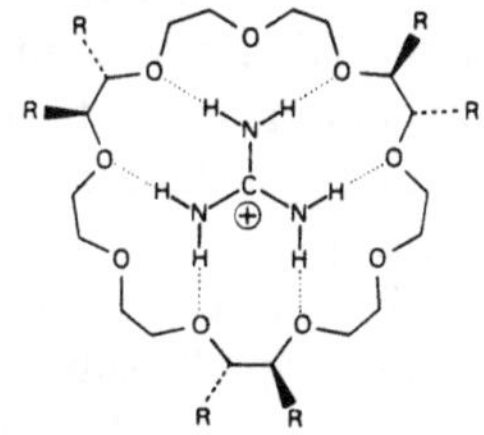

(13)

Varying the side groups X in (12b) allows one to modify the complexation features via interactions (electrostatic, lipophilic, H-bonding, charge transfer) between the lateral units and the R group of the centrally bound R–NH_3^+ substrate, thus bringing about *lateral discrimination* effects. Selective complexation of the guanidinium cation occurs with the larger [27]-O_9 macrocyclic hexacarboxylate forming the complex (13) in which the substrate is bound in an array of six hydrogen bonds to the complementary oxygen sites [14].

Recognition of Linear Triatomic Substrates

In order to further develop the coordination chemistry of anions and to extend recognition of anionic substrates beyond the spherical halides, an ellipsoidal macrobicyclic cryptand Bis-Tren (14) was designed, whose hexaprotonated form was expected to bind various anions [9, 10]. Indeed, potentiometric and spectroscopic measurements showed that (14)-$6H^+$ complexes a number of monovalent and polyvalent anions. The strong and selective binding observed for the linear triatomic anion N_3^- may be attributed to its complementarity to the molecular cavity of (14)-$6H^+$. As confirmed by crystal structure determination, N_3^- forms the cryptate [$N_3^- \subset$ (14)-$6H^+$] (15), in which the substrate is bound inside the cavity by two pyramidal arrays of three $^+N–H \cdots N^-$ hydrogen bonds, which hold the two terminal nitrogen sites.

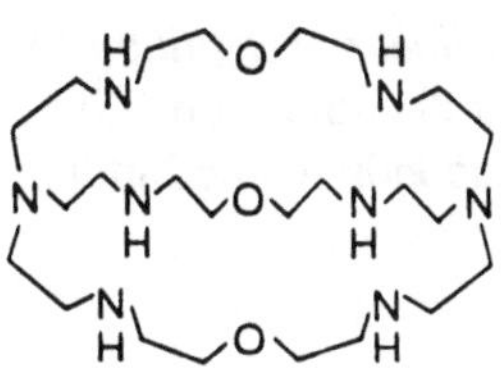

(14)

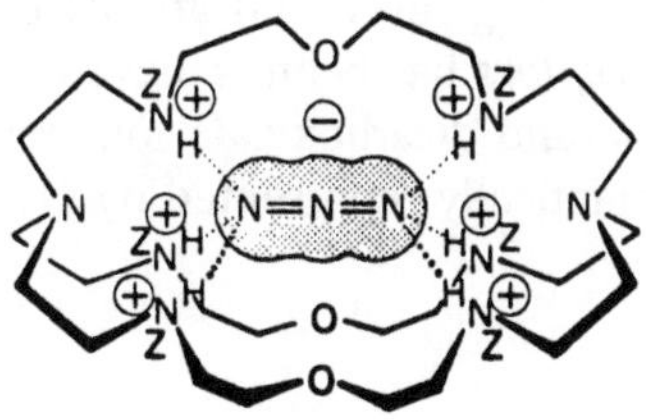

(15)

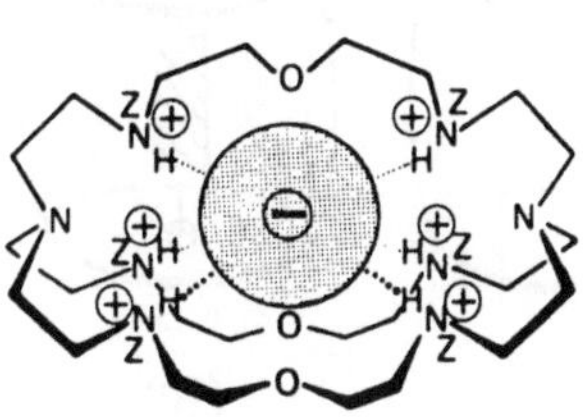

(16)

The non-complementarity between (14)-$6H^+$ and the spherical halide anions results in much weaker binding and in appreciable distortions of the ligand as shown by the crystal structures of the cryptates (16) of (14)-$6H^+$ with F^-, Cl^- and Br^-, the first of these being bound in a tetrahedral array and the other two in an octahedral array of hydrogen bonds [11].

In the supramolecular species, [$N_3^- \subset$ (14)-$6H^+$] (15), the ellipsoidal cryptand is built from two protonated tripodal subunits of the tren type $N(CH_2CH_2NH_2)_3$ located at each pole of the molecule and linked by three bridges. The two subunits are situated at a distance such that each may hold one of the two terminal nitrogens of the N_3^- ion, so that they cooperate in substrate binding. Bis-Tren-$6H^+$ may be considered as a *ditopic coreceptor for linear triatomic substrates* of a size compatible with the size of the molecular cavity.

This system thus brings us from molecular receptors into the field of coreceptors.

Coreceptor Molecules

Coreceptors are polytopic receptor molecules containing two or more discrete binding subunits which may cooperate for the complexation of several singly bound substrates or of a multiply bound polyfunctional species. Their ability, in principle, to perform multiple molecular recognition provides entries into higher forms of molecular behavior, cooperativity, allostery, regulation, and communication (signal transfer). Basic ideas and definitions concerning this area have been presented in more detail elsewhere [4].

Our present aim is to summarize the results obtained in the design of coreceptors capable of performing molecular recognition. Complexation of two or more metal ions by various macropolycyclic ligands to form di- or polynuclear cryptates will not be considered here (see ref. [15]). The discussion will be limited to substrate recognition by ditopic coreceptors for polyatomic molecular cations and anions and to developments towards polytopic coreceptors.

When two binding subunits are located at the poles of a coreceptor molecule, the complexation of a difunctional substrate will depend on the complementarity between the distance of the two binding sites in the receptor and the distance of the two corresponding functional groups of the substrate. Such *linear recognition by ditopic coreceptors* has been achieved for both dicationic and dianionic substrates, diammonium and dicarboxylate ions respectively, and corresponds to the binding modes schematically represented by (17) and (18).

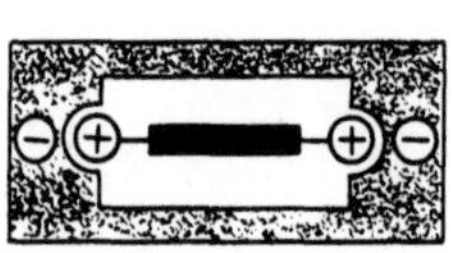

(17)

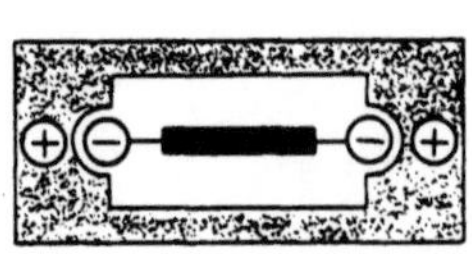

(18)

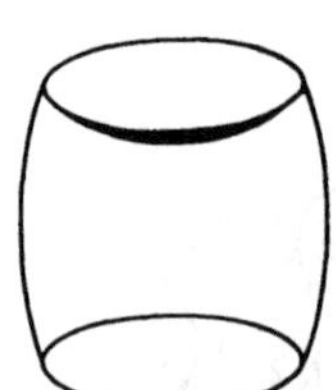

(19)

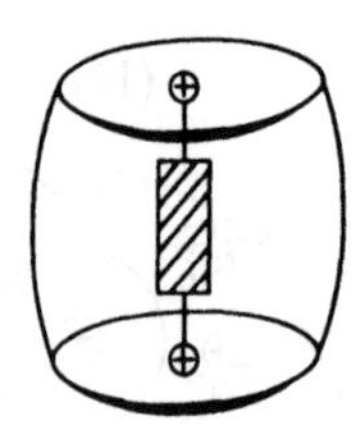

(20)

Linear Recognition of Diammonium Cations

Cylindrical macrotricyclic molecules (19) are ditopic coreceptors constructed on two binding subunits, two macrocycles, linked by two bridges. When the macrocycles chosen are able to bind $-NH_3^+$ groups, diammonium substrates may be expected to form mononuclear dihapto cryptates of type (20) by inclusion into the central molecular cavity of the tricyclic receptor.

This is the case for the macrotricycles (21–24) which yield supramolecular species $[^+H_3N\text{–}(CH_2)_n\text{–}NH_3^+ \subset \text{Receptor}]$ of type (25), where the diammonium substrate is located inside the receptor cavity and held by anchoring of each terminal $-NH_3^+$ group to an [18]-N_2O_4 macrocycle via hydrogen bonds [16]. This was confirmed by the crystal structure of $[^+NH_3\text{–}(CH_2)_5\text{–}NH_3{}^+ \subset (22)]$ which also provided a clear picture of the coreceptor-substrate relationship in the supermolecule [17].

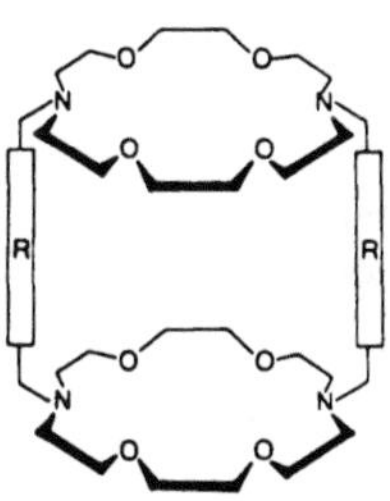

(21)

(22)

(23)

(24)

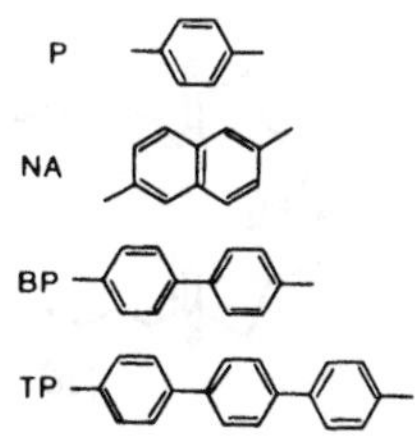

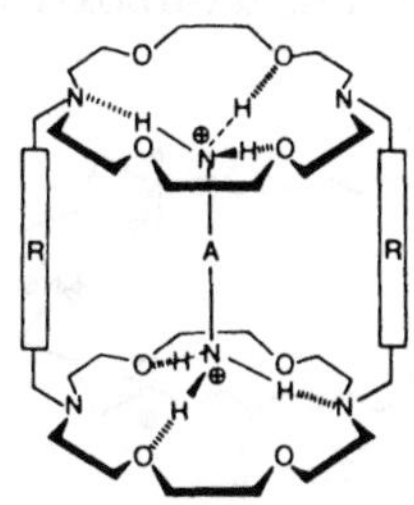

(25)

The stability of these species should depend on the relative length of the substrates and of the bridges R of the macrotricycles (21–24). Although no absolute stability constants are known at present, a study of complexation selectivities by NMR has shown that the $^+H_3N\text{–}(CH_2)_n\text{–}NH_3^+$ substrate preferentially bound was that whose length was complementary to the length of the molecular cavity, with $n = 4, 5$ and 7 for coreceptors (21, 22) and (23) respectively [17–19]. Thus, ligand (22) is able to discriminate between the biologically important cadaverine ($n = 5$) and putrescine ($n = 4$) cations. Similar effects may be expected for other biological diamines or polyamines of various chain lengths (ornithine, lysine, diammonium dipeptides, spermidine, spermine, etc.). Different selectivity sequences have been obtained with other macrotricycles [20] and very high discrimination was found for the diammonium cryptates of cylindrical macrotetracyclic receptors [21].

These results provide a basis for structural receptor-substrate complementarity in *linear molecular recognition.*

^{13}C-NMR determinations of the relaxation and correlation times of supermolecules (25) have indicated that complementary receptor-substrate pairs display similar molecular motions for both partners [19]. This *dynamic fit* may be considered to result from the combination of steric fit with dihapto binding. Thus, complementarity between components of a supramolecular species expresses itself in both its structural and dynamic properties.

Linear Recognition of Dicarboxylate Anions [22]

The two hexaammonium macrocycles (26) and (27) possess the features required for ditopic coreceptor molecules of dianionic substrates, since they contain at the poles of the structure two triammonium chelating subunits which may each serve as a binding site for a negatively charged group, as schematically represented by (28). Indeed, they form strong complexes with dicarboxylates of various types and present a pronounced chain length dependent complexation selectivity among $^{-}O_2C-(CH_2)_m-CO_2^-$ substrates, with preferential binding of $m = 2, 3$ by (26) and $m = 5, 6$ by (27). One notes that this change in selectivity from (26) to (27) corresponds to the same increase in length, by three CH_2 groups, of the polymethylene chains separating both the carboxylate groups of the substrates and the triammonium binding subunits in the receptors. Compounds (26) and (27) also display selective binding of biological dicarboxylates of compatible lengths towards amino acid and dipeptide dicarboxylates respectively.

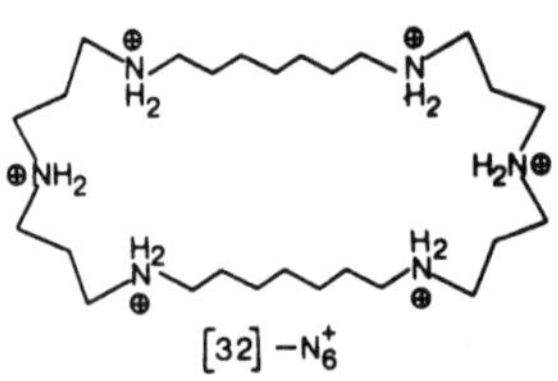

(26)

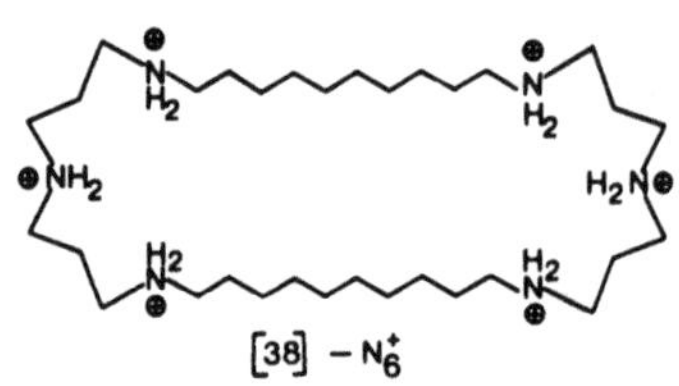

(27)

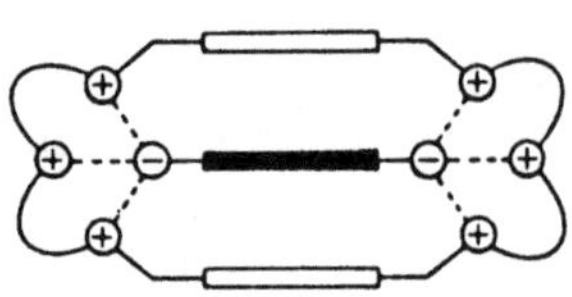

(28)

As in the above case of the binding of diammonium substrates to the macrotricyclic coreceptors, the present chain length selection also describes a *linear recognition* process based on structural complementarity between the dianionic substrates and the coreceptors in a ditopic binding mode (see (28)). In both cases, the receptor molecule acts as a discriminating *sensor of molecular length.*

Molecular Recognition in Speleates

The coreceptor molecules discussed above are built on polar binding subunits. Within the general class of cryptand molecules, the combination of polar binding subunits with more or less rigid, apolar shaping components provides new types of hollow, amphiphilic macropolycyclic coreceptors, termed *speleands,* which form *speleates* by substrate inclusion [4, 23].

The macrocyclic speleand (29) combines two tartaric acid units with two diphenylmethane groups [24]. It strongly binds a whole range of molecular cations by a combination of electrostatic and hydrophobic effects. Thus, primary ammonium ions form complexes which are much more stable than those of the common polyether macrocycles (except (12 a)), providing a novel type of binding unit for $R\text{–}NH_3^+$ substrates. Of special interest is the fact that (29) strongly complexes secondary, tertiary and quaternary ammonium cations as well. Among the latter, complexation of *acetylcholine* is of particular importance, since it provides one answer to the general question of how to bind acetylcholine; furthermore it may shed light on the still poorly understood problem of the nature of the interactions which play a role in biological acetylcholine receptors.

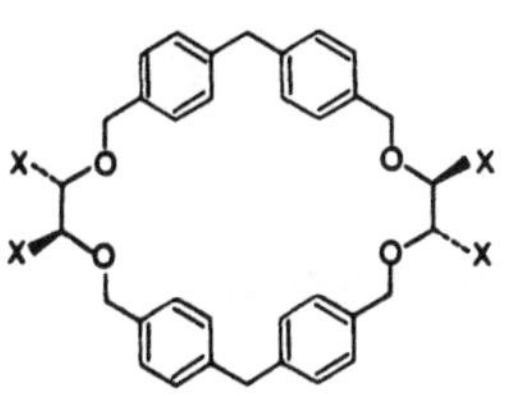

(29)

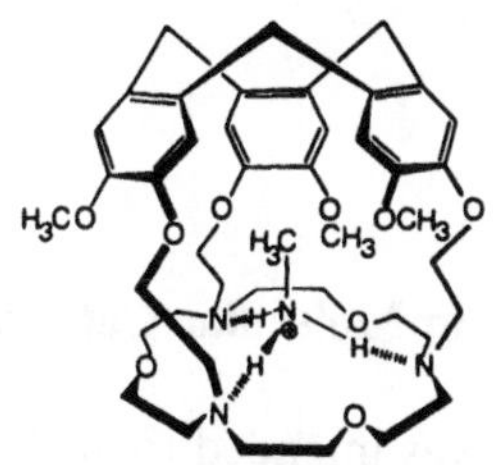

(30 a) ($CH_3\text{–}NH_3^+$ removed)
(30 b)

The macropolycyclic speleand (30 a) combines the [18]-N_3O_3 macrocyclic binding subunit with the cyclotriveratrylene shaping component [23]. Its tight intramolecular cavity allows inclusion of the $CH_3\text{–}NH_3^+$ ion yielding the speleate [$CH_3\text{–}NH_3^+ \subset$ (30 a)] whose structure should be of the type shown schematically by (30 b).

A variety of other combinations of polar binding units and apolar architectural components may be imagined, making speleands a particularly attractive class of macropolycyclic systems. This combination is of prime importance for the design of a variety of novel, efficient and selective molecular receptors of the speleand type, adding another facet to supramolecular chemistry.

Polytopic Coreceptors: Metalloreceptors

Connecting several binding subunits within the same macropolycyclic architecture yields polytopic coreceptors. Some of their general features have been considered elsewhere [4]. We shall only mention briefly the case of metalloreceptors.

These is a special type of polytopic coreceptor which combines different substrate selective binding subunits so as to allow *simultaneous* complexation of metal ions and of organic species within the same superstructure. As a consequence, interaction and reaction (metallo-catalysis) may take place between co-bound substrates.

Metalloreceptors have been obtained by introducing one or two porphyrin or α,α'-bipyridine metal ion binding units as bridges (R) in the macrotricycles of type (21–24) [25]. They separately complex either diammonium ions as in (25) or metal ions. Simultaneous binding of $^+H_3N\text{–}(CH_2)_9\text{–}NH_3^+$ and of two Zn^{2+} cations by such a bis-porphyrin metalloreceptor yields a *mixed-substrate supramolecular species* which may be represented schematically by (31). Numerous variations are possible, involving recognition processes, i. e. selective binding features, for both the organic and inorganic substrates.

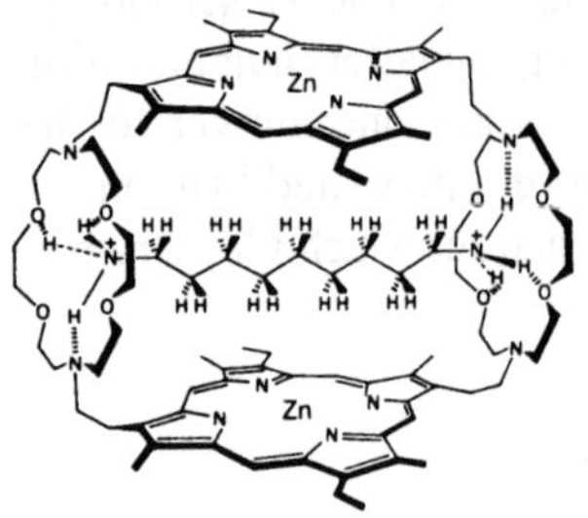

(31)

Molecular Recognition in Molecular Catalysis and Transport Processes

The present discussion was intended to analyze molecular recognition in substrate binding. Of course molecular recognition also plays a major role in molecular catalysis and transport processes where transformation or translocation of the bound substrates is brought about by a suitably functionalized receptor molecule. These most important aspects are however beyond the scope of this presentation. One may just note, for the sake of illustration, that for instance cysteinyl derivatives of receptor (12 b) display marked chiral recognition in transacylation reactions with optically active substrates [26] and that dicarboxylate-dicarboxamide analogs of (12 b) allow pH regulation of the Ca^{2+}/K^{+} selectivity in competitive transport of calcium and potassium ions [27]. Further information about the results obtained in the areas of molecular catalysis and transport may be found respectively in the references [28] and [29] and in the references cited therein.

Conclusion

The chemistry of artificial receptor molecules has produced a great variety of supramolecular structures displaying molecular recognition processes. As a chemistry of the intermolecular bond it has contributed to the fundamental understanding of the elementary interactions on which recognition is based. As a design of efficient and selective receptors for substrates of numerous types it has led to the synthesis of many new molecules and made available a wealth of new properties.

The results obtained may shed light on the interactions responsible for biological specificity and lead to better understanding of the properties of biological receptors. They also point to numerous potential applications in applied fields of science.

Further developments in the chemistry of molecular recognition and in the design of receptor molecules will benefit markedly from computer-assisted modeling, which will allow much faster exploration of the practically infinite number of possible receptor structures and operational evaluation of the supermolecules which they form by substrate binding [30]. One may thus expect that a wealth of new systems will be designed and produced. Suitable modifications will confer new functions – catalysis, transport, regulation, communication – the latter incorporating the *time* dimension into recognition events. As a process of storage and retrieval of information at the molecular level, the chemistry of molecular recognition is also a venture in artificial intelligence, extending still further the domain of supramolecular chemistry.

References

1. J. M. Lehn, *Pure Applied Chem., 50,* 871 (1978); *Accounts Chem. Res., 11,* 49 (1978); *Science, 227,* 849 (1985)
2. J. M. Lehn, *Structure and Bonding, 16,* 1 (1973)
3. J. M. Lehn, Leçon Inaugurale, Collège de France, Paris, 1980
4. J. M. Lehn, in "Biomimetic Chemistry", N. Ise and Z.I. Yoshida, eds., Kodansha, Tokyo, Elsevier, Amsterdam, 1983, p. 163
5. J. M. Lehn and J. P. Sauvage, *J. Am. Chem. Soc., 97,* 6700 (1975)
6. B. Dietrich, J. M. Lehn and J. P. Sauvage, *J. Chem. Soc., Chem. Commun.,* 15 (1973)
7. E. Graf and J. M. Lehn, *J. Am. Chem. Soc., 97,* 5022 (1975); *Helv. Chim. Acta, 64,* 1040 (1981)
8. E. Graf and J. M. Lehn, *J. Am. Chem. Soc., 98,* 6403 (1976)
9. J. M. Lehn, E. Sonveaux and A. Willard, *J. Am. Chem. Soc., 100,* 4914 (1978)
10. B. Dietrich, J. Guilhem, J. M. Lehn, C. Pascard and E. Sonveaux, *Helv. Chim. Acta, 67,* 91 (1984)
11. E. Graf, J. M. Lehn and J. Lemoigne, *J. Am. Chem. Soc., 104,* 1672 (1982)
12. J. P. Kintzinger, J. M. Lehn and E. Graf, unpublished results
13. J. P. Behr, J. M. Lehn and P. Vierling, *J. Chem. Soc., Chem. Commun.,* 621 (1976); *Helv. Chim. Acta, 65,* 1853 (1982)
14. J. M. Lehn, P. Vierling and R. C. Hayward, *J. Chem. Soc., Chem. Commun.,* 296 (1979)
15. J. M. Lehn, *Pure Applied Chem., 52,* 2441 (1980); in IUPAC "Frontiers of Chemistry", K. J. Laidler, ed., Pergamon Press, Oxford, p. 265 (1982)
16. F. Kotzyba-Hibert, J. M. Lehn and P. Vierling, *Tetrahedron Letters,* 941 (1980)
17. C. Pascard, C. Riche, M. Cesario, F. Kotzyba-Hibert and J. M. Lehn, *J. Chem. Soc.,* 557 (1982)
18. F. Kotzyba-Hibert and J. M. Lehn, unpublished results
19. J. P. Kintzinger, F. Kotzyba-Hibert, J. M. Lehn, A. Pagelot and K. Saigo, *J. Chem. Soc., Chem. Commun.,* 833 (1981)
20. N. F. Jones, A. Kumar and I. O. Sutherland, *J. Chem. Soc., Chem. Commun.,* 990 (1981) and references therein
21. F. Kotzyba-Hibert, J. M. Lehn and K. Saigo, *J. Am. Chem. Soc., 103,* 4266 (1981)
22. M. W. Hosseini and J. M. Lehn, *J. Am. Chem. Soc., 104,* 3525 (1982)
23. J. Canceill, A. Collet, J. Gabard, F. Kotzyba-Hibert and J. M. Lehn, *Helv. Chim. Acta, 65,* 1894 (1982)
24. M. Dhaenens, L. Lacombe, J. M. Lehn and J. P. Vigneron, *J. Chem. Soc., Chem. Commun.,* 1097 (1984)
25. A. D. Hamilton, J. M. Lehn and J. L. Sessler, *J. Chem. Soc., Chem. Commun.,* 311 (1984) see also in ref. 4 and 15
26. C. Sirlin and J. M. Lehn, *J. Chem. Soc., Chem. Commun.,* 949 (1978)
27. A. Hriciga and J. M. Lehn, *Proc. Natl. Acad. Sci. USA. 80,* 6426 (1983)
28. J. M. Lehn, *Pure Applied Chem., 51,* 979 (1979); M. W. Hosseini, J. M. Lehn and M. P. Mertes, *Helv. Chim. Acta, 66,* 2454 (1983); M. W. Hosseini and J. M. Lehn, *J. Chem. Soc., Chem. Commun.,* 1155 (1985)
29. J. M. Lehn, in "Physical Chemistry of Transmembrane Ion Motions", G. Spach, ed., Elsevier, Amsterdam, p. 181 (1983)
30. G. Wipff, P. A. Kollman and J. M. Lehn, *J. Mol. Struc. THEOCHEM, 93,* 153 (1983) and references therein

Discussion

List of interventions

1 Intervention of Kellogg

Can the rational basis you describe for designing complementary sites for bonding be extrapolated to design complementary bonding of transition states in order to achieve catalysis?

1.1 Lehn

Yes. But of course you have to know what the structure of the transition state is. This is the major problem.

2 Intervention of Wüthrich

1) In one of your complexes you distinguished an internal water molecule from the bulk water by ^{17}O NMR. How was this done? Is the ^{17}O NMR line of the bound water not broadened? If not, why not?
2) In the "bi-porphyrin sandwich" you conclude from the observed ^{1}H NMR high field ring current shifts that the ligand is located between the two porphyrins. Experience shows that these ring current shifts could also arise if the ligand were attached on the outside of the sandwich near the porphyrin plane. Do you have additional evidence for excluding this latter structure?

2.1 Lehn

1) The ^{17}O NMR was taken with a 400 MHz (for proton) instrument. There is little broadening because the hydrogens exchange quickly. The correlation time is determined by the proton exchange inside the cavity.
2) We have the crystalline structure for an analogous compound [27].

3 Intervention of Luisi

You mentioned that the pK of some ammonium ions in the cage is anomalously high. In order to determine this experimentally you have to carry out pH measurements. If the local pH in the cage is different from the bulk pH – i. e., if the H^+ activity inside is different from outside – your pK determinations are not correct. Generally, in such cases, there is no way of discriminating (on the basis of bulk measurements) between pH and pK changes.

4 Intervention of Breslow

Thermodynamic equilibrium exists between the solution and the cavity, so the activity of H^+, as measured by an electrode, is not different at the cavity.

Biomimetic Control of Chemical Selectivity

R. Breslow

Department of Chemistry, Columbia University, New York, New York 10027, USA

Enzyme mimics and artificial enzymes have been designed to imitate several different aspects of enzyme catalysis. For instance, mimics have been constructed to imitate an enzymatic mechanism, using the same catalytic groups which the enzyme uses and examining the question of whether a proposed mechanism for the enzymatic process is reasonable or duplicable. Another common goal of enzyme mimics is the imitation of the velocity of enzyme catalyzed reactions. Enzymes can increase the rates of ordinary reactions by factors as large as 10^{12} using very ordinary catalytic groups to do so. One of the challenges for enzyme mimics is to duplicate not only the mechanism by which the enzyme operates but also the very large rate of the process. The third area of interest is to imitate the selectivity of enzyme-catalyzed reactions.

Enzymes generally show four kinds of selectivity: substrate selectivity, reaction selectivity, regioselectivity, and stereoselectivity. An enzyme selects one substrate out of a mixture in solution and performs a desired chemical reaction on it. It also directs that reaction to a particular region of the substrate even though there may be others of equal or greater intrinsic chemical reactivity, and achieves the transformation with good stereoselectivity. This can include the formation of a single enantiomer, as in the synthesis of natural amino acids.

We have been pursuing the goal of imitating enzyme-catalyzed reactions for more than 25 years [1]. Very early in our studies we came to realize that selectivity and velocity can be simply two different aspects of the same effect. That is, selectivity may reflect selective acceleration of one reaction rather than another. However, the demands for this selective acceleration are much less than those required to achieve enzymatic rates. If a reaction can be selectively accelerated by as little as 10^2, this can lead to the essentially exclusive production of a single desired product. Even many years ago there were a number of chemical reactions which could achieve this magnitude of acceleration by the use of enzyme-like mechanisms. Thus we set out to try to translate such accelerations into chemical selectivity.

Our first approach to this was to look at chemical reactions, in the intramolecular versions, which were already subject to geometric control. Such reactions are normally very special, requiring starting materials which are set up to perform internal reactions. We sought to generalize this chemistry by attaching rigid reagents to ordinary substrates, so that these reagents could then perform intramolecular functionalization reactions directed by the geometry of the intermediate molecules. This is closely similar to what is done in selective enzyme reactions. A substrate is bound to an enzyme, and there is then selective intracomplex chemistry governed by the geometry of the resulting complex. In this way enzymes are able to perform selective attacks on unactivated carbons in the middle of large hydrocarbon segments, in which the only distinguishing feature of the carbon attacked is its precise geometric position.

The first system we examined involved photochemical functionalization of flexible molecules [2], and later of steroids [3]. We found that flexible molecules could be attacked when they were converted to esters of benzophenonecarboxylic acids, and that the positions attacked were those expected from a detailed consideration of the geometries of the molecules. However, with flexible substrates a number of conformations were available, and the resulting selectivity for any particular position of attack was not high. We will return to this point later, when we discuss ways to immobilize such flexible chains.

When benzophenonecarboxylic acids were attached to steroids, much better selectivity was observed [3]. Now both the reagent and the substrate are rigid, and the only flexibility in the system involves rotation around some of the single bonds where the reagent has been attached to the substrate. The geometric constraints in such systems led to attack on only a few steroid positions, all of them essentially equidistant from the original attachment point. As an example [4] which was particularly well controlled, the ester of 3-α-cholestanol with benzophenone-4-acetic acid underwent hydrogen transfer on photolysis. The ketone carbonyl was reduced, and a double bond was introduced into the steroid substrate. Only one steroid olefin was formed, with the double bond between carbons 14 and 15. This very high selectivity was exactly that predicted from a consideration of the detailed geometry of the system.

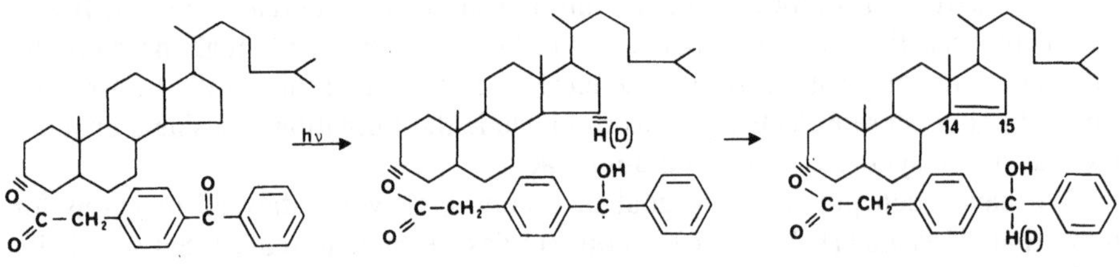

Although this chemistry showed good geometric selectivity in some cases, it was not really practical. Enzymes are catalysts which direct functionalization reactions, while our attached benzophenones were reagents consumed in the reaction. Thus we set out to develop catalytic processes.

An early example [5] was the directed epoxidation of a steroid. It was known that a hydroxyl group next to the double bond in an organic molecule could direct the epoxidation of that double bond. Useful as this reaction is, it is restricted to molecules which happen to have such a nearby hydroxyl group. We set out to generalize the process, by attaching a rigid template molecule which would carry a hydroxyl group into the spacial vicinity of a double bond and direct its epoxidation. An example is seen in the sterol diene shown in Eq. (2). Without our attached template the hydroxyl group of this sterol directs epoxidation of the double bond at the 4,5 position. When we attached a template to the molecule a new hydroxyl group extended in space so as to hold the epoxidizing reagents under the double bond at the far end of the steroid, between carbons 14 and 15. Now epoxidation of this molecule led to a selective attack on this remote double bond, in which both regiochemistry and stereochemistry were completely determined by the geometry of the system. The reaction was quite sensitive to geometry, and related structures in which the template had a slightly different arrangement were not able to direct this process. At the end of the reaction the template had not been consumed, and could be recovered by hydrolysis. Thus this attached hydroxyacid template both catalyzes and directs the epoxidation, so it serves some of the functions of an enzyme.

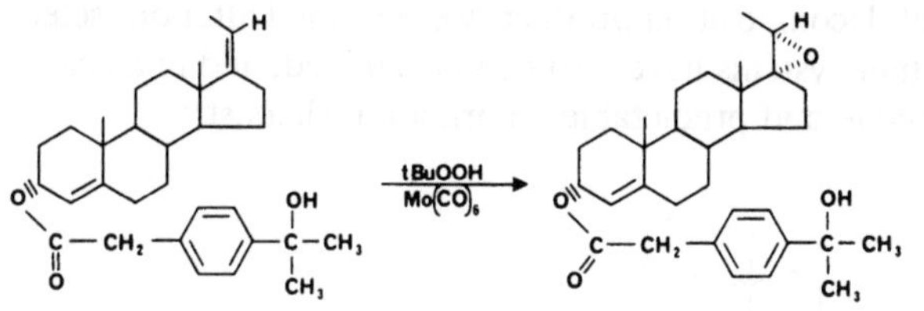

(2)

The most extensive studies we have done in geometrically directed functionalizations have involved the template directed chlorination of unactivated C-H bonds. These have been reviewed in detail [6], so only a few examples will be mentioned. The first system examined [7] involved attachment of an iodoaryl group to a steroid substrate, conversion of the iodine atom to an ICl_2 group, and free radical chain chlorination. After initiation of the chain a radical was produced carrying a single chlorine on the iodine atom; this chlorine then removed a hydrogen atom from the attached substrate.

Selective Chlorination at C-9 of Cholestanol

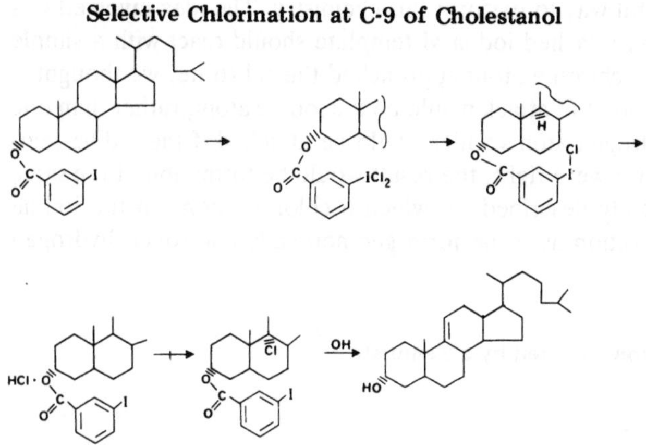

Geometric constraints directed the reaction to a particular region of the substrate, and in this case some chemical selectivity was also helpful. The intermediate iodochloro radical is fairly selective for the removal of tertiary hydrogens if possible. This preference meant that it was quite easy to get a completely selective chlorination process by imposing geometric control on the system. When the attached reagent was a m-iodobenzoate, the iodine was held in such a position that only the tertiary hydrogen at C-9 was accessible during attack on the attached 3-α-cholestanol. Cholestanol chlorinated at C-9 was indeed the only detectable chlorinated product. Other selectivities were imposed by simple geometric changes in the system.

The ester of 3-α-cholestanol with p-iodophenylacetic acid held the iodine in such a position that an attached chlorine atom could reach the hydrogen of C-14, but not the hydrogen of C-9. With this new material, conversion to the corresponding ICl_2 derivative and initiation of the free radical chain indeed led to no chlorination at C-9, but to products exclusively derived from chlorination at C-14. The process was again completely selective, although the product 14-chloro compound is somewhat unstable and other subsequent decomposition products were formed after our selective reaction. A number of other systems have also been examined, and in all cases the geometry resulted in selective and predictable chlorination chemistry.

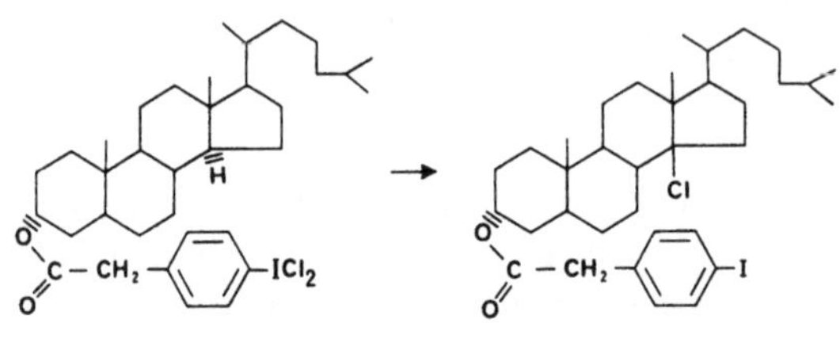

One problem with this reaction was the need to convert the attached iodocompounds to ICl_2 groups, and the fact that the reaction was then stoichiometric in this attached reagent. It occurred to us that it might be possible to invent a new process in which an iodine derivative would act as a template to direct the attack of an external reagent. We devised the radical relay chlorination reaction [8], which has proved to be the most useful way to perform this chemistry. The idea involved was that a substrate carrying an attached iodoaryl template should react with a simple chlorinating agent. When a chlorine atom approached the substrate, we thought it likely that the preferential point of attack would be the iodine atom, rather than one of the many substrate hydrogen atoms which could be attacked if the iodine template were not present. If we were right, the result would be formation of the same intermediate we have already described, in which a chlorine atom on the iodine group is held in such a position as to perform geometrically controlled hydrogen abstraction.

Radical-Relay Chlorination Directed by a Template

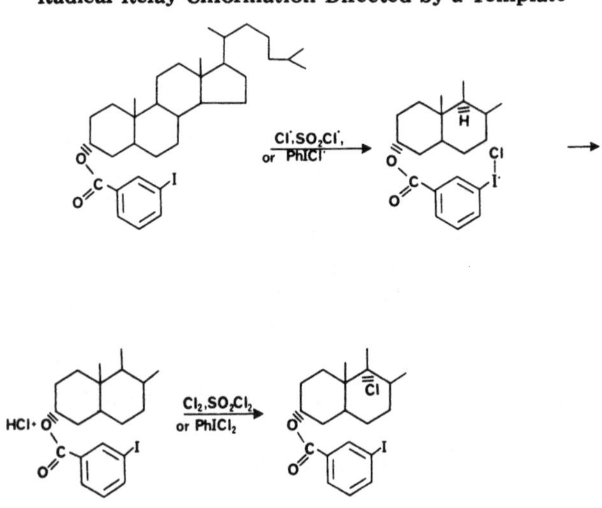

This scheme was successful. In fact, the iodoaryl groups attached as templates to the steroid not only directed the reactions but also catalyzed them. We got geometrically controlled selective chlorination of the steroid when the iodoaryl templates were present, while under the same conditions with attached groups lacking the iodine atom there was no chlorination of the steroid at all. Apparently the process in which iodine captures and delivers a chlorine atom is faster than the decomposition processes which otherwise destroy the reagents before they can attack the steroid. Under more vigorous conditions, in which the steroid is attacked, compounds lacking an iodoaryl template gave quite random chlorination processes.

As we have described elsewhere [9], we were able to use this radical relay process to complete a successful synthesis of cortisone. With another appropriate template we were able to achieve a selective chlorination at C-17 of the steroid, which permitted us to remove the steroid side chain [10]. Recent work in Germany [11] has resulted in a somewhat easier sequence for converting the 17-chloro steroid derivative to a useful intermediate in which the side chain has been removed. It remains to be seen whether the steroid reactions by which we can produce corticosteroids or remove steroid side chains and generate useful intermediates actually lead to practical transformations of industrial interest. These reactions are under active investigation in several companies.

One problem with this chemistry is that the template molecule which is attached to the substrate is what one might call an "honorary" catalyst. That is, it is unchanged at the end of the reaction and can be recovered by hydrolysis, but it is not able to process more than one substrate molecule in the course of a reaction. The catalyst cannot be used in "catalytic" amounts and does not show turnover. Several approaches to turnover catalysts have been made.

In the first of them, we attached three cholestanol molecules to a single template by preparing a silyl triether of the substrate [12]. Two types of templates were examined, one of which had an iodoaryl group while the other had a thiophene ring. In other work we had shown that certain sulfur compounds could also act to capture and transfer chlorine atoms by a radical relay mechanism. We were able to use an excess of chlorinating agent with these triether derivatives and successfully recycle the templates. With the iodoaryl template, for instance, 2.5 out of the 3 steroid units were selectively chlorinated. In this process the template captures an incoming chlorine from an incoming reagent, then geometrically directs that chlorine to abstract the C-9 hydrogen of the steroid. The resulting steroid radical is then chlorinated. The next time the template has a chlorine atom attached on the iodine it can no longer attack this steroid, since the 9-hydrogen is missing and none of the other hydrogens are attacked by this selective species. Thus the template moves on to one of the other unfunctionalized steroids and carries out its conversion. The overall process is good enough that more than 80% of each of the three steroids was chlorinated, while the remainder was simply unfunctionalized material. No other chlorinated products were formed, and the overall process shows turnover catalysis in the sense that one catalyst molecule has now processed three attached substrate species.

In order to achieve large numbers of turnovers, one would need to use a temporary association between catalyst and substrate. The catalyst should bind to the substrate, catalyze and direct a functionalization reaction, then release the product

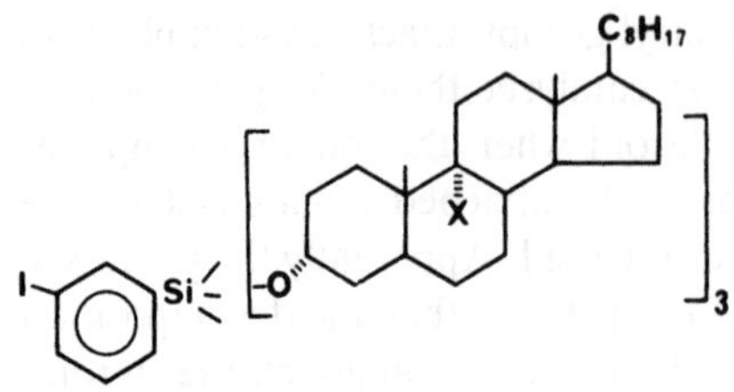

X: H

X: Cl

and go on to perform the reaction again with a new substrate molecule. To achieve this, we have examined temporary associations in a number of systems.

One of the earliest studies was actually done with benzophenone photochemistry. We showed that this photochemistry could be directed if the reagent had a carboxyl group and the substrate was also a carboxylic acid [13]. In non-polar media these two associate by hydrogen bonding; the resulting complex showed reasonably good geometric selectivity. In a related study we recently demonstrated that such hydrogen bonding can be used to solve the problem of flexibility in substrates [14]. We used a benzophenone dicarboxylic acid, which underwent a double interaction with substrate dicarboxylic acids. When the geometry was just right to let the substrate be stretched across the reagent we got very high selectivity for attack on the predicted carbon atom. When systems were geometrically mismatched, the reactions were not selective and interesting product distributions were observed. In this work we also showed that ion pairing was effective. That is, a benzophenone carrying two positively charged groups formed a double ion pair with a substrate dianion. This led to extremely good geometric selectivity, immobilizing the substrate into a single conformation while holding it in a well defined position relative to the benzophenone reagent.

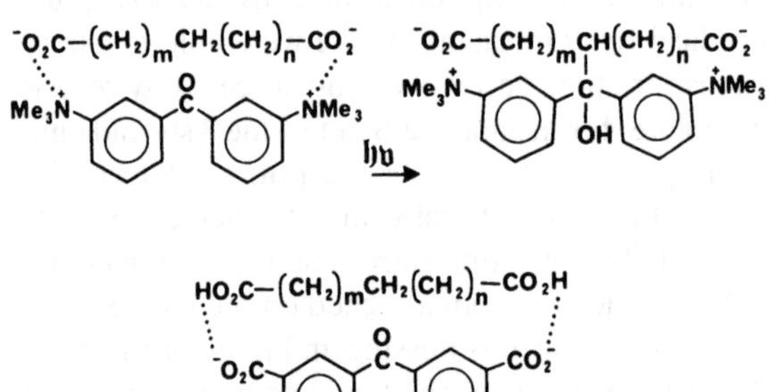

More recently, we have extended ion pairing to template-directed chlorination [15]. Since in the chlorination reactions HCl is produced, we could not use carboxylate anions. However, ion pairs of substrate cations with iodoaryl templates carrying sulfonate anion groups directed chlorination with reasonably good selectivity. Somewhat poorer selectivity, because of flexibility, was achieved when the template carried the cationic group and the substrate was a sulfate anion. The ion paired template shows turnover catalysis by moving on to another substrate molecule after it has finished one reaction. We are currently working on yet other ways to achieve temporary association of substrate and catalyst, so as to continue to improve these systems.

The enzymes which achieve geometrically controlled attack on unactivated hydrogens do not use chlorine atoms for the attack, but instead perform oxygen insertions. They have iron-porphyrin units and are members of a group generally described as cytochromes P-450, named for a particular spectroscopic band. These enzymes have been the subject of very intensive biochemical and model studies.

One of the most interesting model studies was the work of Groves [16], who showed that metal porphyrin derivatives could catalyze oxygen insertion from an oxygen donor iodosobenzene into hydrocarbons such as cyclohexane. It is obviously attractive to try to develop geometrically controlled model systems which would achieve this process; some studies directed toward this goal have been done by Groves and by Chang [17]. In order to achieve turnover catalysis, it would be necessary to produce some sort of temporary association between the porphyrin catalyst and a substrate, with good geometric control within the resulting complex.

We have recently explored a different approach to such systems. Compounds are known, although not well known, in which the oxygen of iodosobenzene is replaced by a nitrogen atom carrying a sulfonyl group. It occurred to us that the extra linkage from this nitrogen atom should make it possible to perform geometrically directed insertion reactions in which the geometry was controlled by the position of the reagent relative to the substrate. Thus a covalent link between reagent and substrate could be used, and catalytic amounts of a porphyrin should be able to promote the overall directed process. This chemistry, still very new, has developed as hoped. We reported recently that various catalysts including metal porphyrins would promote the reaction of $C_6H_5I{=}N$-Ts with cyclohexane to afford N-tosylcyclohexylamine [18]. As expected, it was possible to turn this into an intramolecular reaction with geometric direction, as shown in Eq. (8) [19]. Other work in our laboratory is aimed at developing the same kind of long range geometric control for this process which we had achieved for the reactions discussed earlier. We even have some indication, in collaborative work with Dawson, that rabbit microsomes will catalyze the reaction of our iodine reagent with cyclohexane. If indeed the enzyme cytochrome P-450 can use this reagent and catalyze nitrogen insertions, this could lead to novel selective enzyme-catalyzed processes, producing useful nitrogen compounds from hydrocarbons.

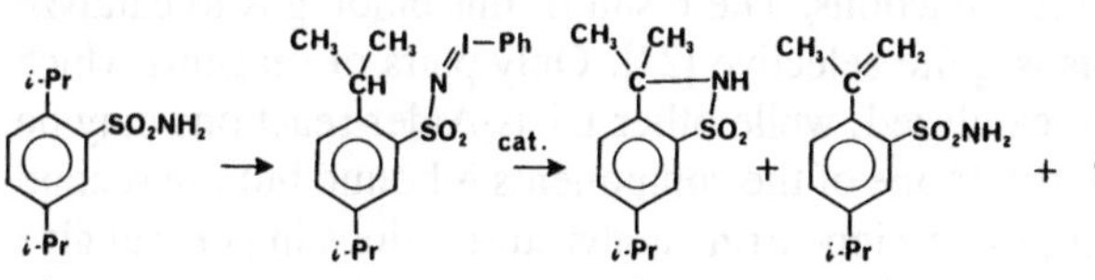

1 2 3 4 (8)

One of the most attractive ways to bind organic molecules to a substrate on a temporary basis is to use hydrophobic binding. Some synthetic hydrophobic cavities have been developed. However, the most generally useful and available materials have been the cyclodextrins, enzymatic degradation products of starch [20]. These simple molecules have glucose units in a ring, and form a hydrophobic cavity into which various substrates will bind. The cavities are highly selective for the size and shape of the substrates.

One of our earliest studies of selectivity using these systems was of aromatic substitution reactions [21]. We found that anisole would bind into α-cyclodextrin (the compound with six glucoses in its ring), and that within this complex aromatic chlorination was completely selective for the para position. By contrast, anisole in free solution is chlorinated rather randomly in both the ortho and para positions. Thus the hydrophobic complex of anisole with cyclodextrin involved temporary association of a template with a substrate, in which the reaction of the substrate with an external reagent was directed by the geometry imposed by the template. The relationship of this to the spirit of our steroid functionalization processes is obvious. Even more interestingly, the reaction turned out to be much more closely related than this general description might imply.

Ionic Relay Chlorination in a Cyclodextrin Complex

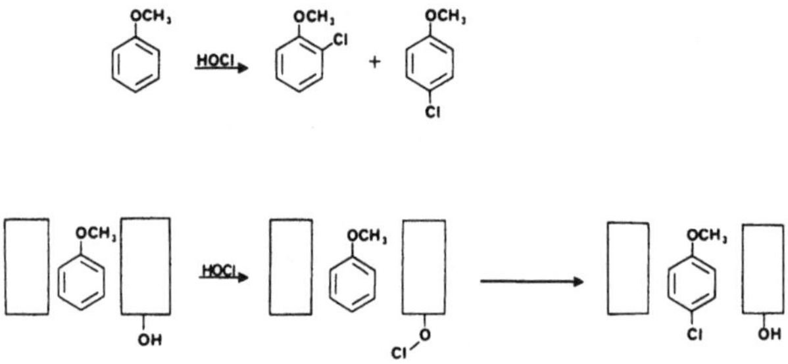

We found that an external chlorinating reagent preferentially passed the chlorine to the template cyclodextrin first, and that the cyclodextrin then relayed the chlorine on to the substrate. Furthermore, this was a catalytic process, and occurred faster than chlorination in the absence of the template. The mechanism involved was established by detailed studies, including reaction kinetics. Modification of the cyclodextrin, and its incorporation into a polymer, have led to the production of highly selective catalysts for this aromatic substitution reaction [22]. In other laboratories an electrochemical adaptation of our reaction has also been made, in which the cyclodextrin molecule is attached to the electrodes [23].

In the larger cavity of β-cyclodextrin it is possible to bind both the diene and the dienophile of certain Diels-Alder reactions. The result of this binding is to catalyze the reaction, and this catalysis is quite selective [24]. Only pairs of reagents which can both fit into the cavity are catalyzed, while other Diels-Alder reactions may be inhibited by the same cyclodextrin if one of the components is bound but the second cannot also fit in. Furthermore, the cyclodextrin-catalyzed reactions in general give a different product distribution from that obtained in the absence of the catalyst, because the geometry in the mixed complex helps to determine the orientation of the two components. There is even a small but real induction of optical activity in the Diels-Alder products, since they are being formed in a chiral environment.

Reactions in cyclodextrins can show chiral inductions because of the chirality of the glucose units. As one example, we have been examining the reaction of certain nitrophenyl esters, bound into the cyclodextrin cavity, with the hydroxyl groups of

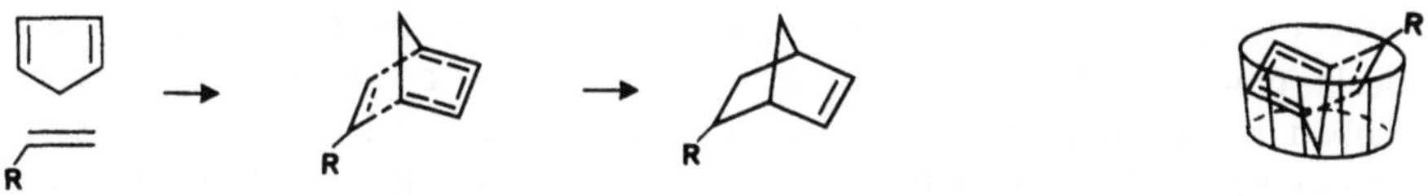

the cyclodextrin. This reaction, which imitates the first step in many serine enzyme hydrolysis processes, has been optimized by appropriate tailoring of the substrate [25]. With the optimal cases it gives extremely high rate accelerations, compared to the reaction of substrate with water alone in the absence of the cyclodextrin. For our current purposes, it should be noted that very high enantioselectivities are also observed in some cases. We have achieved a 62-fold preference with one of the two mirror image substrate molecules in this acylation process, reflecting the chirality of the cyclodextrin cavity [25]. We will return to the influence of cyclodextrin chirality when we discuss the synthesis of amino acids.

A different kind of selectivity has been observed with cyclodextrin bisimidazoles. These species were synthesized as mimics of the enzyme ribonuclease, and indeed they do perform the cleavage of certain bound phosphate esters with a mechanism closely related to that of the natural enzyme [26]. Of course in the natural enzyme there is considerable selectivity, including regioselectivity. For instance, the natural enzyme can cleave a cyclic phosphate ester between C-2 and C-3 of a ribose unit in such a way as to leave the phosphate group entirely on C-3. Ordinary chemical hydrolysis of such a molecule would lead to a mixture of C-2 and C-3 phosphates. Our cyclodextrin bisimidazole is able to imitate such selectivity.

As shown in Eq. (11), when a cyclic phosphate undergoes catalyzed hydrolysis by this molecule the substrate binds to the cavity and is hydrolyzed by the catalytic imidazole groups in such a way that only one of the two possible products is formed. The alternative product is present only in small, but still detectable, amounts. In the absence of the catalyst the hydrolysis is essentially random with a slight preference for the alternative product. Thus we have here an example in which we imitate not only the mechanism of the enzyme but also its selectivity, presumably because of the same geometric control which the enzyme itself exerts.

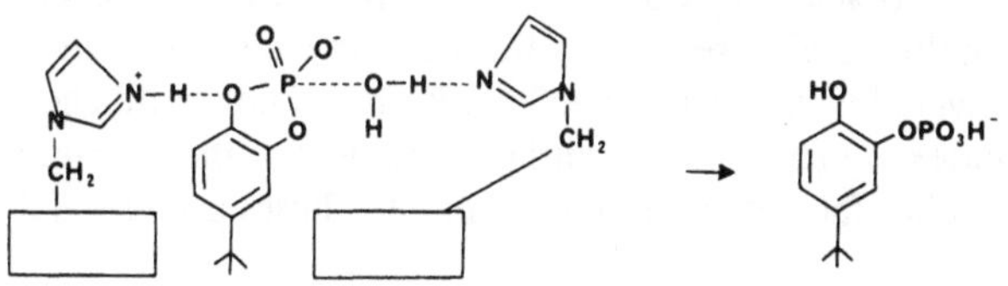

(11)

A great range of interesting biochemical reactions involve coenzymes. These are generally small molecules which are bound to an enzyme in a geometry such that they can interact with a bound substrate. In the chemistry of the synthesis and degradation of amino acids, the coenzyme which plays the important role is pyridoxal phosphate and its derivatives. The classic work of Snell [27] and of Braunstein [28] established the mechanisms of these processes, and demonstrated that the coenzyme can perform many of the reactions by which amino acids are synthesized and degraded even in the absence of the enzyme. However, the reactions are much slower in the absence of the enzyme, and they are much less selective. The enzyme imposes a geometric control on these processes which helps to direct them to a particular path. Thus this was an attractive system for the application of our ideas. We hoped that we could impose geometric selectivity on model systems of this sort, leading to the selective production of particular products with better efficiency than would be available in reactions of the simple coenzyme itself.

In our first work on this problem, we learned how to attach pyridoxamine to β-cyclodextrin at the C-6 position of one of the glucose residues [29]. The resulting compound showed two very interesting kinds of selectivity. First of all, it preferentially reacted with substrates which could use the cyclodextrin binding site. Thus it showed substrate selectivity, preferentially forming aromatic amino acids such as phenylalanine and tryptophan in competition with processes which would form other non-binding amino acids. In the absence of the cyclodextrin unit no such selectivity is observed. The other interesting point was that the product amino acid is being formed in the presence of the asymmetric cyclodextrin unit, and it turns out that it is formed with rather good chiral induction. Thus this system shows stereoselectivity. In the synthesis of phenylalanine, for instance, there was a five to one preference for the formation of the natural L-amino acid by this reagent, although in the absence of the cyclodextrin unit the process gives, as it must, completely racemic products.

At the end of the reaction a keto acid has been converted to an amino acid, and the coenzyme pyridoxamine has been converted to the aldehyde pyridoxal. In enzymatic processes the reaction would then be run in the reverse direction with a different amino acid to regenerate the starting materials, and we found that that was also possible here. Thus this system is truly catalytic, converting a keto acid and a different amino acid into the new amino acid and new keto acid by running the process back and forth. The chiral synthesis of amino acids by this process, so closely related to that in an enzyme, may prove to be of practical interest.

We have also prepared a number of other derivatives in which the pyridoxamine system is attached to binding groups. In one case we attached it to C-3 of β-cyclodextrin, and found that this material also shows a preference for the formation of aromatic amino acids in which the substrate can use the cyclodextrin binding site [30]. More interestingly, the amino acid tryptophan is synthesized with a reversed stereochemical preference from this catalyst relative to the stereochemistry obtained with the previously described cyclodextrin derivative. That is, a derivative attached to the primary carbon of cyclodextrin shows preferential formation of the L-amino acid while that attached to the other face of the cyclodextrin unit shows preferential formation of the D-amino acid.

Certainly one of the most interesting goals in the synthesis of enzyme mimics is to try to incorporate all of the catalytic groups involved in the enzyme. In transamination reactions with pyridoxamine phosphate or pyridoxal phosphate as coenzymes, the enzyme has a catalytic basic group which performs the proton transfers involved in the reaction. In fact, in the enzyme the preferential formation of a chiral amino acid almost certainly is a reflection of the fact that this basic group removes a proton from one part of the intermediate and then transfers it in a stereoselective fashion to the carbon which becomes asymmetric in the product amino acid. We set out to imitate this critical proton transfer process.

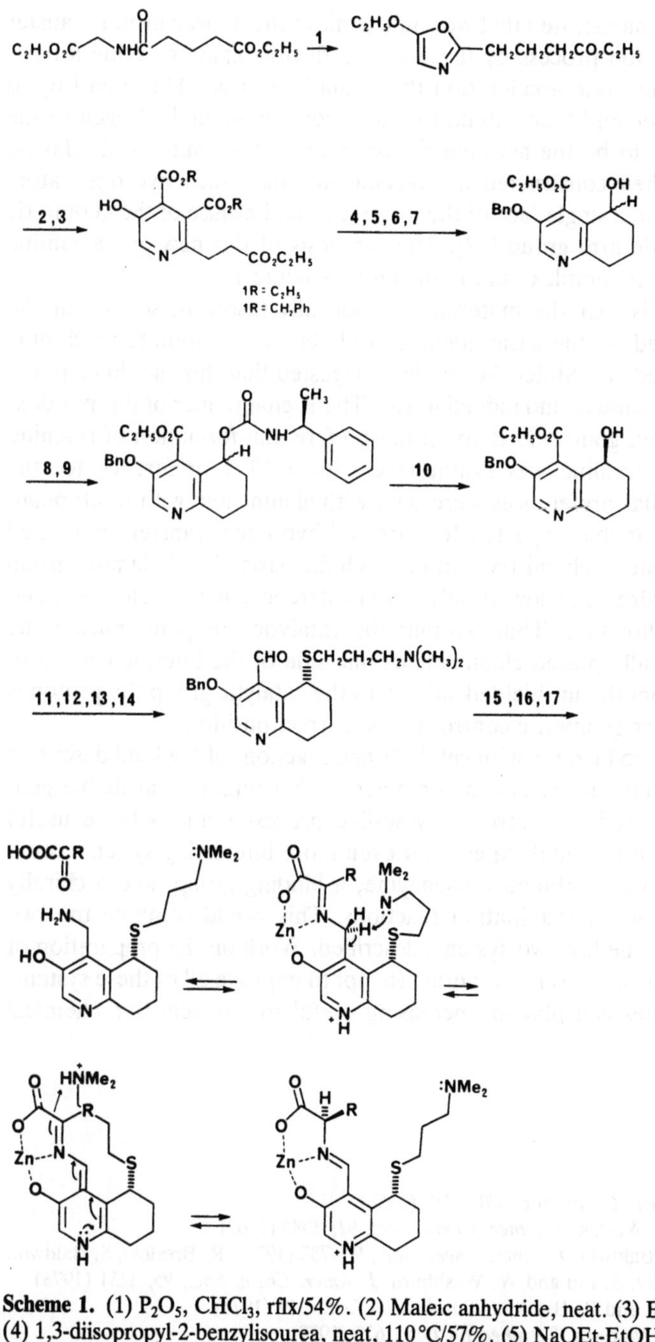

Scheme 1. (1) P_2O_5, $CHCl_3$, rflx/54%. (2) Maleic anhydride, neat. (3) EtOH-sat'd HCl_g, rflx/55%. (4) 1,3-diisopropyl-2-benzylisourea, neat, 110 °C/57%. (5) NaOEt-EtOH, PhH, rflx/99%. (6) 2.4 N HCl, EtOH, rflx/41%. (7) $NaBH_4$, EtOH, 5% $NaOH_{aq}$/99%. (8) carbonyl diimidazole, CH_2Cl_2, R.T. (9) α-1-(−)-Phenethylamine, PhH, $BF_3 \cdot O_2Et$, rflx/58%. (10) MPLC; $HSiCl_3$, Et_3N, PhH, rflx/85%. (11) MsCl, Et_3N, CH_2Cl_2, 0 °C/99%. (12) $HS(CH_2)_3N(CH_3)_2$, NaH, THF, 0 °C-R.T./82%. (13) LAH, THF, R.T./58%. (14) $CrO_3 \cdot Pyr_2$, CH_2Cl_2, R.T. (15) $HONH_2$, NaOAc, H_2O, EtOH. (16) Zn, HOAc; H_2S, CH_3OH; CM 25 sephadex/26% (steps 14→16). (17) 6 N HCl, rflx, 30 mins; CM 25 sephadex/82%

Our early studies demonstrated that we could achieve the desired proton transfer in a model transamination process by the use of a flexible chain carrying a basic group [31]. For optimum rate acceleration this chain had to be able to curl up to reach a nearby hydrogen and then extend to that it could pass the hydrogen to the new carbon which was to be the asymmetric α-carbon of the amino acid. Having established this, we then constructed a molecule in which such hydrogen atom transfers had to occur on a single face of the intermediate, because of the geometric locking of the basic side arm group [32]. The synthesis of this new pyridoxamine derivative was somewhat complex, and is shown in Scheme 1.

A critical element is that the material was optically resolved, so that in the transamination intermediate the basic sidearm would be in a position to reach only one face of the intermediate. Molecular models suggested that this should be possible with the compound shown, and indeed it was. The R enantiomer of this pyridoxamine derivative showed good selectivity in the preferential formation of D-amino acids. For example, norvaline was synthesized with a 19:1 preference for the D-amino acid, and similar preferences were seen with alanine and with tryptophan. The stereochemistry was that expected for directed hydrogen transfer. A related catalyst with the same stereochemistry, but in which the basic dimethylamino group was replaced with a hydrogen, showed only a weak stereochemical preference, and that in the opposite direction. Thus without the catalytic group to transfer the protons an asymmetrically placed chain shields one side of the intermediate, promoting protonation from the unshielded side; with the catalytic group the proton is in fact transferred under geometric control to the correct position.

Much more remains to be done with catalysts and reactions of the kind described in this paper. The selectivities achieved demonstrate that relatively available geometric factors can be used to control enzyme-like processes and achieve useful selectivities. However, one would expect that even more interesting systems could result if, for instance, we combined a coenzyme, a binding group, and a chirally placed catalytic group in transamination reactions. This would combine the two different approaches in the last two systems described. Work on the preparation of such compounds is underway, as is a general attempt to improve all of these systems so that artificial enzymes can play an increasing useful role in selective chemical synthesis [33].

References

1. Cf. R. Breslow, *J. Amer. Chem. Soc., 80,* 3719 (1958)
2. R. Breslow and M. A. Winnik, *J. Amer. Chem. Soc., 91,* 3083 (1969)
3. R. Breslow and S. W. Baldwin, *J. Amer. Chem. Soc., 92,* 732 (1970); R. Breslow, S. Baldwin, T. Flechtner, P. Kalicky, S. Liu and W. Washburn, *J. Amer. Chem. Soc., 95,* 3251 (1973)
4. R. L. Wife, D. Prezant and R. Breslow, *Tetrahedron Lett.,* 517 (1976)
5. R. Breslow and L. M. Maresca, *Tetrahedron Lett.,* 623 (1977)
6. R. Breslow, *Accts. of Chem. Res., 13,* 170 (1980)
7. R. Breslow, R. Corcoran, J. A. Dale, S. Liu and P. Kalicky, *J. Amer. Chem. Soc., 96,* 1973 (1974)
8. R. Breslow, R. J. Corcoran, and B. B. Snider, *J. Amer. Chem. Soc., 96,* 6791 (1974)
9. R. Breslow, B. B. Snider and R. J. Corcoran, *J. Amer. Chem. Soc., 96,* 6792 (1974)
10. B. B. Snider, R. J. Corcoran and R. Breslow, *J. Amer. Chem. Soc., 97,* 6580 (1975)

11. P. Welzel, K. Hobert, A. Ponty, *Tetrahedron Lett., 24,* 3199 (1983)
12. R. Breslow, D. Heyer, *Tetrahedron Lett., 23,* 3451–3454 (1982)
13. R. Breslow and P. C. Scholl, *J. Amer. Chem. Soc., 93,* 2331 (1971)
14. R. Breslow, R. Rajagopalan and J. Schwarz, *J. Amer. Chem. Soc., 103,* 2905 (1981)
15. R. Breslow and D. Heyer, *Tetrahedron Lett.*, 5039 (1983).
16. J. T. Groves, T. E. Nemo and R. S. Myers, *J. Amer. Chem. Soc., 101,* 1032 (1979)
17. D. K. Chang, M. Kuo, *J. Amer. Chem. Soc., 101,* 3413 (1979)
18. R. Breslow, S. H. Gellman, *J. Chem. Soc. Chem. Comm.*, 1400 (1982)
19. R. Breslow and S. Gellmann, *J. Amer. Chem. Soc., 105,* 6729 (1983)
20. For a review, cf. M. L. Bender and M. Komiyama, "Cyclodextrin Chemistry", Springer-Verlag, New York, 1977.
21. R. Breslow and P. Campbell, *J. Amer. Chem. Soc., 91,* 3085 (1969); R. Breslow and P. Campbell, *Bioorg. Chem., 1,* 140 (1971)
22. R. Breslow, H. Kohn and B. Siegel, *Tetrahedron Lett.*, 1645 (1976)
23. T. Matsue, M. Fajihira and T. Osa, *J. Electrochem. Soc., 126,* 500 (1979)
24. R. Breslow and D. Rideout, *J. Amer. Chem. Soc., 102,* 7816 (1980)
25. R. Breslow, G. Trainor and A. Ueno, *J. Amer. Chem. Soc., 105,* 2739 (1983)
26. R. Breslow, J. Doherty, G. Guillot and C. Lipsey, *J. Amer. Chem. Soc., 100,* 3227 (1978)
27. E. E. Snell, *J. Biol. Chem., 157,* 491 (1945)
28. Cf. A. E. Braunstein, in "The Enzymes", 2nd edition, P. Boyer, H. Lardy and K. Myrbäck, eds., Academic Press, New York, 1960, vol. 2, ch. 6.
29. R. Breslow, M. Hammond, M. Lauer, *J. Amer. Chem. Soc., 102,* 421 (1980)
30. R. Breslow, A. W. Czarnik, *J. Amer. Chem. Soc., 105,* 1390 (1983)
31. S. C. Zimmerman, A. W. Czarnik and R. Breslow, *J. Amer. Chem. Soc., 105,* 1694–1695 (1983).
32. R. Breslow and S. Zimmerman, *J. Amer. Chem. Soc., 106,* 1490 (1984)
33. Support of this work by the National Science Foundation and the National Institutes of Health, and the contributions of my coworkers, are gratefully acknowledged

Biopolymers in Reverse Micelles as Guest – Host Molecular Systems

Pier Luigi Luisi[1] and Linda Magid[2]

[1] Institut für Polymere, ETH-Zentrum, CH-8092 Zürich, Switzerland.
[2] Dept. of Chemistry, University of Tennessee, 37996-1600 Knoxville, USA.

General properties of reverse micelles

Surfactant molecules are amphiphilic, possessing nonpolar regions (frequently *n*-alkyl group(s)) as well as polar regions (generally referred to as the head groups) which may or may not bear charges. In solvents having low dielectric constants, for example aromatic, aliphatic and halogenated hydrocarbons, surfactant aggregation is dominated by solvophobic interactions involving the head groups. These groups are sequestered in the interior of the aggregates, largely shielded from interactions with bulk solvent; hence the name reverse micelles, since these aggregates are inside out compared to normal micelles, which are formed in water and have their head groups exposed to the solvent (see Fig. 1). Important contributions to micelle stability in apolar solvents come from hydrogen bonding, dipole-dipole, dipole-induced dipole and dipersion interactions. Several classes of surfactants which aggregate in

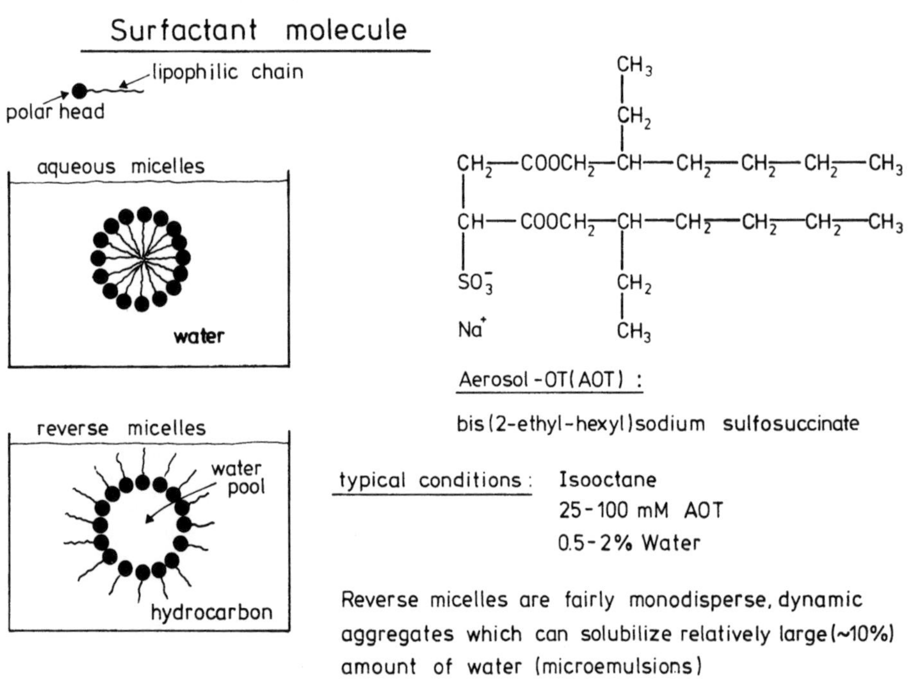

Fig. 1. Schematic representation of aqueous and reverse micelles (cross-section); and the structure of the surfactant AOT, with the typical conditions to which most of the work presented in this paper refer

Table 1.

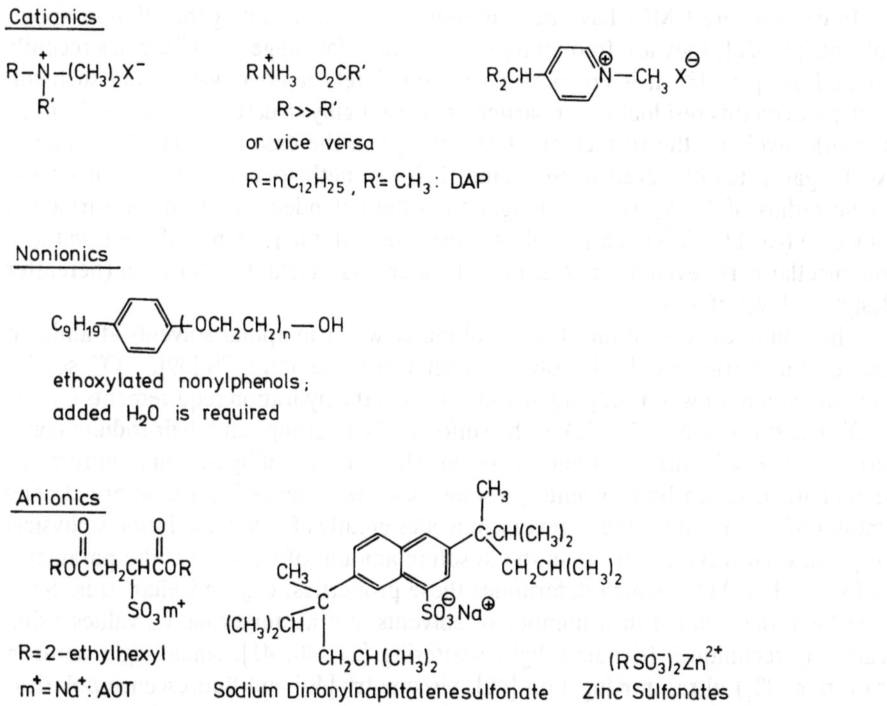

apolar solvents are shown in Table 1. The formation and characterization of reverse micelles have been reviewed recently [1, 2], so we present here only a brief introduction.

Unlike surfactant aggregation in aqueous solutions, which is often characterized by a well-defined critical micelle concentration (CMC) and monomer ⇄ $\bar{n}$-mer association (n frequently taking values of ca. 50–100), surfactants in nonpolar solvents often display indefinite self association [3–13]. In several cases, the stepwise equilibrium constants for the aggregation monomer ⇄ dimer ⇄ trimer ⇄ . . . ⇄ $\bar{n}$-mer have been found to be equal; average aggregation numbers are frequently as low as three to seven at moderate total surfactant concentrations. Mono-di- and trialkylammonium salts and many non-ionic surfactants display this aggregation behaviour.

There are also surfactants which display monomer ⇄ $\bar{n}$-mer association in nonpolar solvents [12, 14]; they often contain two hydrocarbon tails. Dialkyl sulfosuccinates [15–22], dinonylnaphthalenesulfonates [23, 24] and phosphatidylcholines [25–31] are perhaps the most extensively studied. Aggregation numbers typically range from 12 to about 30, but dodecylammonium benzoate forms a trimer in a monomer ⇄ $\bar{n}$-mer association [32]. Even with these surfactants, the definition of a CMC is not without ambiguity. Thus in the case of Aerosol OT (sodium di(2-ethylhexyl) sulfosuccinate, AOT) [33–36], a premicellar linear trimer which undergoes a conformational transition to a closed trimeric unit has been implicated.

In certain solvents, the aggregation of AOT to 6-mers and 14-mers has also been inferred [37].

In cases where CMCs have been measured for surfactant aggregation in apolar solvents [15–22], they are frequently in the millimolar range. As Eicke has recently pointed out [21, 38], it is not possible to completely remove water from such solvents; indeed this residual water participates in a highly structured hydrogen-bonded network involving the surfactant's head groups in the case of AOT. The smallest AOT aggregates observed in isooctane (2,2,4-trimethylpentane) have a hydrodynamic radius of 15 Å, which is longer than the extended length of the surfactant molecule (ca. 11–12 Å). This result is consistent with the presence of some water in the micellar core, even at an operational water-to-surfactant molar ratio (hereafter designated w_0) of zero.

The ability of reverse micelles to solubilize water in apolar solvents at ambient temperature varies widely. In solvents such as toluene and CCl_4 [39], AOT solubilizes an amount of water only slightly in excess of the hydration requirements (ca. six H_2O's per surfactant molecule) of the sulfonate head groups and their sodium counterions, while solubilizing amounts substantially in excess of hydration requirements in aliphatic hydrocarbon solvents (in some cases w_0 exceeds 70) are observed. The amount of water solubilized in reverse micelles greatly affects the solution's physical properties. However, more than the absolute amount of water, it is the parameter w_0 ($w_0 = H_2O/AOT$) which determines these properties, e. g., micellar sizes. Sizes have been determined in a number of solvents and at numerous w_0 values using scattering techniques (dynamic light scattering [21, 40, 41], small-angle neutron scattering [42]) ultracentrifugation [40], viscometry [40] and fluorescence polarization [43, 44] measurements. As w_0 increases, the diameters of the central water pools increase and the number of water and AOT molecules per aggregate also increase. Table 2 shows radii obtained using dynamic light scattering. There is some penetration of the continuous phase into the surfactant tails, as the small-angle neutron scattering (SANS) data in Table 3 indicate (note that $R_2 - R_1$ is less than 12 Å). It is also widely accepted that reverse micelles, in contrast to normal micelles in water, have a very moderate size distribution.

The appearance of free water ($w_0 \geq 10$–15), which coincides with the aggregates' acquiring a water pool encased by a curved surfactant monolayer having a constant

Table 2. The dependence of hydrodynamic radius on w_0 for AOT reverse micelles in isoocatane at 30 °C (light scattering)

w_0	R_H, Å	R_W, Å*
0	15	–
5	31	19
9	35	23
15	40	28
27	61	49
38	105	93
50	145	133

* $R_W = R_H - 12$ (i. e. assuming no solvent penetration) (from Ref. [20])

Table 3. The dependence of overall and water pool radius on w_0 for AOT reverse micelles in heptane at 25 °C obtained from small angle neutron scattering (SANS) measurements

w_0	R_O, Å	R_W, Å
10	33	26
15	36	29
20	42	35
30	–	48
40	–	60
50	–	90
60	–	120

(from Ref. [42])

area per head group, marks the transition to the water-in-oil (w/o) microemulsion domain. Once this condition obtains, the aggregate sizes (with further addition of water) show a very strong dependence on temperature and pressure [21] (size is virtually T and P independent below $w_0 = 15$, although of course dependent on w_0 itself). It should be noted that the substantial solubility of AOT in both water and hydrocarbon solvents (which is sometimes expressed by saying that AOT has a favorable hydrophile-lipophile balance) makes the attainment of large w_0's possible in certain hydrocarbon solvents without the requirement of an additional component (a cosurfactant, generally a medium chain length alcohol). Many single-tailed ionic surfactants do require a cosurfactant because of their marked hydrophilicity. It is advantageous to be able to avoid this fourth component when studying the properties of the water pools and surfactant interphase (see below), since its distribution among the various environments must first be elucidated.

The w_0-dependent conformational properties of AOT molecules in the surfactant interphase of reverse micelles have been investigated using ^1H and ^{13}C NMR [45–48]. The two hydrocarbon tails are conformationally distinct, making AOT in some respects a rather good model for phosphatidylcholines, which are im portant constituents of cell membranes. The w_0-dependence of the rotational correlation times (evaluated from ^{13}C T_1 values) for carbon atoms in the hydrocarbon tails indicates increased overall mobility of the AOT molecule but a decrease in the extent of segmental mobility as w_0 increases [47, 48]. The decrease in segmental mobility parallels the decrease in area per hydrocarbon tail which occurs at the surfactant/continuous phase boundary as w_0 increases. The extent of solvent penetration into the surfactant layer (benzene and CCl_4 penetrate to a greater extent than cyclohexane) may also be inferred from the NMR data. Depending on the time scale of experimental observation, micelles can be seen as static spheres or as dynamic entities participating in several rapid exchange processes among themselves and with the solvents. We will come back to this feature later on.

Several groups of investigators have looked at the effect of reverse micelles on reactions involving small molecules or ions; a rationalization of the kinetic data obtained requires some knowledge about a) the solubilization sites for the small molecules within the micelles; b) the nature of the water pool and c) the dynamics of solubilizate exchange among micelles. From a practical point of view, carrying out chemical reactions at the polar/apolar interfaces found within the reverse micelles in homogeneous solution offer certain advantages: the catalytic transformation of lipophilic substrates by enzymes, for example, is possible in such a system.

A number of studies have indicated that while exchange of small solubilizate molecules among reverse micelles is generally much more rapid than any chemical reactions the molecules may participate in (this may not be true for enzymes as solubilizates), exchange is a factor of ca. 10^3 below the diffusion-controlled rate [42, 49]. Exchange processes on time scales ranging from 100 μs to 10 msec have been observed for the Ni^{2+}/murexide ion pairs [50] and for the $TbCl_3$/phenylacetic acid pair [51] (monitored by enhancement of Tb^{3+} luminescence intensity) in AOT reverse micelles. Schelly and coworkers [37, 52] have investigated the kinetics of picric acid and 7,7,8,8-tetracyanoquinodimethane solubilization by AOT and other reverse micelles in benzene (using concentration jump experiments with absorption detection), identifying a rapid rate process (submillisecond) associated with adsorption at the AOT carbonyl groups and additional processes corresponding to further penetration of the picric acid to the level of the surfactant's sulfonate head groups.

w_0-Dependent Properties of the Water Pool Water and Sites of Solubilization for Small Molecules

Having described some of the general aspects of reverse micelles, let us consider now in more detail some of their physical properties. As already mentioned, these depend upon the amount of water solubilized, i. e. upon the parameter w_0. At values of w_0 less than about 10 in the case of AOT reverse micelles (the numerical value cited depends on the hydration requirements of the surfactant), the water pool water has properties which are substantially different from those of normal bulk water. Strong water-head group interactions cause dielectric constant decreases (often monitored by using a solubilizate which has solvent-sensitive spectral properties) and immobilization of the water molecules (inferred from ^1H NMR T_1s for the water [53, 54] and from microviscosities obtained by fluorescence polarization measurements employing various solubilizates) [43, 44, 55, 56]. The water's freezing point can be lowered by as much as 30 °C, which has interesting applications in cryoenzymology [57].

Above a w_0 of 10, the water activity [58] in AOT reverse micelles in isooctane approaches 1.0, the area per sulfonate head group at the surfactant water interface levels off at 55 $Å^2$ and the water molecules' rotational correlation times [53] (from ^1H NMR T_1s) reach their minimum value. The w_0-dependence of water pool polarity (often designated micropolarity) has been investigated using solubilizated which have solvent-sensitive UV-VIS absorption spectra [59, 60] and/or solvent-sensitive fluorescent lifetimes and quantum yields [61, 62]. Fluorescence polarization decay has been used to assess microviscosities. In such studies, it is important to determine by an independent technique the solubilizate's location within the reverse micelle, so that one is certain that the probe molecule is indeed located within the water pool rather than adsorbed at the interface (i. e. at the micelle boundary). At low w_0, contact between probe and interface is of course unavoidable. Even molecules which are quite soluble in water, such as phenols, are found to bind at the interface

[59, 60], with the hydrocarbon portion of the molecule intercalated among the surfactant tails. Electrostatic repulsion between surfactant head groups and the solubilizate may be used to ensure a water pool solubilization site for the probe. Thus perylene-tetracarboxylate (PTC) and pyranine [56] have been used to good advantage in studying AOT systems.

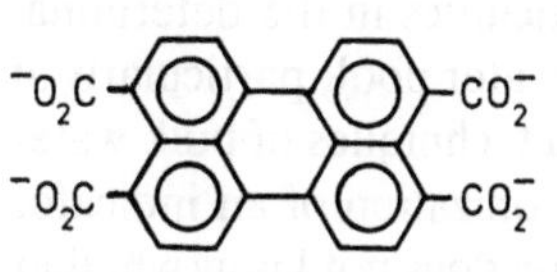

PTC

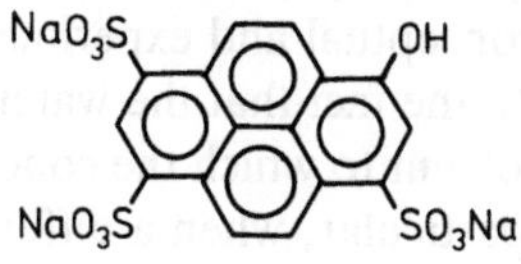

Pyranine

As expected, water pool microviscosity decreased and micropolarity increases as w_0 increases. With some probes, bulk water values are not obtained until w_0 is much greater than 10; these may be cases where the probe reports on interfacial water as well as on the free bulk water in the pool. Indeed, the fluorescence polarization measurements of Zinsli [55] using 1-amino-naphthalene-4-sulfonic acid (a water pool probe) and 2-(N-tetradecyl)amino-naphthalene-6-sulfonic acid (an interfacial probe) and the pK_a measurements of Sunamoto and co-workers [56] on excited state pyranine have been interpreted in terms of a radially inhomogeneous microviscosity and micropolarity, with a viscous boundary layer of hydration water (type I) at the pool periphery and an inner core of water (type II) whose properties rapidly approach those of ordinary bulk water as w_0 increases.

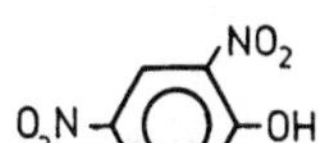

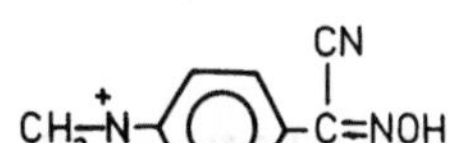

in water	7.14	4.1	4.6
in AOT (interfacial)	11.5	5.8	5.15
(water pool)	7.6 - 7.9		

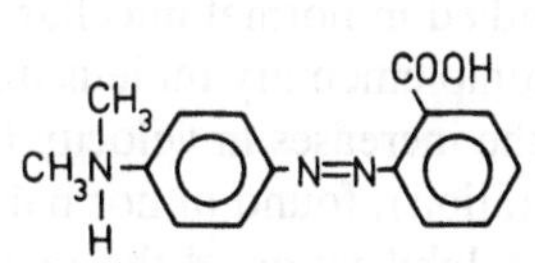

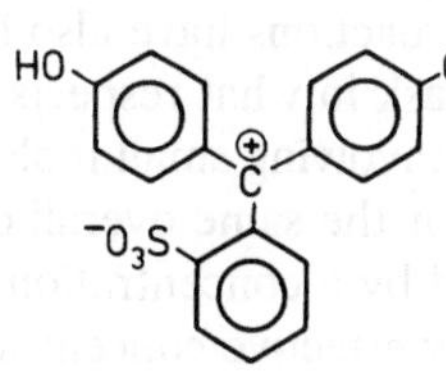

in water	4.95	7.8
in AOT	8.2	8.6 - 11.0

Solubilizate binding constants have been determined for a few reverse micelle – small molecule (phenols, imidazole, simple alcohols) combinations [60, 63, 64]. In the case of imidazole binding to dodecylammonium propionate (DAP) reverse micelles, the use of the chemical shift dependence on DAP concentration for imidazole protons led to nonunique (i. e., proton-dependent) binding constants, a physically unreasonable situation [64]. When one is studying chemical reactions involving reactants which partition among the water pools, with the surfactant layer on the continouous phase, one must know the effective concentration of each reactant at the reaction locus. This question of local versus overall reactant concentration is discussed further in the next section.

A knowledge of water pool pH(pH_{wp}) is extremely useful when one attempts to rationalize the effects of the water pool microenvironments on the conformations of a reaction participation by biopolymers in reverse micelles.

These are, however, conceptual and experimental difficulties in the determination of pH_{wp}, due mostly to the fact that the water of the water pool, particularly at low w_0 values, is a novel solvent to which the concepts and techniques of bulk water do not apply directly. In particular, when a difference in ionization of an indicator arises between bulk water and reverse micellar solution, one does not know whether it should be ascribed to differences in pH or pK. Also one does not always know the binding site of the indicator (what portion of the aggregate is it reporting on?) and whether the species involved in the indicator equilibria in bulk water and in the water pools are the same [70–72]. If one makes the assumption that it is possible to evaluate the local pH by using solubilized indicators, one observes substantial shifts in their pK_as. For example, interfacially solubilized *p*-nitrophenol in AOT reverse micelles has a pK_a of 11.5, while material in the water pool (primarily in the type II water) has a pK_a of 7.5–7.9 vs. 7.14 in bulk water [59]. Generally pK_a shifts are quite small (often less than one pH unit) for indicators residing in the water pools; with anionic reverse micelles one may select the phosphate buffer system (studied using ^{31}P NMR) as suitable for monitoring pH_{wp}, since the various anionic species are very hydrophilic and experience electrostatic repulsions with the surfactant head groups [69].

Reverse Micelles as Novel Microenvironments for Chemical Reactions

The effects of reverse micelles on a wide variety of chemical reactions (metal ion-ligand exchange, ester hydrolysis and aminolysis, proton transfers, etc.) have been investigated and the topic has been reviewed a number of times in recent years [2, 3, 63, 73, 74]. Chemical reactions have also been studied in normal micellar systems, and it is legitimate to ask in what respects these complementary milieus differ. For bimolecular reactions involving small molecules, the increases in velocity (over the values in bulk water for the same overall concentration), found in normal micellar systems are dominated by a concentration effect: solubilization of the reactants by micelles increases their effective concentrations. Of course, for the evaluation of a real "micellar catalysis", this concentration effect must be taken care of [84]. Unfortunately, this useful and trivial correction of the rate enhancement is seldom done. For some reactions, such as unimolecular decarboxylations [75], it has also been verified that generalized medium effects (the dielectric constant of the surface of normal micelles being lower than that of bulk water) also affect the rate constant. One would expect reverse micelles to be analogous to normal micelles in these two respects; for many reactions this is the case.

The dependence of the reactivity in aminolysis of a series of aliphatic *p*-nitrophenyl esters on hydrocarbon tail length in AOT reverse micelles (acetate fastest, dodecanoate slowest) is the reverse (as expected based on partitioning considerations) of what one finds in normal micellar systems [61]. Increasing the amount of water (and therefore the pool size) increases the aminolysis velocity because the effective concentration of the ester in the reverse micelles increases. No unusual effects due to the properties of the water pool of the reverse micelles could be discerned.

Sunamoto and coworkers [76] have investigated in AOT reverse micelles the reaction given in Eq. (1). They have found a substantial rate decrease for the ring

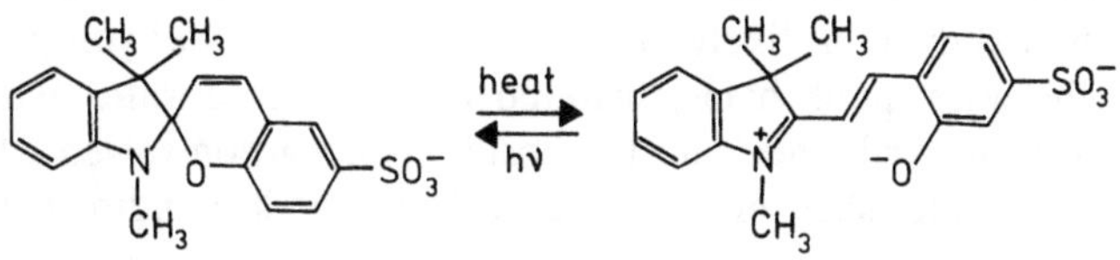

(1)

opening and subsequent cis-trans isomerization. This decrease is interpreted to arise from the particular microenvironment of the reverse micelles. In several cases where water pool water is a reactant, such large rate enhancements have been observed [77] (for example the aquation of Cr^{3+} complexes), that one can speak of real micellar catalysis, i. e. of marked increase of the bimolecular rate constant. In certain other cases, for example the hydration of 1,3-dichloroacetone [78], there is a mechanistic change (revealed in this case by a proton inventory study) upon going from bulk water to water pools in AOT reverse micelles, i. e., two different transition states for $w_0 = 1.3$ and for $w_0 = 11.1$ have been inferred. Attempts have been made to exploit the idea that micelle-reactant interactions should be more specific (in the sense of one reactant orientation being markedly favored over others and/or chiral discrimination by surfactants of optically active reactants occuring in reverse rather than in normal micelles). Thus Jaeger and Ippoliti [79] attempted to carry out lactonization of long chain ω-hydroxyacids in reverse micelles; however, the production of polyesters was not significantly hindered by the micellar environment. Similarly, only modest chiral discrimination has been observed in reactions carried out in reverse micelles: Kitahara and coworkers [80, 81] observed an enantioselectivity of 2.5 in the aminolysis reaction of (+) or (−)-α-phenethylamine with (+) or (−) -*p*-nitrophenyl-α-methoxyphenyl acetates in cationic reverse micelles. From the examples of stereoselective catalysis in normal micellar and vesicular systems, it is likely that providing for multiple hydrogen bonding and/or hydrophobic interaction sites for surfactant-reactant interactions will be necessary in order to produce significant stereoselectivity [74].

Enzymes in Reverse Micelles. The Water-Shell Model

We have discussed until now some of the basic properties of reverse micelles, and indicated how the particular environment of the water pool influences some of the chemical properties of small guest molecules. We will consider now the case of biopolymers (in particular enzymes and nucleic acids) hosted in reverse micelles, discussing first the structural and later the reactivity aspects.

Hydrophilic enzymes, such as lysozyme, α-chymotrypsin, ribonuclease, horse liver alcohol dehydrogenase, urease; as well as hydrophobic enzymes, such as lypo-oxygenase, can be readily solubilized in the system Aerosol-AOT/isooctane/water [85–90]. From the above list it is clear that the range of molecular weight of the proteins which can be solubilized is rather large, extending from about 10,000 to 500,000 Dalton. Considering that even a plasmid with molecular weight of 2.8 million Dalton has been solubilized [91], it appears that the large size of biopolymers is not a limiting factor, and actually it still remains to be clarified where the upper limit is. What is the structure of these enzyme-containing aggregates? We proposed in our earlier papers [88] that the most likely situation is one in which the hydrophilic biopolymer resides in the water pool, being surrounded by a shell of water, which protects it from the surfactant wall and from the hostile hydrocarbon solvent. We refer to this model as the water-shell model. The evidence for it is indirect, but rather convincing.

First of all, all enzymes investigated to date remain active, as much as in aqueous solution, and their reactivity and mechanism do not seem to differ from those in bulk water. Also, the enzyme activity and conformation depend on the water content of the system, i. e. on the micelle size – which is an indication that the enzyme recognizes the amount of water around it; the spectroscopic properties, although somewhat changed, are typical of aqueous solution, e. g. the fluorescence properties of α-chymotrypsin [90] and ribonuclease [92] (although this finding may not necessarily be valid for all proteins). Alternative models are possible (for example, one in which the protein sticks to the negatively charged inner wall of the AOT micelle; or one in which a partly hydrophobic protein extends from the water pool into the lipophilic part of the surfactant layer or into the solvent). Although we do not exclude in principle the existence of such alternative forms of organization, we have found no experimental indications for them until now and we would like to remain, for the time being, with the water-shell model.

This does not mean that all is clear with the water-shell model. First of all, as already mentioned, water in the water pools, particularly at low water content, behaves anomalously. Secondly, it is not yet clear what happens to the geometry of the micelle when the protein is accepted in the water pool. Originally [88], we proposed a model which is based on the assumption that the water content and w_0 of the host micelle does not change upon acceptance of the guest biopolymer. This model demands that at low water content, when micelles are small, the insertion of a protein produces a large enhancement of the size, whereas at high water content a small protein like chymotrypsin or ribonuclease brings about an enlargement which is barely appreciable. Preliminary ultracentrifugation experiments appeared to confirm this model, and later experiments in the Martinek group supported this view [93]. A later extension of this model, to correct for the dependence of f_H (the size occupied by a surfactant head) from w_0, gave essentially the same results [94]. However, more refined neutron scattering experiments [100] seem to show that the situation is more complex. In fact, we have observed that lysozyme at $w_0 = 8$ produces an enlargement of size, from a pool with radius 20 Å to one with radius 33 Å (both species can be observed experimentally). In the case of chymotrypsin however, under otherwise the same conditions, we could not detect any increase in size. This last observation was made in the group of B. Robinson [95].

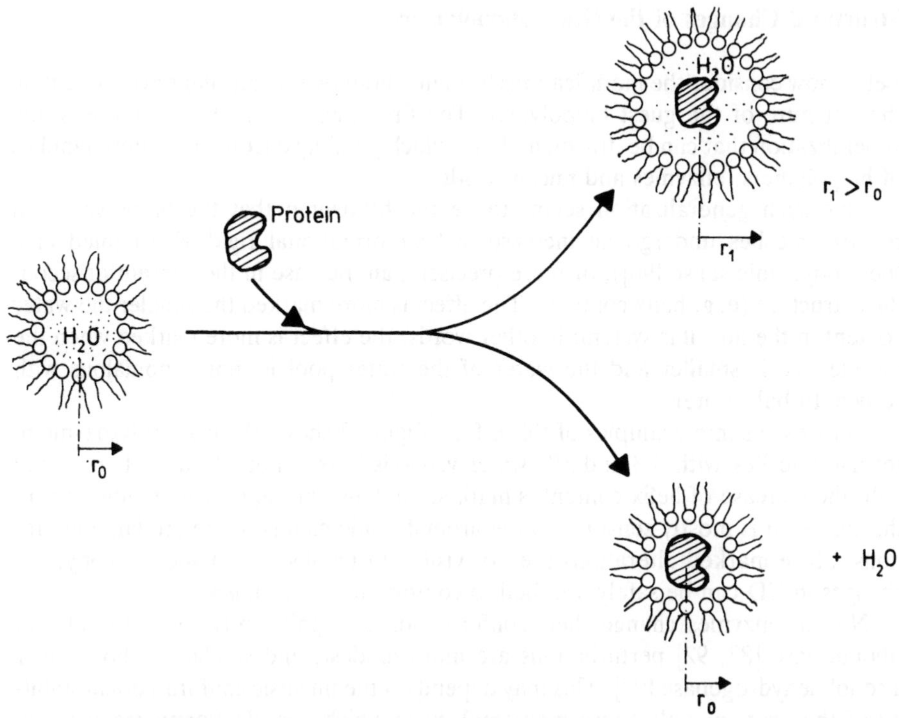

Fig. 2. Representation of two possible models of protein uptake into a reverse micelle. In the first case the protein produces an increase of the size of the micelle (see Ref. [88] for detailed information); in the second, the size of the micelle remains the same.

Whereas it is too early to draw conclusions from these sparse data, it appears possible that the incorporation of an enzyme into a reverse micelle obeys one of the following two extreme models: (Fig. 2)

In the first case, the total water volume which pre-existed in the "empty" micelle is maintained in the enzyme-containing micelle, and there is an enlargement of the size of the micelle (with the necessary reequilibration of the system to respect mass balance). In the second case, it is the original micellar size which is maintained. This latter model demands that water is expelled from the water pool when the protein comes in, this water being reequilibrated in the overall system.

The observation that α-chymotrypsin may follow the second pathway is countered by the fluorescence data, which suggest a highly aqueous environment. The situation is even more complex in the case of DNA, where it is not yet established whether the micelles remain spherical when DNA is hosted, or whether they assume a more cylindrical form, respecting the rigid elongated structure of the DNA-duplex. Thus, the situation concerning the form and size of biopolymer-containing reverse micelles is at present uncertain. Several groups are now actively working to elucidate this point by means of various techniques, and there is confidence that a clearer picture will emerge shortly.

Structural Changes of the Guest Biopolymers

Let us now consider the modifications brought about by the micellar environment on the structure of the guest biopolymer. The first question is whether there is any generalization that can be drawn, and one which possibly encompasses both families of biopolymers, enzymes and nucleic acids.

One such generalization seems to be the following: that the biopolymers in reverse micelles undergo an increase in "conformational rigidity" (defined in a thermodynamic sense [96]), or more precisely, an increase in the percent of secondary structure (e. g. helix content). The effect is more marked the smaller the water content in the micellar system: in other words, the effect is more marked when the micelle size is smaller and the water of the water pool is more anomalous with respect to bulk water.

Let us see some examples of this effect. Figure 3 shows the case of lysozyme in reverse micelles with 0.3 and 3% water vis-à-vis bulk water. As already reported [89], the increase of helix content is marked, and also in the near uv-region, where the spectrum is mostly sensitive to the aromatic side chains, there are large alterations. Since marked alterations are not visible in uv-absorption spectroscopy, the changes in CD can be safely ascribed to conformational changes.

Not all enzymes change their conformation so significantly. For example, in ribonuclease [87, 92] perturbations are more modest, and similary in horse liver alcohol dehydrogenase [97]. This may depend on the intrinsic conformational stability of the protein: only those macromolecules which contain floppy regions may show a significant increase in secondary structure, whereas those which are already highly rigidified by internal constraints (S–S bonds or other kinds of forces) may change little. Chymotrypsin is an intermediate case, giving practically no change in the near uv-region and marked small but very significant changes in the 230 nm region [90] (see Fig. 3).

As shown in Fig. 4, the UV-absorption of RNA in micellar solution shows an ipochromic effect, which is the more marked the lower the water content. The ellipticity shows an increase in intensity, and the two pieces of data put together indicate a higher degree of hydrogen bonding, which causes an increase of stacking in the single stranded RNA molecule [91]. In the double-stranded, more rigid DNA

Fig. 3 a–c. Circular dichroic spectra of proteins in reverse micelles. **a)** CD spectra of lysozyme. The ▶ CD spectra of lysozyme (0,1 μM) were recorded in water (——) and in 50 mM bis(2-ethylhexyl) sodium sulfosuccinate (AOT) in isooctane containing 0.3% water (v/v) (.–.) and 0.3% water (v/v) (---) at 23 °C. In addition, the CD spectra of lysozyme in reverse micelles were recorded at – 30 °C and no change in the CD spectra (compared to 23 °C) was observed. The ordinate gives the molar ellipticity. The bars represent the experimental error (from Ref. [89]). **b)** Circular dichroic spectra of ribonuclease in Tris buffer 0.05 M, pH 7.13 (---) and in n-octane/AOT (50 mM)/H_2O (0.94 M) (——) obtained by mixing 50 μl of a stock solution of the enzyme (14 mg/ml) in borate buffer 0.05 M pH 9.8 with 3 ml of the n-octane/AOT solution. The final concentration of the enzyme is the same in the two cases (15 μM). Practically the same CD spectrum is obtained in micelles prepared from pH 8.0 phosphate buffer (from Ref. [87]). **c)** Circular dichroism spectra of α-chymotrypsin (overall concentration 11 μM) in water and micellar solutions at different w_0: ——, water 7.9; ---, water, pH 9.8; –.–, $w_0 = 22.5$; –..–, $w_0 = 18$; ..., $w_0 = 13.5$. The CD spectrum at $w_0 = 6.6$ (not shown) is very similar to the spectrum = 13.5. $[\theta]$ is the molar ellipticity per residue (from Ref. [90])

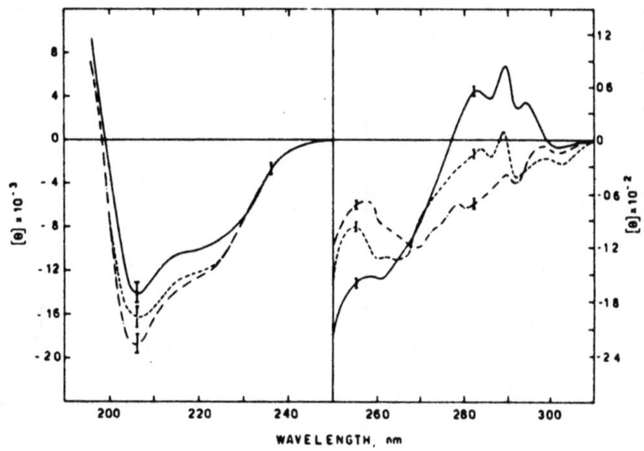

a)

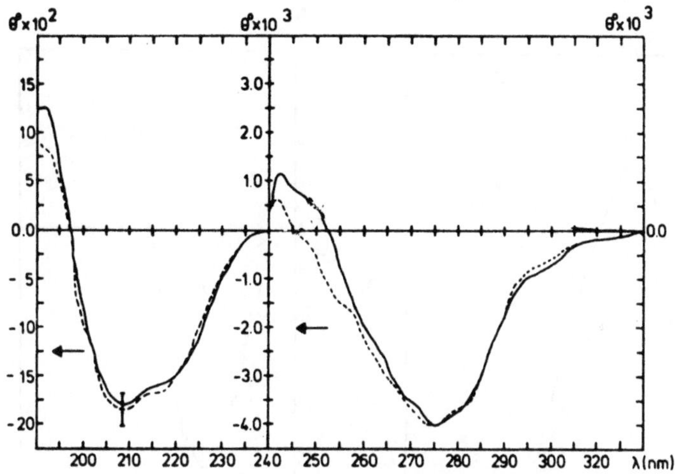

b)

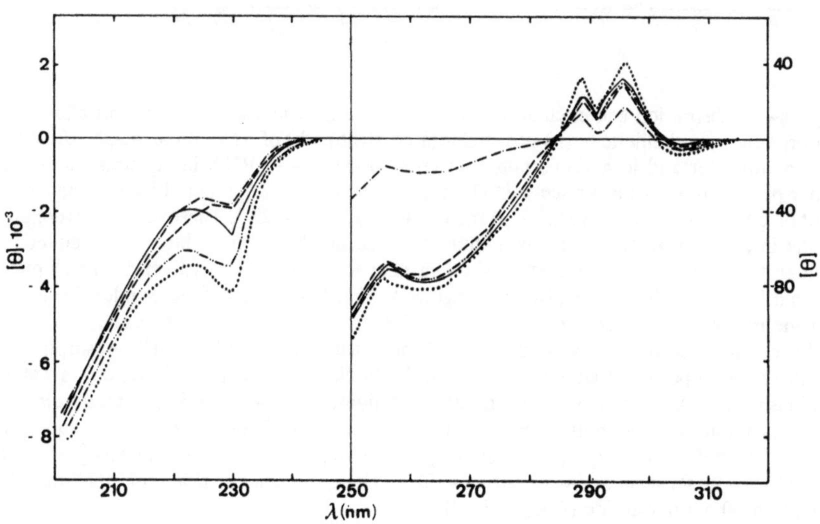

c)

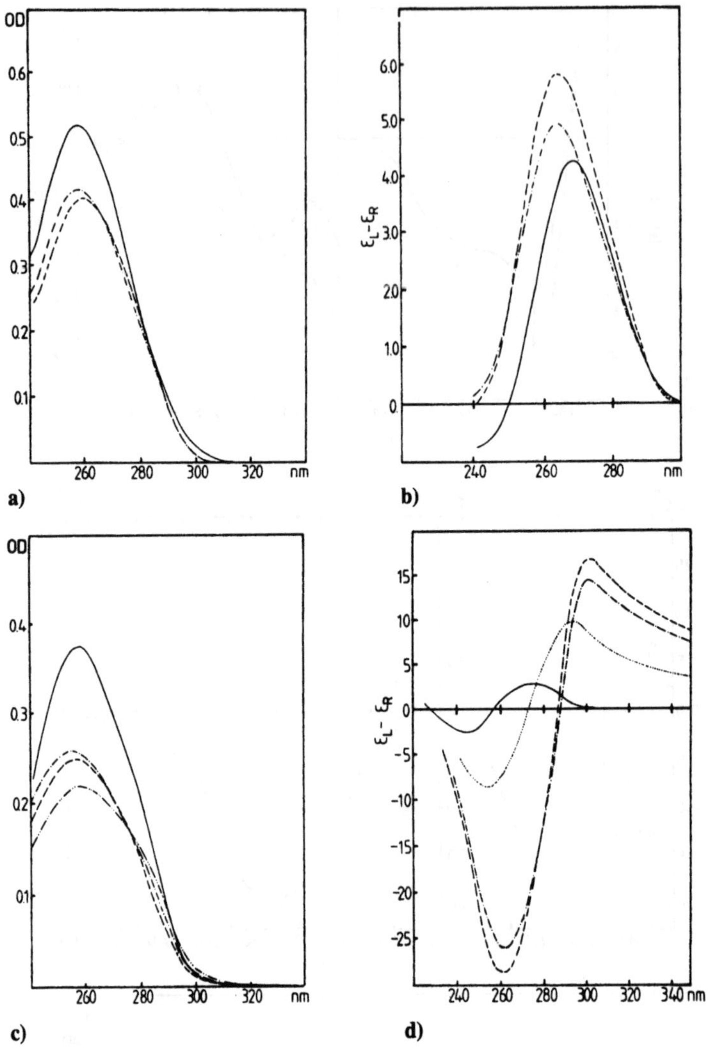

Fig. 4a–d. Absorption and circular dichroic spectra of nucleic acids in reverse micelles. Absorption (**a**) and ellipticity (**b**) properties (1 cm light path) of RNA (mol. weight 20,000–30,000 Dalton) in water and in hydrocarbon micellar solutions. (—) RNA in aqueous solution borate buffer pH = 9.0; (---) in isooctane/AOT solution at $w_0 = 14.8$ (prepared from an aqueous stock solution borate buffer pH = 9.0); (-.-) same as (---) but $w_0 = 22.2$. These solutions were prepared by the injection procedure, i. e. by adding the same number of microliters of an aqueous stock solution to 3 ml of water or of hydrocarbon micellar solution. [AOT] = 50 mM. The ellipticities in this and in the next figure are given in (degrees × 1/nucleotide-mol × cm), under the assumption that the molar concentration in the micellar solutions and in water are the same. The extinction coefficient in water for RNA was taken as 7000 per nucleotide (from Ref. [91]). Absorption (**c**) and ellipticity (**d**) properties (1 cm light path) of DNA (mol. weight ca. 250,000 Dalton) in water (—) and in isooctane/AOT (50 mM)/water micellar solutions: (---) $w_0 = 18.5$ (prepared from an aqueous stock solution borate buffer pH = 9.0): (-.-..) $w_0 = 18.5$ (prepared from an aqueous stock solution Tris buffer pH = 7.5; (...) same as (---) but $w_0 = 25.9$; (-.-.-) same as (---) but $w_0 = 14.8$; (-.-.- in a)) same as (---) but $w_0 = 22.2$. The extinction coefficient of DNA in water was taken as 6450 per nucleotide (from Ref. [91])

molecules changes are less marked, which is in agreement with the picture drawn above for proteins. However, when plasmids with very large molecular weight are incorporated in reverse micelles, drastic changes in the CD spectrum are observed (Fig. 4). We have interpreted this as being due to a superpacking of DNA (as in the head of certain viruses or in chromatin). The nature of the supercondensed structure of DNA is not known, but it appears safe to propose that it must be based on a higher degree of intramolecular interactions, with H-bonds playing an important role.

Why should the micellar environment act as a medium in which H-bonds are strengthened?

A decrease of the dielectric constant can be thought of as the most obvious factor to increase the probability and strength of H-bonds. In fact, as already mentioned, indirect fluorescent probing has suggested that the dielectric constant in the water pool may be considerably smaller than that of bulk water [98].

New Working Hypothesis Concerning Nucleic Acids in Reverse Micelles

The increase in the H-bond capability of the micellar environment may be important not only to modify the conformation of the biopolymers but also to favor intermolecular associations. Actually, the micellar environment appears to be the ideal medium for the reconstitution experiments whenever electrostatic interactions are important. In fact, hydrophilic macromolecules introduced in a hydrocarbon micellar solution will necessarily tend to concentrate in the water pools; and acting upon the relative concentration, it should be possible to have different biopolymers in the same water pool, even if they do not have a great affinity for each other. Due to the increased strength of electrostatic interaction they may tend to form complexes more efficiently than in bulk water.

Recognizing this, we have thought of a series of experiments with nucleic acid-containing micelles. Taking advantage of the informal format of this meeting, we would like to briefly mention them, and expect to receive comments and/or criticism from the participants.

1) *Interaction between DNA and histones.* In order to reconstitute chromatin like material, it would be particularly interesting to see whether such complexes can be obtained by mixing two micellar solutions, one containing the histones, the other containing DNA and to compare the results with analogous reconstitution experiments in aqueous solution [99]. These studies would also be useful to clarify the rate processes by which large molecules can be exchanged among micelles.

2) *Template-induced polymerization of nucleotides.* The idea is to have one single-stranded polynucleotide (e. g. poly A) as guest in reverse micelles, as well as an excess of mononucleotide (e. g. U or of short oligomers thereof). The binding A=U should produce an array of U lined up to the poly A matrix, and the polymerization should then take place by chemical activation. The process, if successful, can lead to any series of block copolymers or random copolymers.

3) A more ambitious process is an attempt to carry out a *ribosome-free protein biosynthesis.* We need a polynucleotide, e. g. poly-U, as a simple m-RNA and the binding to it of charged Phe-t-RNA, followed by polymerization of the aligned

aminoacids via chemical activation (or enzymatic coupling). This type of experiment is also related, as one can easily understand, to prebiotic chemistry. Preliminary binding studies between poly-A and uncharged Lys-t-RNA have already been successfully accomplished by Ergin Imre in our group in Zürich.

4) The *replication of DNA* is the most challenging experiment. The two basic elements are a few micelles containing DNA and an excess of micelles containing free nucleotides; the aim is to multiply the number of DNA-containing micelles at the expense of the other. In this respect, the DNA-containing micelles would work like a virus.

To a certain extent, all such systems can be considered as models for the corresponding biological systems *in vivo*. A few words should be said to clarify the term "biological model", since it is often used in quite different connotations by different authors. This is done in Appendix 1.

Enzymes in Reverse Micelles, General Considerations

To date, several enzymes have been investigated in reverse micelles. Not only our group [87–92, 97] but that of Martinek's in Moscow [85, 93], Robinson's in Kent, U.K. [95], Douzou's in France [57], Laane's in the Netherlands [107], Sumabramanian's in India [101], Menger's [86] and Montal's [102] in the United States have widely contributed to the field. A few generalizations can be made on the basis of these papers, and are discussed below.

Enzymes retain their activities and tend to follow a Michaelis-Menten kinetics. In addition, the activity of the enzymes depends on the water content of the hydrocarbon micellar system; interestingly, the activity is not the highest at the maximal water content. Rather, it seems to reach its maximal value at w_0 values of 10–15. Figure 5, composed from the work of Sumabramabianan and Martinek, together with our work on lysozyme, illustrates this point.

How does the enzyme steady state cycle function in reverse micelles? Just as in water, enzyme and substrate must encounter, and product must leave the enzyme. In hydrocarbon micellar solutions these events are subordinate to the encounter of enzyme-containing micelles with substrate-containing micelles, whereas the "empty" micelles may play the important role of incorporating the product (this holds in the case of hydrophilic substrate and products which are soluble only in the water pools and not in the hydrocarbon bulk – otherwise partition coefficients must be included, a problem which for the sake of simplicity we neglect here). In an earlier paper [88], we propose the steady state cycle as shown in Fig. 6.

It is easy to see from this Figure how the Michaelis-Menten scheme could be violated (for example when there are too few empty micelles, or when the concentration of reagents is so low that collisions among micelles become the rate limiting step). Until now, however no deviations from the Michaelis-Menten kinetics have been reported.

An interesting conceptual problem arises when one asks which concentration should be used to express the kinetic parameters. (Let us remain still in the simple frame that only water-soluble substrates and products are present, so that all material is gathered in the water pools.)

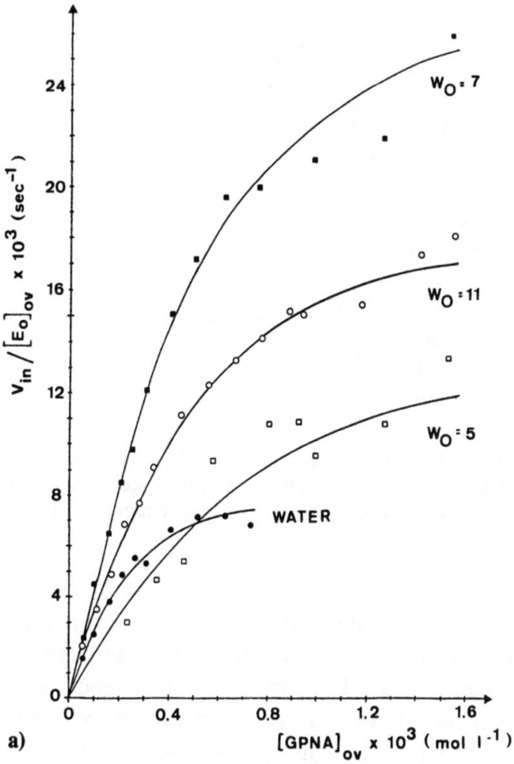

a)

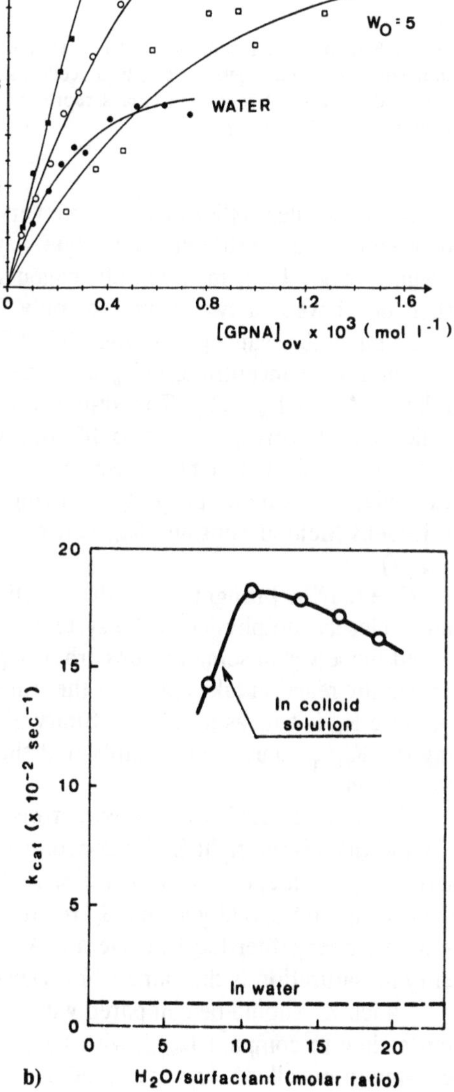

b)

Fig. 5. a, b Influence of water in the reverse micellar system on enzyme activity. **a)** Enzyme activity profile in the water pool of chymotrypsin. The substrate is glutatyl-phenyl alanine p-nitro anilide. **b)** Dependence of the first-order rate constant for peroxidase oxidation of pyrogallol on the content of water solubilized in 0.1 M Aerosol OT plus aqueous buffer (0.02 M phosphate-borate-acetate, pH 7.0) plus octane at 26 °C. For comparison, the value of k_{cat} in the same buffer (dashed line) is shown. The initial steady-state rate of formation of purpurogallin was followed spectrophotometrically (420 nm) (from Ref. [105])

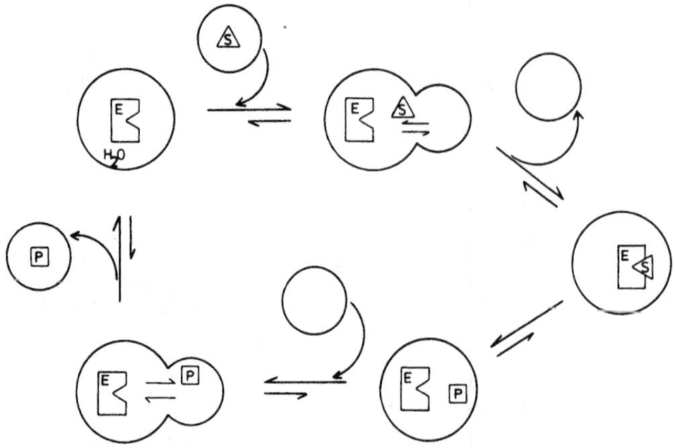

Fig. 6. Schematic representation of an enzymatic reaction in a reverse micellar system under steady state conditions and excess of "empty" micelles. E,S, and P represent the enzyme the substrate, and the product, respectively. The circles represent the water pools of the reverse micelles (cross-section) (from Ref. [88])

Since micelles collide and coalesce very rapidly, the system from the kinetic point of view can be considered as a single water bubble with a fraction F_w of the total volume (e.g., 1%), in which all reagents are sequestered. The concentration can then be viewed in two ways, not only referring to overall volume C_{ov} but also referring to the water pools alone (C_{wp}).

The local concentration C_{wp} is related to C_{ov}, in our simple case, by the simple relation $C_{ov} = F_w \cdot C_{wp}$. The system then has two different concentration scales, which simply correspond to two different ways as looking at the system – not to two different physical systems. However, this duality may lead to some confusion. For example, all kinetic properties having dimensions of concentration, like the Michaelis-Menten constant K_m, can be expressed by two different numbers (see Fig. 7).

One may be brought to ask the question, which of the two K_m values represents more closely the physics of the system. In fact, this is the question which we have asked ourselves in some of our earlier papers, generally opting for $K_{m, wp}$, since the enzymatic reaction takes place in the water microphase. The question is important, because K_m is a measure of the affinity of the enzyme for the substrate, and accepting the $K_{m, wp}$ value, would imply that the enzyme in reverse micelles has a rather low affinity.

Upon closer analysis however, it becomes apparent that the very question of asking which is the right K_m? is conceptually wrong. In fact, when one considers the large $K_{m, wp}$ value, one must by the same token also acknowledge that the substrate concentration S_{wp} is larger than S_{ov} by the same factor F_w^{-1}. In other words $K_{m/S}$ is the same number either local or overall. And this shows the obvious, namely that the enzyme saturation is the same when viewed overall or local (see Fig. 7).

Which K_m should be compared with that of the bulk water solution? It would be misleading to compare $K_{m, wp}$ with $K_{m, aq}$ since this would convey the idea that the enzyme in micelles is less saturated at a given overall concentration. Thus, it now

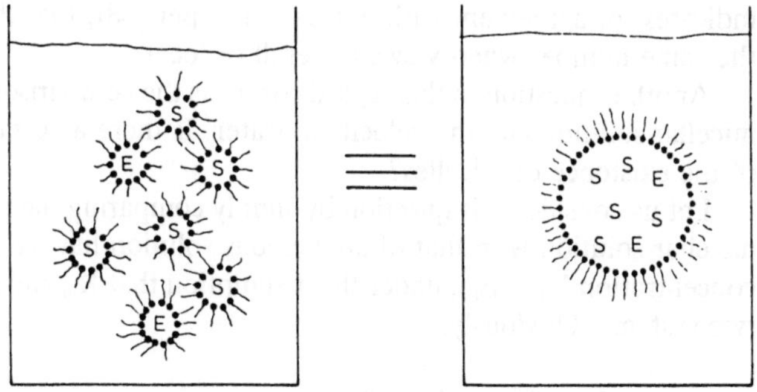

Fig. 7a. An equivalent model of reverse micelle in the case of E and S soluble only in water

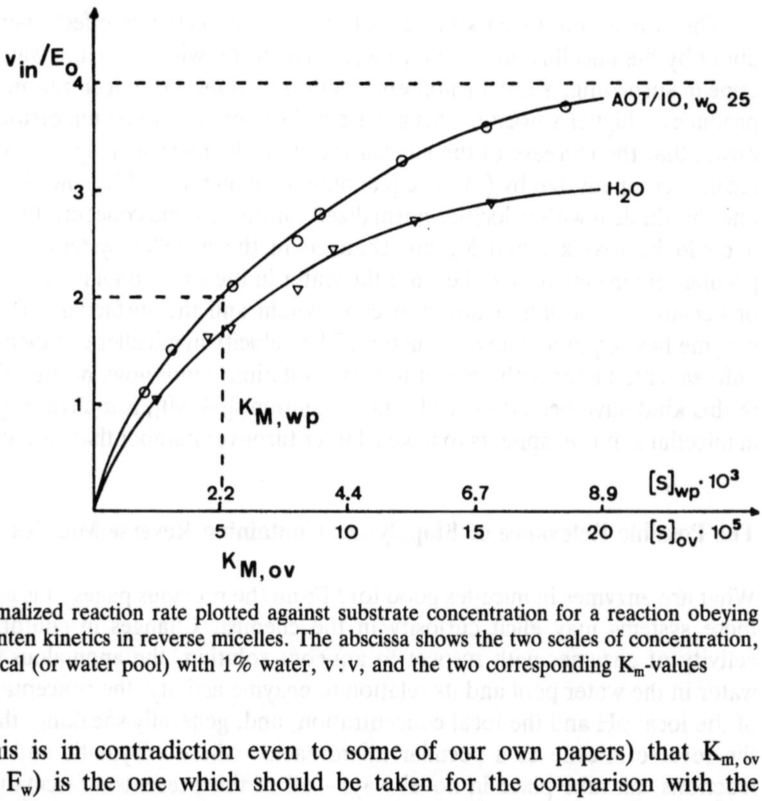

Fig. 7b. Normalized reaction rate plotted against substrate concentration for a reaction obeying Michaelis-Menten kinetics in reverse micelles. The abscissa shows the two scales of concentration, overall and local (or water pool) with 1% water, v : v, and the two corresponding K_m-values

appears (this is in contradiction even to some of our own papers) that $K_{m, ov}$ ($= K_{m, wp} \cdot F_w$) is the one which should be taken for the comparison with the aqueous solution.

Let us consider now the problem of reaction velocity. How does the overall reaction velocity v_{ov} compare with the water pool velocity v_{wp}? Is there an increase of velocity because of the concentration of reagents in the water pool?

Clearly, the number of molecules of product produced per time unit per unit of volume in the water microphase is numerically larger, by a factor F_w^{-1}, than that evaluated in the overall volume. If, however, one normalizes for the respective enzyme concentrations (which eventually one needs to do anyway in order to obtain the turnover number having the dimensions s^{-1}), then obviously $v_{wp/E_{wp}} = v_{ov/E_{ov}}$. This indicates, in agreement with our earlier paper [88], that the normalized velocity is the same number when viewed overall or local.

Another question in this regard concerns the comparison between the velocity in micellar solution and the velocity in water. Is there a "catalytic" effect just because of the existence of micelles?

Let us consider this question by simply comparing the normalized velocities of a micellar solution with that of an aqueous solution having the same overall enzyme concentration $E_0 = E_{ov}$, under the assumption that K_m and k_{cat} are unchanged in the two systems. Obviously,

$$\frac{v_{aq}}{E_0} = \frac{k_{cat} \cdot S}{K_m + S} = \frac{k_{cat} \cdot S_{ov}}{K_m + S_{ov}} = v_{ov/E_{ov}}$$

The conclusion appears to be that there is no "catalytic effect" per se brought about by the micellization of enzyme and substrates with respect to water. This may appear surprising, since in non-enzymatic reactions the micellization of reagents produces a higher velocity. This apparent discrepancy can be understood by recognizing that the increase of the concentration of the micellar reagents, when seen as local, is compensated by (a) the equal numerical increase of K_m and (b) the fact that one should deal with velocities normalized for the enzyme concentration. Of course, it could be that k_{cat} and K_m are changed in the micellar system because of the peculiar environment, i. e. because the water in the water pool is not normal water, or because of the interaction between protein and the surfactant, or because the enzyme has acquired a novel structure. The velocity in micelles can then differ by an unforseeable factor with respect to water solution. Until now, no significant effects of this kind have been observed. One exception, [68, 90], is α-chymotrypsin, which in micellar solution appears to have a larger turnover number than in water solution.

The Possible Relevance of Biopolymer-Containing Reverse Micelles

What are enzymes in micelles good for? From the previous pages, it is apparent that these systems may elicit curiosity in the chemist. Changes in conformation and activity of enzymes with respect to aqueous solution, the anomalous structure of water in the water pool and its relation to enzyme activity, the conceptual problems of the local pH and the local concentration, and, generally speaking, the picture of the reverse micelle as a peculiar microreactor where enzymatic and nucleic acid reactions can take place in novel ways – all of this presents a fascinating research enterprise.

On the other hand, all this may appear as a closed system with its own intrinsic and quite specific laws, so that one may consider the whole story as lacking in general relevance (particularly at this time, when the scientists' amusement in pursuing a certain line of research is not a sufficient justification, certainly not in the eyes of the financing agencies). The question of "relevance" must then be posed. What can you do with enzymes in micelles, or nucleic acid in micelles? How do you come out from the closed mini-world to impinge upon themes which have a greater relevance?

Recently, a meeting on reverse micelles took place in Switzerland with the title *Technological and biological relevance of reverse micelles* [104]. A few contributions at this meeting were devoted to biopolymers in reverse micelles, and on this basis a couple of general observations can be made. Concerning the technical relevance, some research effort is channelled into the utilization of enzymes in reverse micelles for the catalytic transformation of water insoluble, hydrocarbon-soluble, substrates. Lipooxygenase, which has a water insoluble substrate, could also be easily investigated in reverse micelles in isooctane, where clear solutions are formed which are viable for spectroscopic and mechanistic studies [106]. In the case of α-chymotrypsin, it has been shown that two water soluble substrates, which are soluble in the water pools, can be coupled to yield a water insoluble product which is expelled out of the microreactor (the micelle) into the bulk hydrocarbon solvent [106]. Certain steroids or long chain aldehydes can be readily oxidized by alcohol dehydrogenase in reverse micelles [97, 105, 106]; Laane and coworkers reported the reduction of a chetosteroid utilizing a chain of three different enzymes in reverse micelles, and speculate that the system can be interesting for biotechnological applications [107]. Indeed, an active industrial screening research is now under way, mostly in the United States, to test the viability of this method.

With regard to the practical importance of nucleic acids in reverse micelles, things are even less mature, and no serious hypothesis can be made at this point. Of course, of the research projects outlined before, two (the one on protein biosynthesis and the one on template-induced copolymerization of nucleotides) may in principle be interesting technologically. The future will tell whether the difficulties afforded by trivial parameters such as yields, costs, reaction times, etc., can be successfully overcome.

Concerning the biological relevance, some papers have appeared which discuss the possible presence of reverse micelles *in vivo*, i. e. in the lipid membranes [102, 108]. To what extent this is so, and what the relevance is of such structures, remains to be clarified. But this also appears to be an area in which the world of proteins in reverse micelles would grow beyond the narrow limits of the synthetic AOT-reverse micelles.

In conclusion: the world of reverse micelles, in addition to having its own internal coherence (and fun), appears to extend into more general fields of biotechnology and biochemistry – an indication of vitality, which will certainly stimulate the activity of other researchers.

Appendix 1: About the Meaning of the Word "Model"

One meaning of the word is connected with the argument that since biological systems are very complex, one should attempt to elucidate them by utilizing simpler synthetic or semisynthetic systems. Eventually, from the understanding of such artificial systems, a better understanding of the biological function will follow. We claim that such a meaning of the term model is generally bound to increase confusion. The construction of a synthetic model always imposes structural constraints which are not present in the biological systems and the fact that it contains fewer structural parameters does not signify at all that the information can be transferred to the more complex *in vivo* systems. In fact, up to now almost no instance can be cited in which the biologist has learned something about the *function* of a real biological system on the basis of synthetic models (it is perhaps different in the case of *structure,* in particular conformation).

But the term "biological model" has simpler and cleaner connotation, which actually corresponds more closely to the philological meaning of the word "model". The paradigm is now to look at the biological system in order to obtain a hint for a new synthetic system. Thus, the construction of synthetic macromolecules as synthetic catalysts based on model enzymes is very interesting and useful from the point of view of polymer chemistry, but one should not try to use them the other way round, i. e. to obtain information about the catalytic and biological activity of enzymes.

In connection with micelles and their biological similarities the term "biomimetic systems" has been used [2], and it is certainly more proper than the term "biological models". The term "bio-analogous systems" may be even more appropriate.

In addition to the above discussed meanings, the term "model" has acquired in the literature yet another broader meaning, to signify simply a scheme, design, pattern (see for example the water-shell model, or the two models of Fig. 6). Statisticians use the word in the same sense to signify an equation or a set of equations which describe a given physical system. The use of this term here is harmless – one should perhaps merely be aware of the difference in emphasis.

The authors thank Prof. I. Klotz for a helpful discussion about this argument.

References

1. H.-F. Eicke, *Topics Curr. Chem., 87,* 85 (1980)
2. J. H. Fendler, "Membrane Mimetic Chemistry", Ch. 3, John Wiley & Sons, New York, 1982
3. J. H. Fendler, *Acc. Chem. Res., 9,* 153 (1976)
4. A. S. Kertes and H. Gutmann, *Surf. Colloid Sci., 8,* 193 (1975)
5. F. Y. Low, B. M. Escott, E. J. Fendler, E. T. Adams, Jr., R. D. Larsen and P. W. Smith, *J. Phys. Chem., 79,* 2609 (1975)
6. K. Tsujii, J. Sunamoto, F. Nome and J. H. Fendler, ibid., *82,* 423 (1978)
7. P. S. Sheih and J. H. Fendler, *J. Chem. Soc., Faraday Trans. 1, 73,* 1480 (1977)
8. P. Debye and H. Coll, *J. Colloid Sci., 17,* 220 (1962)
9. I. Paligoric and I. J. Gal, *J. Inorg. Nucl. Chem., 40,* 285 (1978)
10. P. Mukerjee, *Ber. Bunsenges. Phys. Chem., 82,* 931 (1978)
11. U. Herrmann and Z. A. Schelly, *J. Amer. Chem. Soc., 101,* 2665 (1979)
12. N. W. Fadnavis and J. B. F. N. Engberts, *J. Org. Chem., 47,* 2923 (1982)
13. E. Ruckenstein and R. Nagarajan, *J. Phys. Chem., 84,* 1349 (1980)
14. S. M. F. Tavernier and R. Gijbels, *Bull. Soc. Chim. Belg., 89,* 579 (1980)
15. K. Kon-no and A. Kitahara, *J. Colloid Interface Sci., 35,* 636 (1971)
16. S. G. Frank and G. Zografi, *J. Pharm. Sci., 58,* 993 (1969)
17. M. Ueno, H. Kishimoto and Y. Kyogoku, *J. Colloid Interface Sci., 63,* 113 (1978)
18. H.-F. Eicke and H. Christen, *ibid., 46,* 417 (1974)
19. H.-F. Eicke and V. Arnold, *ibid., 46,* 101 (1974)
20. H.-F. Eicke and J. Rehak, *Helv. Chim. Acta, 59,* 2883 (1976)
21. H.-F. Eicke and M. Zulauf, *J. Phys. Chem., 83,* 480 (1979)
22. H.-F. Eicke, *Pure Appl. Chem., 53,* 1417 (1981)
23. R. C. Little and C. R. Singleterry, *J. Phys. Chem., 68,* 3453 (1964)
24. R. C. Little, *J. Phys. Chem., 74,* 1817 (1970)
25. P. H. Elworthy and D. S. McIntosh, *Kolloid Z.-Z. Polymere, 195,* 27 (1964)
26. P. H. Elworthy and D. S. McIntosh, *J. Phys. Chem., 68,* 3448 (1964)

27. R. J. Braedley, D. H. Grant, V. C. Reinsborough and P. A. Ross, *Canad. J. Chem., 54,* 3070 (1976)
28. J. P. M. Janson, M. Kunst, A. Rip and P. Bordewijk, *Chem. Phys. Lipids, 9,* 147 (1972)
29. G. Klose, G. Hempel and Th. Zglinicki, *ibid., 21,* 261 (1978)
30. R. Haque, I. J. Tinsley and D. Schmedding, *J. Biol. Chem., 247,* 157 (1972)
31. P. H. Poon and M. A. Wells, *Biochemistry, 13,* 4928 (1974)
32. H.-F. Eicke and A. Denss, *J. Colloid Interface Sci., 64,* 386 (1978)
33. H.-F. Eicke and H. Christen, *J. Colloid Interface Sci., 48,* 281 (1974)
34. H.-F. Eicke, R. F. W. Hopmann and H. Christen, *Ber. Bunsenges. Phys. Chem., 79,* 667 (1975)
35. H.-F. Eicke in "Micellization, Solubilization and Microemulsions", K. L. Mittal, Editor, Vol. 1, p. 429, Plenum Press, New York, 1977
36. Y. Jean and H. J. Ache, *J. Am. Chem. Soc., 100,* 984 (1978)
37. K. Tamura and Z. A. Schelly, *J. Am. Chem. Soc., 103,* 1018 (1981)
38. H.-F. Eicke and H. Christen, *Helv. Chim. Acta, 61,* 2258 (1978)
39. S. G. Frank and G. Zografi, *J. Colloid Interface Sci., 29,* 27 (1969)
40. R. A. Day, B. H. Robinson, J. H. R. Clarke and J. V. Doherty, *J. Chem. Soc., Faraday Trans. 1, 75,* 132 (1979)
41. E. Gulario, B. Bedwell and S. Alkhafaji, *J. Colloid Interface Sci., 77,* 202 (1980)
42. P. D. I. Fletcher and B. H. Robinson, *Ber. Bunsenges Phys. Chem., 85,* 863 (1981)
43. B. Valeur and E. Keh, *J. Phys. Chem., 83,* 3305 (1979)
44. E. Keh and B. Valeur, *J. Colloid Interface Sci., 79,* 465 (1981)
45. M. Ueno, H. Kishimoto and Y. Kyogoku, *J. Colloid Interface Sci., 63,* 113 (1978)
46. H. N. Maitra and H.-F. Eicke, *J. Phys. Chem., 85,* 2687 (1981)
47. C. A. Martin and L. J. Magid, *J. Phys. Chem., 85,* 3938 (1981)
48. L. J. Magid and C. A. Martin, Proceedings of the Fourth ESF Workshop on Polymer Sciences, Rigi-Kaltbad, Switzerland, in press
49. P. D. J. Fletscher, A. M. Howe, B. H. Robinson and D. C. Steytler, Proceedings of the Fourth ESF Workshop on Polymer Sciences, Rigi-Kaltbad, Switzerland, in press
50. B. H. Robinson, D. C. Steytler and R. D. Tack, *J. Chem. Soc., Faraday Trans. 1, 75,* 481 (1979)
51. H. F. Eicke, T. C. W. Shepherd and A. Steinemann, *J. Colloid Interface Sci., 56,* 168 (1976)
52. S. Harada and Z. A. Schelly, *J. Phys. Chem., 86,* 2098 (1982)
53. M. Wong, J. K. Thomas and T. Nowak, *J. Am. Chem. Soc., 99,* 4730 (1977)
54. G. Bakale, G. Beck and J. K. Thomas, *J. Phys. Chem., 85,* 1062 (1981)
55. P. E. Zinsly, *J. Phys. Chem., 83,* 3223 (1979)
56. H. Kondö, I. Miwa and J. Sunamoto, *J. Phys. Chem., 86,* 4826 (1982)
57. P. Douzou, E. Keh and C. Balny, *Proc. Natl. Acad. Sci., 76,* 681 (1979)
58. R. Kubik, H.-F. Eicke and B. Jönsson, *Helv. Chim. Acta, 65,* 170 (1982)
59. F. M. Menger and G. Saito, *J. Am. Chem. Soc., 100,* 4376 (1978)
60. L. J. Magid, K. Kon-no and C. A. Martin, *J. Phys. Chem., 85,* 1434 (1981)
61. F. M. Menger, J. A. Donohue and R. F. Williams, *J. Am. Chem. Soc., 95,* 286 (1983)
62. M. Wong, M. Grätzel and J. K. Thomas, *Chem. Phys. Lett., 30,* 329 (1975)
63. A. Kitahara, *Adv. Colloid Interface Sci., 12,* 109 (1980)
64. A. O. El Seoud and J. H. Fendler, *J. Chem. Soc., Faraday Trans. 1, 71,* 452 (1975)
65. L. J. Magid, K. Kon-no and C. Martin, *J. Colloid Interface Sci., 83,* 307 (1981)
66. A. V. Levashov, V. I. Pantin and K. Martinek, *Colloid J. USSR, 41,* 453 (1979)
67. A. T. Terpko, R. J. Serafin and M. L. Bucholtz, *J. Colloid Interface Sci., 84,* 202 (1981)
68. F. M. Menger and K. Yamada, *J. Am. Chem. Soc., 101,* 6731 (1979)
69. R. E. Smith and P. L. Luisi, *Helv. Chim. Acta, 63,* 2302 (1980)
70. O. A. El Seoud and R. C. Vieira, *J. Colloid Interface Sci., 93,* 289 (1983)
71. O. A. El Seoud and M. R. Shimizu, *Colloid Polymer Sci., 260,* 794 (1982)
72. O. A. El Seoud, A. M. Chinelatto, M. R. Shimizu (1982) *J. Colloid Interface Sci., 88,* 420
73. J. H. Fendler and E. J. Fendler, "Catalysis in Micellar and Macromolecular Systems", Ch. 10, Academic Press, New York, 1975
74. L. J. Magid, Proceedings of the Fourth ESF Workshop on Polymer Sciences, Rigi-Kaltbad, Switzerland, in press

75. C. A. Bunton, A. Kamego, M. J. Minch and J. L. Wright, *J. Org. Chem.*, *40,* 1321 (1975)
76. J. Sunamoto, K. Iwamoto, M. Akutagawa, M. Nagase and H. Kondo, *J. Am. Chem. Soc.*, *104,* 4904 (1982)
77. C. J. O'Connor, E. J. Fendler and J. H. Fendler, *J. Chem. Soc., Dalton Trans.*, 625 (1974)
78. O. A. El Seoud, R. C. Vieira and J. P. S. Farah, *J. Org. Chem.*, *46,* 1231 (1981)
79. D. A. Jaeger and J. T. Ippoliti, *J. Org. Chem.*, *46,* 4964 (1981)
80. K. Kon-no, M. Tosaka and A. Kitahara, *J. Colloid Interface Sci.*, *79,* 581 (1981)
81. K. Kon-no, M. Tosaka and A. Kitahara, *J. Colloid Interface Sci.*, *86,* 288 (1982)
82. O. A. El Seoud, in Reverse Micelles, – technological and biological relevance, P. L. Luisi and B. Straub editors, Acad. Press, in press
83. B. Robinson, in Reverse Micelles, – technological and biological relevance, P. L. Luisi and B. Straub editors, Acad. Press, in press
84. C. A. Bunton, in Solution chemistry of surfactants, K. L. Mittal editor, Plenum Press, 1979
85. K. Martinek, A. V. Levashov, N. L. Klyachko and I. V. Berezin, *Dokl. Akad. Nauk SSSR* (Engl. edition), *236,* 951 (1978)
86. F. M. Menger and K. Yamada, *J. Am. Chem. Soc.*, *101,* 6731 (1979)
87. R. Wolf and P. L. Luisi, *Biochem. Biophys. Res. Comm.*, *89,* 209 (1979)
88. F. J. Bonner, R. Wolf and P. L. Luisi, *J. Solid-Phase Biochem.*, *5* (4), 225 (1980)
89. C. Grandi, R. E. Smith and P. L. Luisi, *J. Biol. Chem.*, *256* (2), 837 (1981)
90. S. Barbaric and P. L. Luisi, *J. Am. Chem. Soc.*, *103,* 4239 (1981)
91. V. E. Imre and P. L. Luisi, *Biochem. Biophys. Res. Comm.*, *107,* 538 (1982)
92. A. Pande and P. L. Luisi, in preparation
93. A. V. Levashov, Y. L. Khmelnitzky, N. L. Klyachko, V. Y. Chernyak and K. Martinek, *Anal. Biochem.*, *118,* 42 (1981)
94. R. Wolf, Dissertation (ETH No. 7027) (1982)
95. P. D. I. Fletscher and B. H. Robinson, manuscript in preparation
96. P. Meier, P. L. Luisi, *Naturwissenschaften, 64,* 569 (1983)
97. P. Meier, P. L. Luisi, *J. of Solid-Phase Biochemistry, 5 (4),* 269–282 (1980)
98. M. Wong, J. K. Thomas and T. Nowak, *J. Am. Chem. Soc.*, *99,* 4730 (1977)
99. R. Sasi, P. E. Hüvös and G. D. Fasman, *J. Biol. Chem.*, *257,* 11448 (1982)
100. L. J. Magid, E. Imre, H. Jäckle, P. L. Luisi, in preparation
101. C. Kumar and D. Balasubramanian, *J. Phys. Chem.*, *84,* 1895 (1980)
102. V. Ramakrishnan, A. Darszon and M. Montal, *J. Biol. Chem.*, *258,* 4857, 1983
103. P. D. I. Fletscher, A. H. Howe, B. H. Robinson and D. C. Steytler in "Reverse Micelles", editors P. L. Luisi and B. Straub, Plenum Press, in press
104. "Reverse Micelles", editors P. L. Luisi and B. Straub, Plenum Press, N.Y. (1983)
105. K. Martinek, A. V. Levashov, Y. L. Khmelnitsky, N. L. Klyackko, I. V. Berezin, *Science, 218,* 889 (1982)
106. P. L. Luisi, P. Lüthi, A. Pande, J. Prenosil, I. Tomka, Proceedings of the VII. Intern. Enzyme Engineering Conference, in press
107. R. Hilhorst, C. Laane and C. Veeger, FEBS Lett., in press
108. B. de Kruijff. P. R. Cullis, A. J. Verkleij, *TIBS,* 79 (1980)

Discussion

List of interventions

1 Intervention of Franklin

1) Is it possible that your reverse micelles may offer a model for intracellular organelles such as lysosomes?
2) I want to express some scepticism about your proposal to use the reverse micelles as artificial ribosomes. The ribosomes have about 80 proteins, many of which are involved in protein biosynthesis. It seems unlikely to me that you will be able to mimic this complicated biological system with your simple reverse micelles.

2 Intervention of Breslow

Since the treatments are equivalent, it is easier to consider a change in pK's rather than a change in H^+ activity.

3 Intervention of Scheraga

I have two questions.

1) First, to reconcile your lack of correspondence of CD spectra and enzymatic activity, shouldn't you look at the CD spectra of enzyme-substrate complexes rather than of enzyme alone?

2) Second, since the water in your reverse micelles does not freeze at $-40\,°C$, have you examined the temperature dependence of enzymatic activity in this supercooled water?

3.1 Luisi

You are right. In fact, preliminary experiments show that lysozyme in reverse micelles is conformationally much more stable in the presence of NAG_3 (see B. Sternmann, M. Jackle, P. L. Luisi, Biopolymers, in press – *added in the proofs*).

3.2 Luisi

No, we have not.

4 Intervention of Wüthrich

You mentioned that the NMR signal of the water in the inverted micelle is different from the NMR signals of bulk water.

4.1 Luisi

In about 0.1 or 0.2% of water you have a resonance at 4.1 ppm instead of at 4.8 ppm.

Communications

Intimate Ion Pair Intermediates in the Solvolysis of Thio Addition Products of NAD(P)$^+$ Analogs and Their Relevance to the Chemistry of 3-Phosphoglyceraldehyde Dehydrogenase

Richard M. Kellogg and Berend J. Van Keulen

Department of Organic Chemistry, University of Groningen (NL)

The enzyme 3-phosphoglyceraldehyde dehydrogenase has at the active site an active cysteine (Cys-149) and a histidine residue (His-176). In the presence of the coenzyme NAD$^+$ the substrate 3-phosphoglyceraldehyde is oxidized to an acyl phosphate (Eq. (1)). NADH is produced. A thioacylderivative of Cys-149 is first formed and is

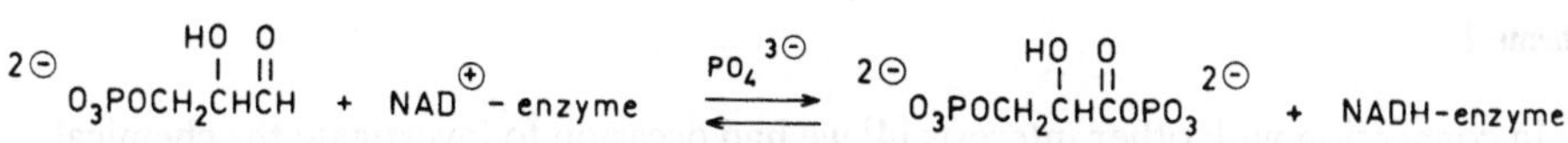
(1)

subsequently converted to the acyl phosphate at another site on the enzyme. A hemithioacetal intermediate with Cys-149 is involved in the oxidative step. The overall sequence is summarized in Scheme 1 [1].

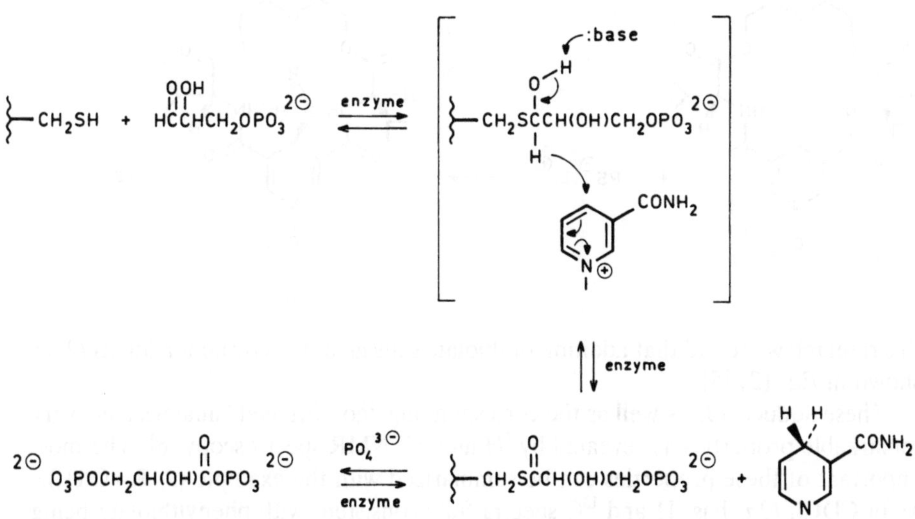

Scheme 1

This enzyme has the unusual property of developing a yellow color on incorporation of NAD$^+$. An explanation offered in the early fifties by Krimsky and Racker [2] was that cysteine adds to NAD^+ to form a 4-thiosubstituted-1,4-dihydropyridine intermediate. Such compounds have spectral properties consistent with the observed color of the enzyme [3]. However, because of the subsequent establishment of an obligatory hemithioacetal intermediate with Cys-149, postulation of other causes of the yellow color, and the absence of an obvious mechanism to convert a NAD^+-cysteine-149 addition product to the obligatory hemithioacetal, this suggestion fell by the wayside. Were such a cysteine-NAD^+ to be involved the general sequence of Scheme 1 would have to be modified in the sense indicated in Scheme 2.

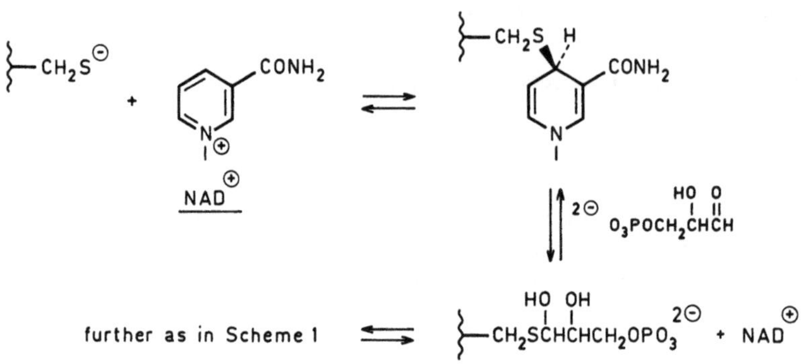

Scheme 2

In connection with other interests [4] we had occasion to investigate the chemical behavior of some 4-thiosubstituted-1,4-dihydropyridines. We demonstrated that (*1*, ClO_4^-) and the corresponding 1,4-dihydropyridine form (not illustrated) are remarkably good surrogates for the NAD^+/NADH coenzyme system. In an extension of

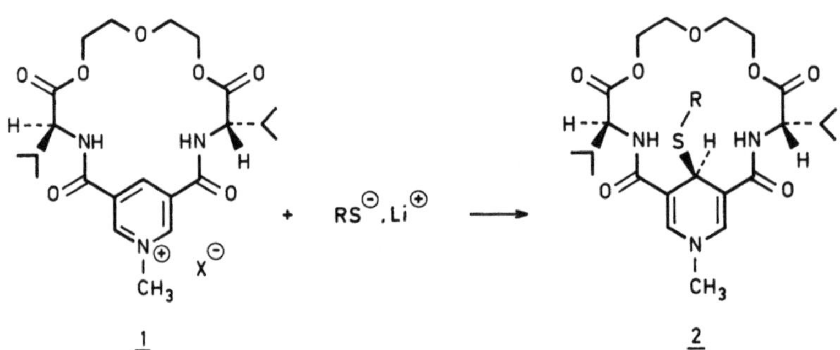

(2)

this research we found that addition of thiolate salts gives the covalent adducts *(2)* as shown in Eq. (2) [5].

These adducts *(2)* as well as the corresponding "non-bridged" analogs have very remarkable properties as revealed by ^1H and ^{13}C NMR spectroscopy [6]. The most important of these properties can be summarized with the example (*2*, $R=C_6H_5$).

a) In $CDCl_3$ *(2a)* has ^1H and ^{13}C spectra fully consistent with phenylthiolate being bonded *covalently* to the 4-position. In CD_3OD as solvent (CD_3CN, CD_3COCD_3, and CD_3SOCD_3 lead to similar effects) *H4* (see drawing of *2* for numbering) shifts 1.3 ppm downfield and in the ^{13}C spectrum *C4* shifts 29 ppm downfield. No other significant changes occur in the spectra.

b) From temperature dependent ^1H NMR measurements in CD_3OD it was determined that *(2a)* is in equilibrium with an isomeric form *(2a')*; the ratio *2a'/2a* is approximately 60 : 40 at – 55°C. This isomeric form differs *only* in the chemical shift of *H4*, which is shifted downfield relative to covalent *(2a)*. At room temperature an averaged spectrum is observed for *2a/2a'*.

c) There is no measurable difference in $J_{^{13}C\text{-}H}$ coupling constants for the C4/H4 coupling in CD_3OD compared to $CDCl_3$. In either solvent the observed $J_{^{13}C\text{-}H}$ is 149 Hz. This means that the intermediate *(2a')* cannot have undergone significant rehybridization at C4.

d) By means of "cross-over" experiments, i.e. mixing structurally different 4-thiosubstituted-1,4-dihydropyridines in CD_3OD under conditions under which intermediates analogous to *(2a')* are formed, it was established that the thiolate substituents *remain bonded* to the pyridine ring. There is no observable dissociation/recombination.

These data, as well as other information accumulated [6], are not in agreement with the most obvious interpretation of these phenomena, namely that partial ionization to (1^+, SC_6H_5) occurs on dissolution in apolar solvent. The NMR chemical shift behavior (i.e. effects are localized at C4/H4), lack of apparent rehybridization (a pyridinium salt has $J_{^{13}C\text{-}H}$ = 180 Hz in contrast to the measured 149 Hz), and absence of cross-over all indicate that a pyridinium thiolate (1^+, SR) *is not* formed. A clue to the cause of these unusual effects is found in the observation that:

Dialkyl substituted amides derived from diethyl amine or proline otherwise analogous in structure to *(2a)* ionize immediately to pyridinium thiolates in CD_3OD. *No* intermediate is observed. The spectra are fully consistent with pyridinium ions [6]. It is concluded that an amide proton is necessary for formation of the intermediate species.

An explanation consistent with these observations is that an *intimate ion pair (3)* [7] is formed wherein departing thiolate is held by a hydrogen bond to the amide. The degree of association between C4 and thiolate is so strong that little rehybridization has occurred (C4 is therefore *not* a pyridinium carbonium ion) and the possibility of exchange is eliminated.

This intimate ion pair has special properties of reactivity. A mixture of *(2b)* and salicylaldehyde (as model for 3-phosphoglyceraldehyde) in CH_3OH (but *not* $CHCl_3$) with a proton donor, NH_4Cl, reacts within seconds to give a hemithioacetal and pyridinium salt *(1)* as shown in Eq. 3. This is the step indicated in Scheme 2 as amendment to the mechanism of Scheme 1 were a Cys-149/NAD^+ adduct to be involved in the enzymic mechanism.

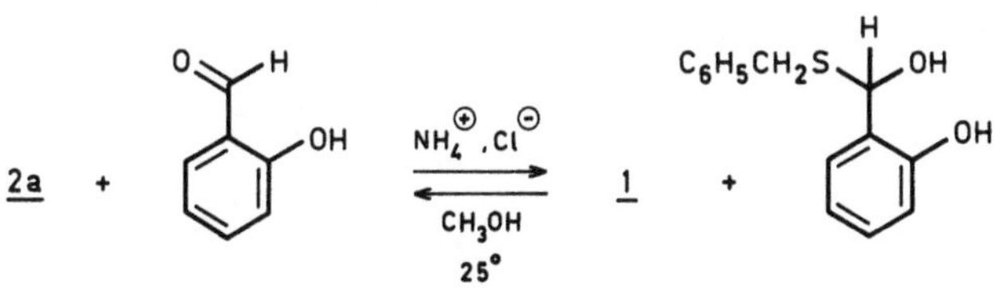

(3)

A corollary derived from these observations is that the carboxamide side arm of NAD^+/NADH may have a conformational role in adjusting reactivity in some cases. It could, by virtue of the potential for hydrogen bond formation through NH_2 but

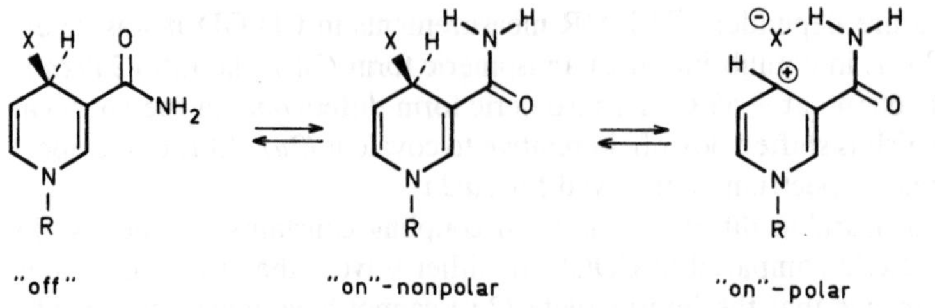

(4)

not =O, act literally as an "on-off" switch as illustrated rather crudely in Eq. 4. This would offer the opportunity to make use of the full scale of reactivity available as a covalent bond (in this case C–S) undergoes a graded transition to ionic. The role of ion pairs is cardinal and is not restricted to the example provided here.

References

1. a. Rossman, M. G., Liljas, A., Brändén, C. I., Banazak, L. J. (1975) in "The Enzymes", Vol. IX, ed. Boyer, P. D., Academic Press, New York, pp. 61–102
   b. Walsh, C. (1979) "Enzymatic Reaction Mechanisms", W. H. Freeman and Co., San Francisco, pp. 322–330
2. a. Krimsky, I., Racker, E. (1955) Science *122,* 319
   b. Racker, E., Krimsky, I. (1952) Nature *169,* 1043
3. Wallenfels, K., Hofmann, D., Schüly, H. (1959) Justus Liebig's Ann. Chem. 621, 188
4. We have made extensive use of bridged chiral pyridinium salts like *(1)*, and the derived 1,4-dihydropyridines for enantioselective reductions of carbonyl groups of activated ketones: Jouin, P., Troostwijk, C. B., Kellogg, R. M. (1981) J. Am. Chem. Soc. *103,* 2091; Kellogg, R. M., Talma, A. G., Jouin, P., de Vries, J. G., Troostwijk, C. B., Werumens Buning, G. H., Waninge, J. K., Visscher, J. (1985) J. Am. Chem. Soc. *107,* 3981
5. Piepers, O., Kellogg, R. M. (1980) J. Chem. Soc., Chem. Commun 1147
6. The complete discussion of the results has been submitted for publication: Van Keulen, B. J., Kellogg, R. M. (1984) J. Am. Chem. Soc. *106,* 6029
7. We imply here a species with the general properties postulated by Sneen, R. A. (1973) Accts. Chem. Res. *6,* 46. The intimate ion pair has in particular not yet undergone rehybridization of the carbon atom bearing the leaving group

Stereoselective Guest-Host Relationships

G. Blaschke

Pharmaceutical Institute, University of Münster (FRG)

This is a short contribution concerning stereoselective guest-host relationships, which were presented by Prof. Cram in his lecture.

Chiral recognitions are not restricted to specially designed synthetic models of hosts but are commonly observed in optically active polymeric systems.

Structure 1 shows the structure of such a synthetic optically active polymer derived from the amino acid L-phenylalanine.

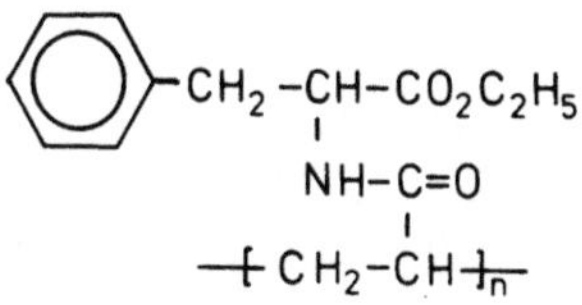

(1)

By radical polymerization in the presence of crosslinking agents, clefts or cavities are formed. These completely insoluble polymers adsorb for example by hydrogen bonding with the polyamide backbone the enantiomers of a racemate, the binding sites being surrounded by asymmetry. This asymmetry enables chiral recognitions to occur. An application is the use of these polymers as adsorbents for the direct chromatographic resolution of racemates. This resolution of chiral structures including drugs is very important in our field of Medicinal Chemistry.

Another optically active adsorbent, microcrystalline cellulose triacetate, also permits the chromatographic resolution of many racemates including drugs.

Examples of complete resolutions on a preparative as well on an analytical scale are thalidomide (2), ifosfamide (3), praziquantel (4) mianserin (5) and barbiturates like (6) and (7).

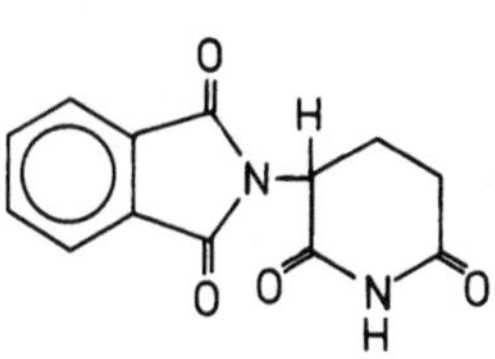

(2)

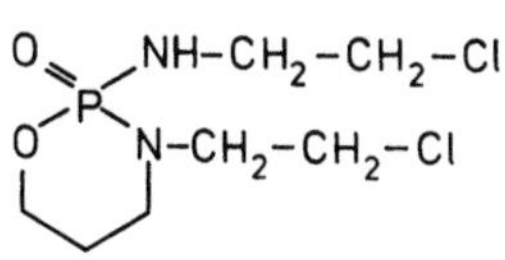

(3)

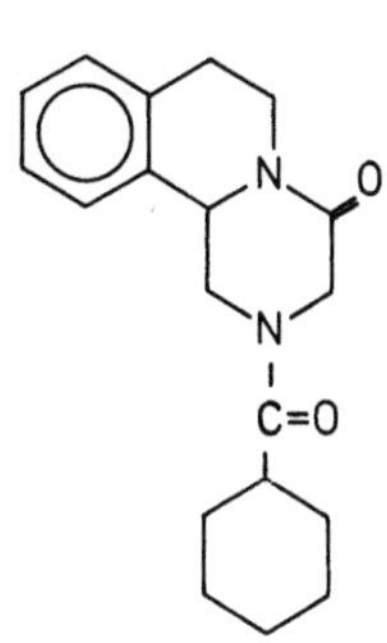

(4)

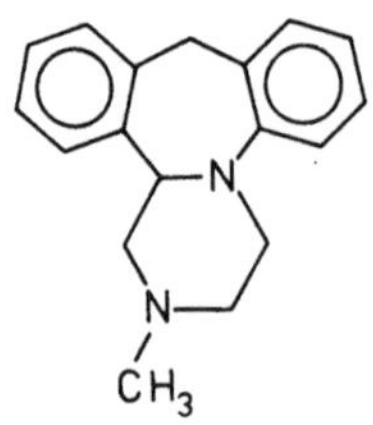

(5)

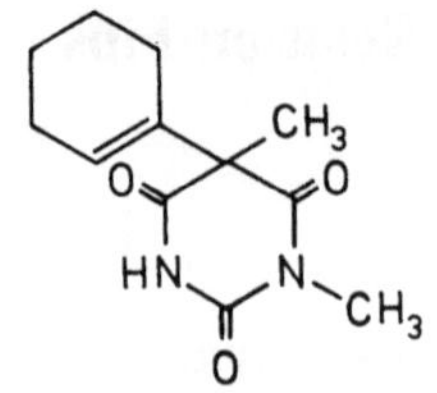

(6)

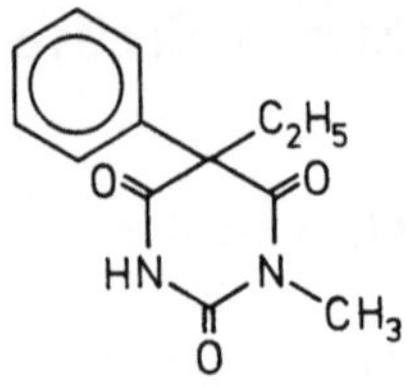

(7)

Discussion

List of interventions

1 Intervention of Breslow

Are these better than other commercially available materials for chiral separation? Is your material commercially available?

1.1 Blaschke

I think so. By the optically active acids or bases normally used for separation the racemates mentioned in the slides cannot be separated. Also columns of optically active adsorbents which are already commercially available do not resolve these racemates. Our materials will soon be commercially available.

On the Dynamics of Embedding in Imprinted Polymers

G. Wulff, D. Oberkobusch and M. Minárik

Inst. Org. Chemistry, University of Düsseldorf (FRG)

In recent years a number of low-molecular-weight catalysts were designed containing a cavity for binding as well as functional groups attached in a defined vicinity to it [1]. The use of macromolecular substances would offer some advantages. One possibility would be to attach the substances of low molecular weight to a polymer. Another more interesting possibility would be to use the whole polymer as the carrier of the stereochemical information. Several years ago we introduced a new approach on these lines [2]. For the construction of a binding site, suitable binding groups were bound to a template molecule in the form of polymerizable vinyl derivatives (see Scheme 1). This monomer was then copolymerized in the presence of a high amount of a crosslinking agent in order to obtain a rigid polymer with chains in a fixed arrangement. Afterwards the templates were split off and the remaining cavities possessed a shape and an arrangement of functional groups that corresponded to that of the template. The functional groups in this polymer were located at fixed points of the polymer chain; they are held in a spatial relationship by the crosslinking of the polymer. It is a type of arrangement that in enzyme chemistry has been called "discontinuate word" by R. Schwyzer [3].

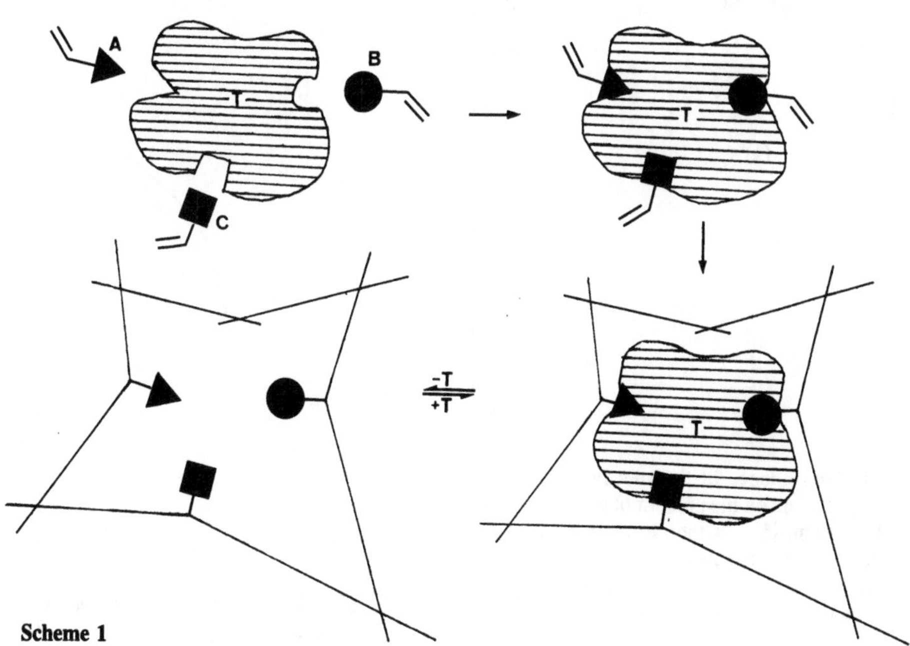

Scheme 1

This approach has been used with quite a number of different templates and binding groups [4]. Some aspects of the dynamics of the embedding of substances to these polymers can now be discussed more particularly. Appropriate investigations have been performed with monomer (1) in which two molecules 4-vinylphenylboronic acids have been bound to phenyl-α-D-mannopyranoside as the template by ester linkages to each of two hydroxyl groups. Monomer (1) was copolymerized to a macroporous polymer from which 80% of the templates could be split off. The D-mannoside was a chiral template and therefore the accuracy of the steric arrangement in the cavity could be tested by the ability of the polymer to resolve the racemate of the template. This has been done mostly in the simple batch procedure. Figure 1 shows the dependence of the separation factor α on the amount of crosslinking in the polymers [5]. Below 10% crosslinking, no selectivity is observed. Between 50–65%, a very strong increase occurs with maximal values at 95% crosslinking of $\alpha = 3.66$. The polymers used in the following investigations were prepared with 95% crosslinking agent glycoldimethacrylate and 5% monomer (1) in the presence of the same volume of benzene/acetonitrile as the inert solvent.

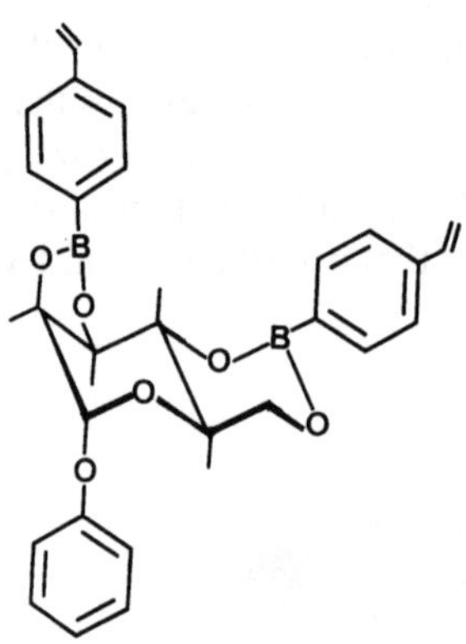

(1)

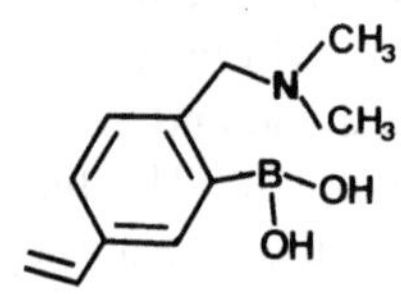

(2)

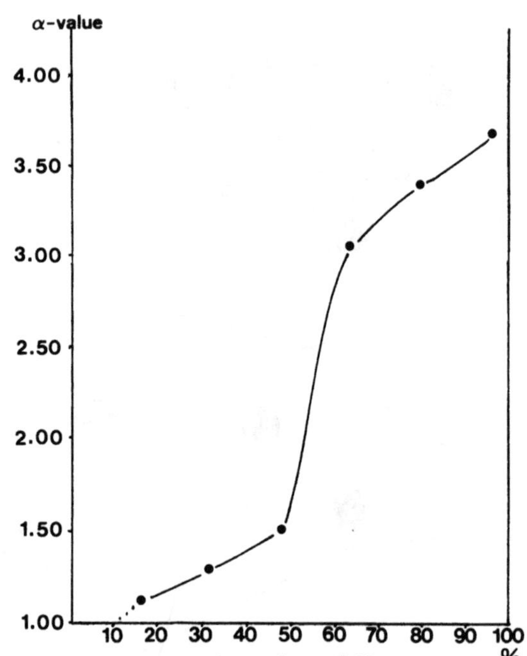

Fig. 1. Dependence of the specificity of the polymers on the amount of crosslinking agent [5] α is the ratio of the distribution coefficients of the D- and the L-form in the batch procedure. Crosslinking agent: Glycoldimethacrylate

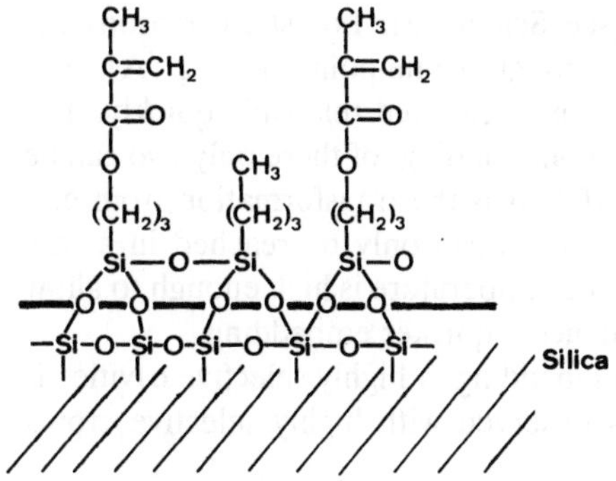

Scheme 2. Silica (500 Å) containing 2–4 μmoles methacrylate per g on the surface was used to prepare a polymer layer of 20–40 Å thickness with monomer (1) and glycoldimethacrylate

To obtain more information on the mechanism of separation, the polymers were investigated by using them as packings for columns in liquid chromatography (HPLC) of the racemate. Contrary to the relatively good efficiency for nonspecifically interacting solutes (e.g. benzene), both sugar enantiomers exhibited strong peak broadening. This phenomenon indicates a slow mass-transfer. It could be explained as follows:

1) The diffusion of the substrate within the polymer matrix is restricted.
2) Reversible covalent binding (i.e. the boronic diester formation) is too slow.
3) The embedding of the substrate into highly selective cavities is restricted.

In order to test whether diffusion through the polymer chains effects the mass-transfer, we prepared a thin layer (20–40 Å) of the same type of polymer on the surface of silica (see Scheme 2). This material had a very good selectivity for racemic resolution (α-value around 3.5), but in chromatography the mass-transfer was not much improved compared to bulk polymers.

To clarify the second topic, boronic acids with strongly enhanced reactivity were prepared [6]. It was found that by the introduction of e.g. an o-dimethylaminomethylene group, an enhancement resulted in the kinetics of the transesterification as well as of the esterification and hydrolysis by 6–8 orders of magnitude. Therefore in monomer (1) the 4-vinylphenylboronic acid moiety was replaced by that of (2) [7] and crosslinked polymers were prepared from this monomer as described above. After splitting off the template, these polymers were also investigated by liquid chromatography. However, there was only some enhancement in mass transfer [8]. Similarly, chromatographic results were somewhat improved by the addition of piperidine or ammonia to the mobile phase.

The investigation of the chromatographic behavior under different conditions provided more information. If during the chromatographic separation the temperature is increased or the flow rate is decreased, the chromatographic behavior of the less retarded enantiomer is only slightly affected. The peak width and the retention time remain nearly constant. By contrast, with the original template (the more retarded enantiomer) the retention time increases when the flow rate is reduced. At very low flow-rates it is difficult to elute the second enantiomer at all. At the same time stronger peak broadening is observed. Stronger retardation of this enantiomer together with less peak broadening is observed at higher temperature. Thus, at a higher temperature combined with very low flow rates, extremely high selectivities can be observed.

The explanation might be as follows (see Scheme 3): For steric reasons the wrong enantiomer (L) can only be bound through a one-point-binding (A). The equilibrium of the one-point binding is apparently reached relatively quickly. The D-enantiomer has four possibilities for a one-point binding; of these only two can be transformed to the right two-point binding (C). It is the transformation from e.g. B to C that is apparently kinetically restricted. C can only be reached in a true equilibrium if enough time is available, or if the temperature is high enough to allow some flexibility of the polymer chains and hence a quicker embedding.
Thus in these experiments it is shown that embedding in highly selective cavities is kinetically restricted. Similar behavior was observed with highly selective crown

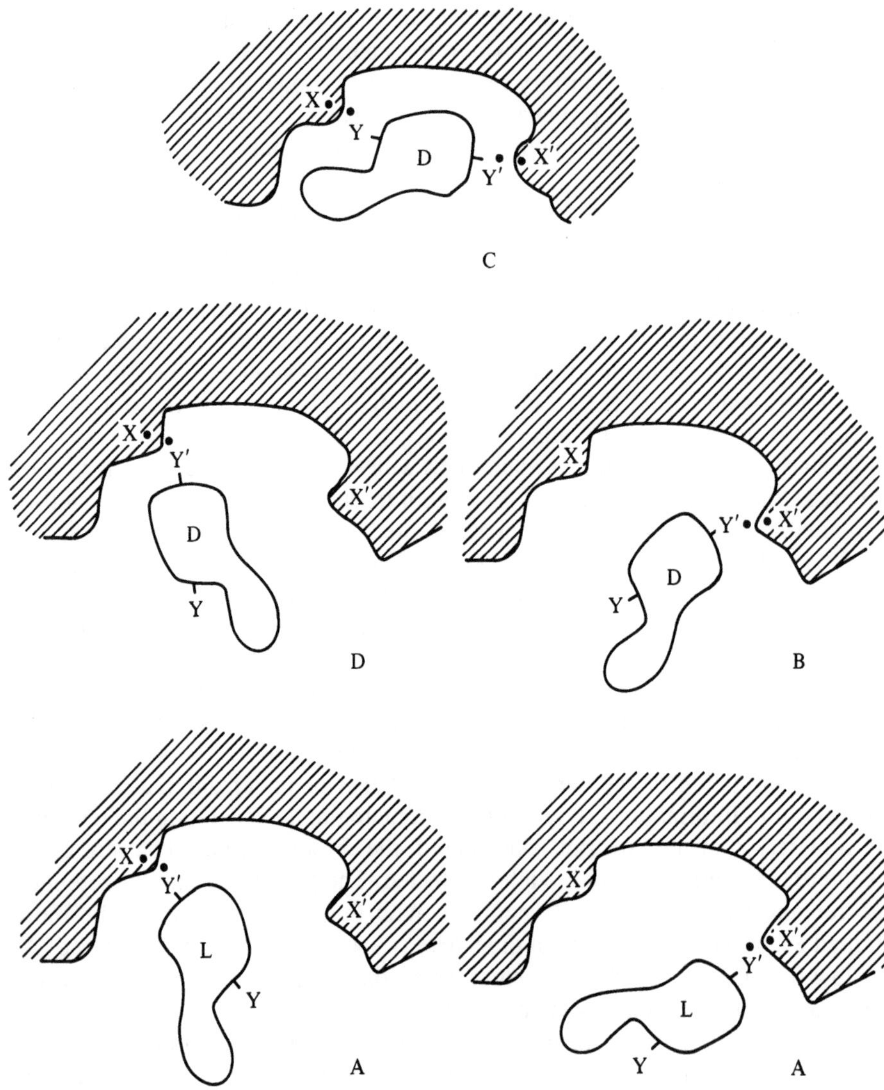

Scheme 3

ethers or cryptates. The release of substrates from those complexes can be extremely slow [1]. It should be borne in mind that the polymers used in our experiments were not completely rigid. In the solvent used in chromatography they show a swellability of about 60% after splitting off the template. On the addition of templates to the swollen polymer the original volume is nearly restored. This swelling and deswelling should preferably occur in the cavities by the solvation of the functional groups. Thus, since there is an equilibration of uptake and release of substances from the cavities, one would expect a quick fluctuation in the size of these cavities.

Acknowledgements

Thanks are due to Deutsche Forschungsgemeinschaft, Fonds der Chemischen Industrie, and the Minister für Wissenschaft und Forschung des Landes Nordrhein-Westfalen for financial support.

References

1. See e.g. the contributions of Cram, D. J., Breslow, R. and Lehn, J.-M. in this book
2. Wulff, G. and Sarhan, A. (1972) Angew. Chem. Int. Ed. Engl. *11,* 341; Wulff, G., Sarhan, A., Zabrocki, K. Tetrahedron Lett. *1973,* 4329; Wulff, G., Vesper, W., Grobe-Einsler, R., Sarhan, A. (1977) Makromol. Chem. *178,* 2799
3. Schwyzer, R. (1970) Proc. Fourth Int. Congr. on Pharmacology *5,* 196
4. See e.g. Wulff, G. in Laskin, A. J. (Ed.): "Enzyme Engineering VII", Annals New York Acad. Sciences *434*, New York 1984, p. 327; Wulff, G., Sarhan, A. in Green, B. S., Ashani, Y., Chipman, D. (Eds.): "Chemical Approaches to Understanding Enzyme Catalysis", Elsevier, Amsterdam 1982, p. 106
5. Wulff, G., Kemmerer, R., Vietmeier, J., Poll, H.-G. (1982) Nouveau J. Chim. *6,* 681
6. Lauer, M., Böhnke, H., Grotstollen, R., Salehnia, M., Wulff, G. (1985) Chem. Ber. *118,* 246
7. Wulff, G., Dederichs, W., Grotstollen, R., Jupe, C. in Gribnau, T. C. J., Visser, J., Nivard, R. J. F. (Eds.): "Affinity chromatography and Related Techniques", Elsevier, Amsterdam 1982, p. 207
8. Wulff, G., Grotstollen, R., unpublished results

Discussion

List of interventions

1 Lehn
2 Hamers

1 Intervention of Lehn

I should like to come back to Professor Luisi's question regarding the contribution made by organic receptor chemistry to the understanding of biological receptors. I feel that much depends on the question asked: if the question is *how,* then one should study the biological material itself; if the question is *why,* then the chemistry of artificial receptors should be studied not only for its own sake but also because it provides the elements for understanding the basis of binding and selectivity.

2 Intervention of Hamers

When Professor Patchornik mentioned GOD in relation to antibody specificity, he was obviously referring to the Generator Of Diversity. In immune systems a vast number of cells, each equivalent to an experimenter tries hard to make stereospecific sites by a statistical diversification process and the successful ones are selected. My suggestion is that the technology of monoclonal antibodies can now be used to generate antibodies with specific binding sites against any chosen molecule and that our present mastery of genetic engineering allows us to obtain the sequence of this immunoglobulin molecule quite readily. The time is not far off when these sequences can be used to computer-generate the chemical geometry of the binding site. This might then very well be used as an economical method of designing and constructing stereospecific sites.

Photo Reversible Affinity Labeling

A. Patchornik, K. Jacobson and M. P. Strub

Department of Organic Chemistry, Weizmann Institute of Science, Rehovot, Israel

We propose a novel concept of photo reversible affinity labeling (ReAL) in which the molecule is labeled by the formation of a photosensitive linkage and is later cleaved under mild physiological conditions by the action of light.

Ring-substituted derivatives of the o-nitrobenzyl group were used to modify enzymes and other proteins selected to serve as models for this method. The modified proteins were irradiated with filtered light at wavelengths (350–400 nm) known not to cause photo-damage to proteins.

ReAL reagents, 4-bromomethyl-3-nitrobenzene and derivatives were synthesized. These compounds, upon binding to the protein, alkylate via the reactive bromomethyl group at a nucleophilic site.

Results with chymotrypsin, ribonuclease A, bovine serum albumin and model compounds show that the method has potential usefulness, but practical problems remain to be solved.

The advantages of photolyzable linkages over bonds broken by chemical reagents are:

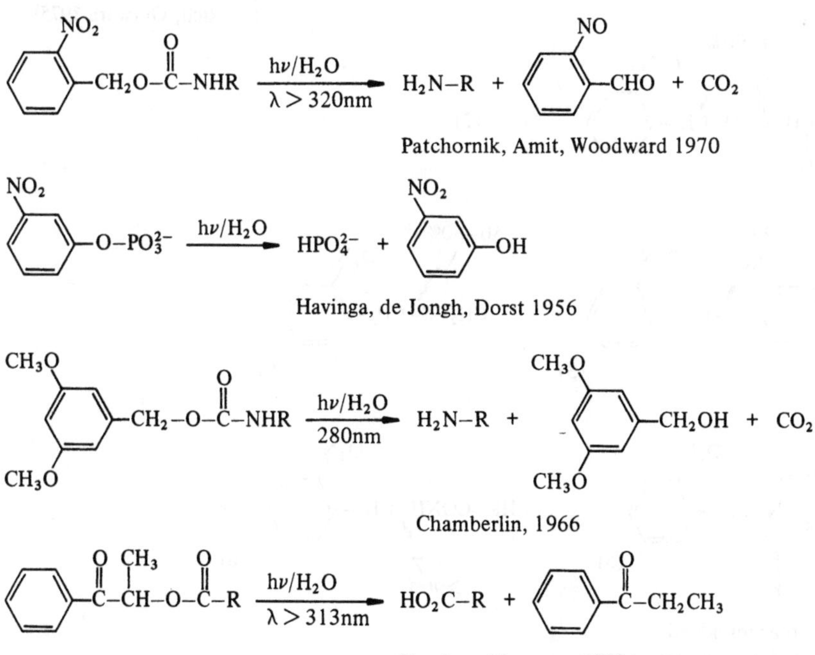

Fig. 1. Representative photocleavable groups

- *Physiological conditions;* the photocleavable group can be eliminated under mild conditions (aqueous medium, neutral pH and low temperatures).
- *Specificity;* at sufficiently long wavelengths only the bond designated for cleavage is affected by irradiation; wavelengths of 350 nm or longer are generally thought not to damage biological systems [1] (cells, membranes, proteins).

Various photo-removable groups described in the literature have been used as protecting groups in organic synthesis [2]. A protecting group to be removed by light must absorb in a region that is not damaging to the rest of the molecule and must be stable to all chemical manipulations before the removal step.

Our class of photocleavable groups is based on the photolability of nitroaromatics having a benzylic H atom (Fig. 1). This group was used in the synthesis of peptides [2, 3], sugars [2] and nucleotides [2, 4]. Other photolyzable groups that have been investigated as protecting groups are: the m-nitrophenyl phosphate [5], the 3–5 dimethyl benzyloxycarbonyl group [6], the phenacyl group [7], and the 5-bromo-7-nitroindolinyl group (BNI) [11] (Fig. 1).

Our study was based initially on 2-nitrobenzyl derivatives. Cleavage of this group occurs at 350 nm. The mechanism involves an intramolecular redox reaction; i.e., the transfer of an oxygen from the nitro group to the benzylic carbon atom. This is

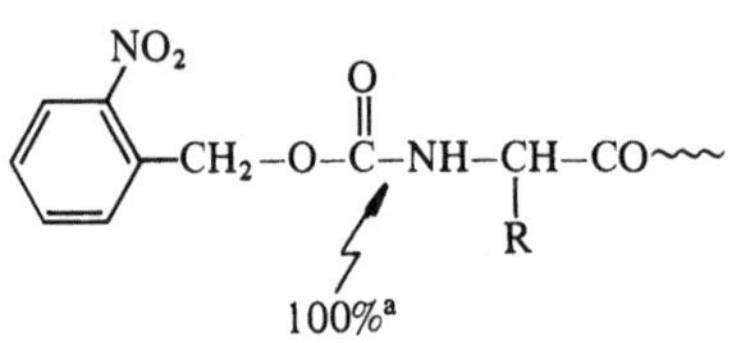

(1)

(1) Patchornik et al. 1970
(2) Amit, PhD. thesis 1974
(3) Amit et al. 1977
(4) Hazum et al. 1980
(5) Kalbag, Roeske 1975
(6) Rich, Gurwara 1975

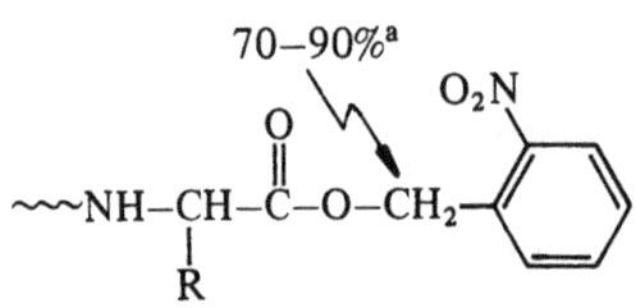

(2)

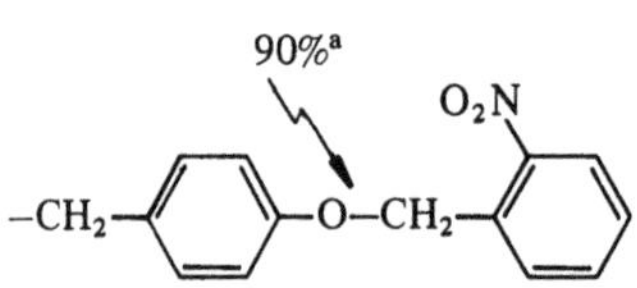

(3)

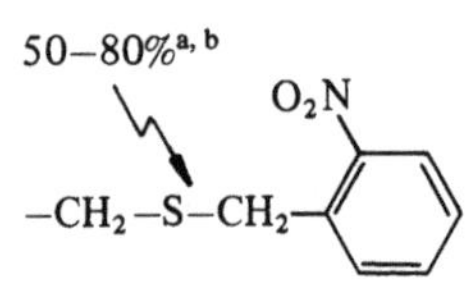

(4)

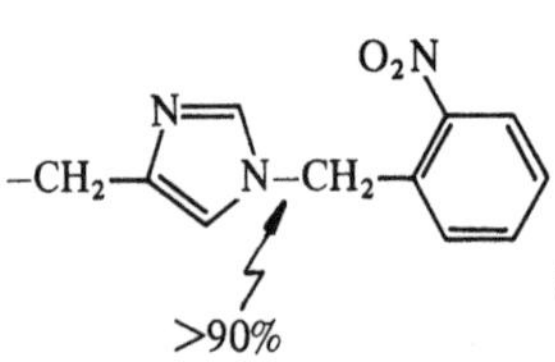

(5)

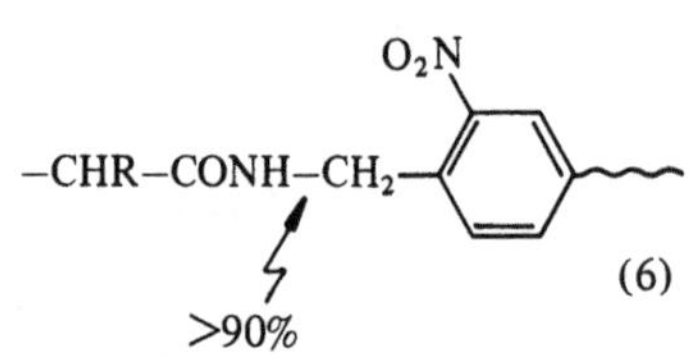

(6)

[a]-aldehyde reagent added
[b]-photooxidation quencher added

Fig. 2. Photocleavage of o-nitrobenzyl derivatives of amino acids and peptides followed by release of the labeled moiety through the formation of the 2-nitrosobenzaldehyde derivative [2]. This reaction occurs rapidly and with high yield.

In our laboratory various o-nitrobenzyl derivatives have been studied as light-sensitive protecting groups for amino acids. Derivatives of 6-nitroveratryloxycarbonyl and 2-nitrobenzoyloxycarbonyl were used as blocking groups for amino and carboxyl functions in peptides and were released with 100% yield. Photo deprotection of tyrosine, cysteine and histidine residues, in high yields, was also demonstrated (Fig. 2).

In this communication we would like to describe some preliminary experiments that indicate the potential application of the method to various proteins. ReAL reagents, are constructed to include "ligands" (substrate, inhibitor, hormone agonist or antagonist, or antigen) that are directed by complementary structures to the specific binding sites in the desired protein (enzyme, receptor or immunoprotein). Labeling of the protein will then occur by alkylation at nucleophilic residues such as Ser, Tyr, Cys, Lys, Asp, Glu, Met, Trp, His or the N-terminal amino group. Covalent binding may reduce biological activity if the binding site occupied by the reagent is essential for function. When the modified protein is subjected to irradiation, the free protein released by photocleavage can be expected to demonstrate regeneration of its biological activity.

The ReAL reagents of Fig. 3 must include specific ligands for the various proteins to be modified in addition to the functions that form the photolabile covalent derivative. A family of such specific reagents can be synthesized easily by derivatizing ligands with a parent compound, a nitrobenzylbromide bearing an acylating group *4* (Fig. 4). This basic reagent may be radioactive to permit trace amounts of reaction to be followed.

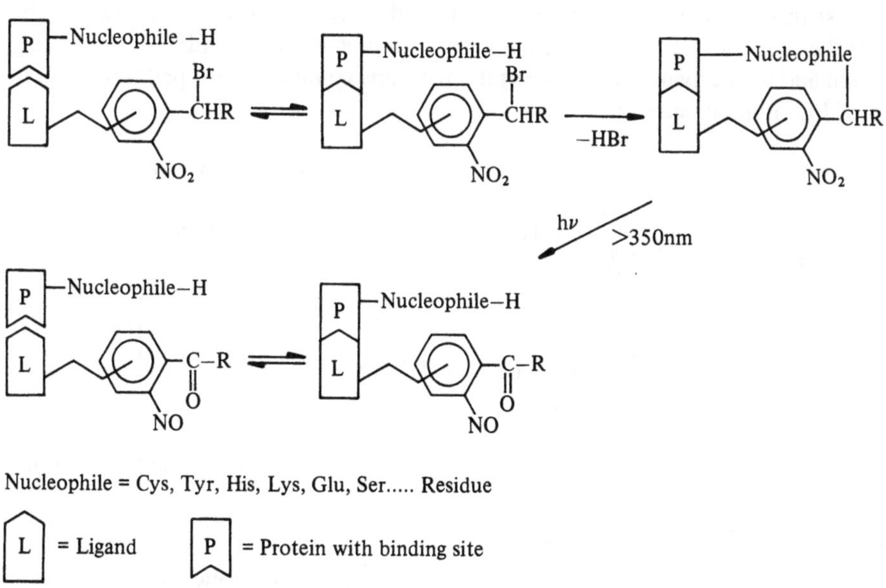

Fig. 3

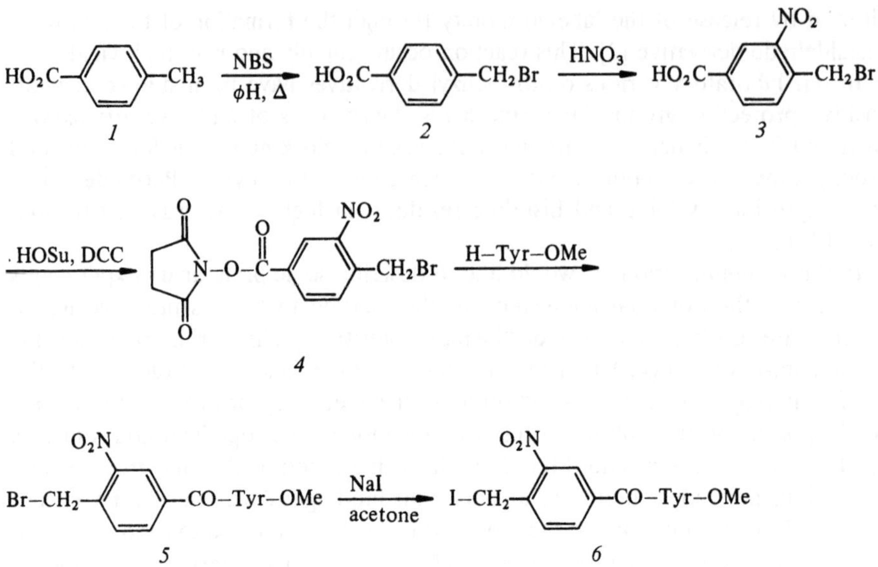

Pass, Jacobson, Strub, Patchornik 1982

Fig. 4. Synthesis and reaction of bifunctional ReAL reagent

The first application of ReAL reagents to proteins showed that there are unanticipated difficulties. Compounds *3a* and *3b* (Fig. 5) were designed to react with chymotrypsin. The tyrosine derivative *3a* did not alter the catalytic activity significantly after 2 hours of incubation although 80% of the activity was lost after 5 days. The acid *4b*, however, inactivates chymotrypsin very efficiently; 88% of the activity was lost in one minute. Attempts to restore the activity were not successful; the control enzyme lost 28% of its activity when irradiated for 2 h with light of roughly 350 nm and the derivatized enzyme under the same conditions dropped from 12% to 7% of the original activity.

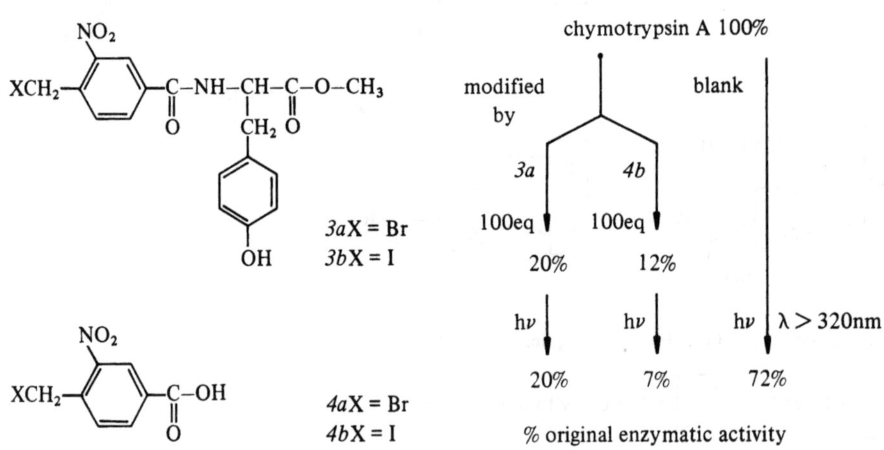

Fig. 5. Inhibition of chymotrypsin by ReAL reagents

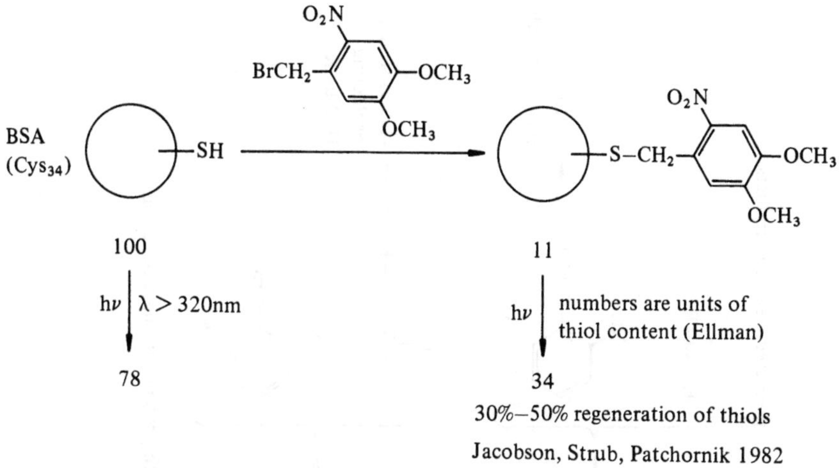

Fig. 6. Photo-regeneration of thiol group of bovine serum albumin

More encouraging results were obtained with bovine serum albumin. The alkylating reagent (Fig. 6) masked all of the sulfhydryl groups that reacted with Ellman's reagent [14]. Irradiation at pH 7 for 10 h with 345 nm light filtered through a copper sulfate solution restored 1/3 of the titratable SH groups.

Improvement of the method resulted from synthesis of a new set of photocleavable inhibitors (Fig. 7). Compound *1* acts as acylating agent and was expected to modify lysine residues. The bond formed is a photocleavable urethane [2]. The cleavage is a two-stage process. The first step involves formation of 2-nitrosobenzaldehyde, leaving a carbamic acid. The second step is the loss of carbon dioxide to give the amine. The derivatives of amino acids are cleaved 100%. A nitrogen laser beam was used as a source of monochromatic light in order to have maximum photons absorbed in minimum time.

Ribonuclease partially inhibited by compound *1* was laser-irradiated for 1/2 h at 365.5 nm and regained 8% (73%–81%) of the total activity.

The 2-nitrobenzyl group is prone to side reactions, and chemical interactions of cleavage products with the enzyme may interfere with the recovery of the catalytic

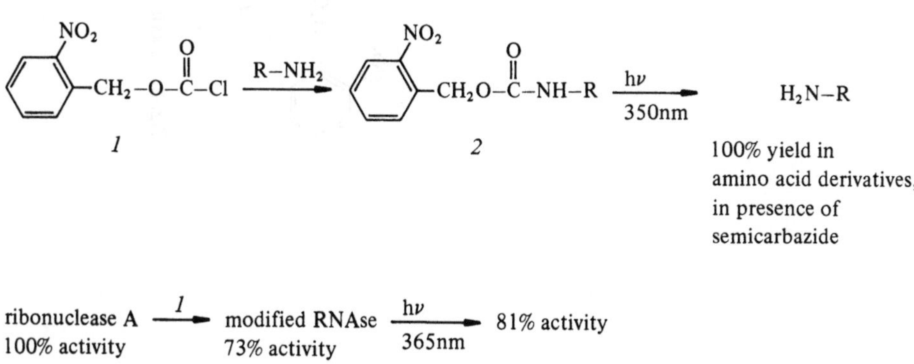

Fig. 7. Formation of photocleavable urethane bonds

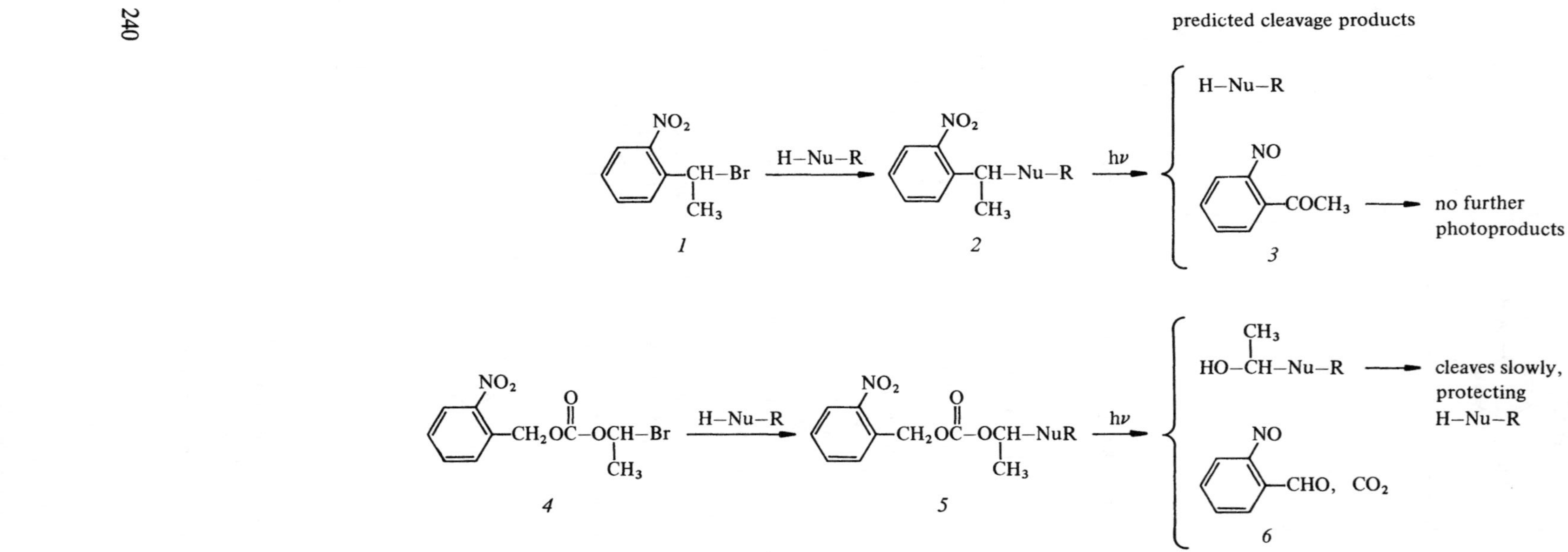

Fig. 8. Alkylating agents for ReAL concept designed to avoid side reactions with proteins during irradiation

activity. We will approach the problem of side reactions by synthesizing o-nitrobenzyl derivatives *1* having substitutions at the benzylic carbon (Fig. 8) or by incorporation of a two step photolyzable linkage *4*. Compound *1* is expected to have fewer photocleavage side products than the unsubstituted o-nitrobenzyl group because the transfer of oxygens from the nitrogen atom cannot proceed further than o-nitroacetophenone [2–3]. The o-nitrobenzaldehyde can repeat the oxygen transfer process to form benzoic acid derivatives.

References

1. Lester, H. A. and Nerbonne, J. M. (1982) Ann. Rev. Biophys. Bioeng. *4,* 151–175
2. Patchornik, A., Amit, B. and Woodward, R. B. (1970) J. Amer. Chem. Soc. *92,* 6333–6335

2a. Amit, B., Zehavi, U. and Patchornik, A. (1974) Isr. J. Chem. *12,* 103–113
2b. Amit, B. (1974) PhD Thesis
2c. Pillai, V. N. R. (1980) Synthesis 1–26

3. Rich, D. H. and Gurwara, S. K. (1973) J. Chem. Soc. Chem. Comm. 610–611
4. Rubinstein, M., Amit, B. and Patchornik, A. (1975) Tetrahedron Lett. 1445–1448
5. Havinga, E., de Jong, R. O. and Dorst, W. (1956) Rec. Trav. Chim. *75,* 378–383
6. Chamberlin, J. W. (1966) J. Org. Chem. *31,* 1658–1660
7. Sheehan, J. C. and Umezawa, K. (1973) J. Org. Chem. *38,* 3771–3774
8. Birr, C., Flor, F., Fleckenstein, P. and Wieland, T., Peptides, proc. 12th Eur. Pept. Symp.
9. Amit, B., Ben-Efraim, D. and Patchornik, A. (1976) J. Amer. Chem. Soc. *98,* 843–844
10. Amit, B., Hazum, E., Fridkin, M. and Patchornik, A. (1977) Int. J. Peptides Prot. Res. *9,* 91–96
11. Hazum, E., Gottlieb, P., Amit, B., Patchornik, A. and Fridkin, M., In "Peptides 1980", p. 105–110
12. Kalbag, S. M. and Roeske, W. (1974) J. Amer. Chem. Soc. *96,* 440–441
13. Rich, D. H. and Gurwara, S. K. (1975) J. Amer. Chem. Soc. *97* (6), 1575–1579
14. Ellman, G. L. (1958) Arch. Biochem. Biophys. *82,* 70–72

Discussion

List of interventions

1 Intervention of Breslow

Does irradiation cause damage to the enzyme which has been photosensitized by your reagents?

1.1 Patchornik

The wavelength used in our photo-irradiation experiments is 3500 Å, known not to cause any damage to a protein. However, the presence of photosensitizers from the reagent might cause some damage to the activity of the enzyme due to the formation of singlet oxygen which might modify histidine residues.

Topic 4

Design of New “Guests”

Molecular Discrimination in Crystal Growth in the Presence of "Tailor-Made" Inhibitors

L. Addadi, Z. Berkovitch-Yellin, I. Weissbuch, J. Van Mil, M. Lahav and L. Leiserowitz

Department of Structural Chemistry, The Weizmann Institute of Science, Rehovot, 76100, Israel

1 Introduction

The relation between the morphology of a crystal and its internal symmetry at a molecular level attracted the attention of scientists a long time ago. The most remarkable experiment in this context was that of Louis Pasteur [1], who in 1848 separated for the first time the two enantiomers of sodium ammonium tartrate, by utilizing the asymmetric habit of their crystals. However, it became evident over the years that the morphological characteristics of the crystals of a given compound depend not only on its crystal structure but also on external parameters of the overall crystallizing system such as solvent, supersaturation, temperature and, specifically, impurities present in the system [2].

In 1931 Miles [3] demonstrated that crystals of $PbCl_2$, of centrosymmetric point symmetry $2/m2/m2/m$, when grown in the presence of dextrin, assume a chiral morphology of symmetry 222. With the advent of new and efficient methods of crystal structure determination, most efforts were devoted to such determinations and less interest was focused on correlating crystal structure with morphology. As a result many important properties of materials, which depend on this correlation, were not exploited in a systematic way. On the other hand, the use of dopants for inducing desired morphological changes is a widespread technological tool, even though the mechanism of the effect is not always well understood [4]. Thus, for example, small amounts of ions such as Pb^{2+} or SO_4^{2-} have long been routinely used as crystallization moderators in the industrial production of large NaCl single crystals for IR and Raman spectroscopy. Similarly, the sugar technologists [5] recognized and studied the critical role played by oligosaccharides in determining the rate of crystallization and the habit of sucrose crystals on extraction from molasses.

In the course of investigating the packing and growth of organic molecular crystals we initiated a study on the systematics of the interactions between a crystallizing substrate and its environment in general, and between the solvent and stereospecific "tailor-made" growth inhibitors in particular. We shall demonstrate that the approach used in the present study should be of relevance to the field of receptors interacting with agonists and antagonists. The growing sites of the various crystal surfaces may be regarded as structured receptors, and the substrate and the inhibitors as agonist and antagonists respectively. The ability to assign the structure of the receptor sites with a very high degree of accuracy by using diffraction techniques should make the present studies most attractive as models for the investigation of the nature of the receptors. We shall illustrate here some mechanistic aspects of the above interactions as they have emerged during these studies and some applications thereof to amplification of chirality by crystallization, resolution of conglomerates, direct and relative assignment of absolute configuration of chiral molecules, study of molecular interactions in crystals, and engineering of crystals with desired morphologies.

2 Amplification of Chirality by Crystallization. A New Method for Resolution of Conglomerates

Synthesis of chiral enantiomerically pure materials from non-chiral reagents has been accomplished by crystallization of the symmetrical (in solution) substrate in appropriately packed chiral single crystals, followed by a lattice controlled reaction [6]. This concept is illustrated in Scheme 1 for the generation of chiral cyclobutane polymers from non-chiral dienes packing in "engineered" chiral crystals [7].

Subsequently, with the aim of amplifying the chirality created in the system in this first generation step, the effect of the chiral products on the crystallization of the two enantiomeric parent crystals, A and $\overline{A}$, was investigated. A number of disubstituted phenylene-diacrylates packing in chiral crystals with the structural motif of Scheme 1 were crystallized, either from the melt or from solution, in the presence of enantiomerically pure dimers, trimers or oligomers generated photochemically from the same phases in a previous step. It was observed in all experiments carried out

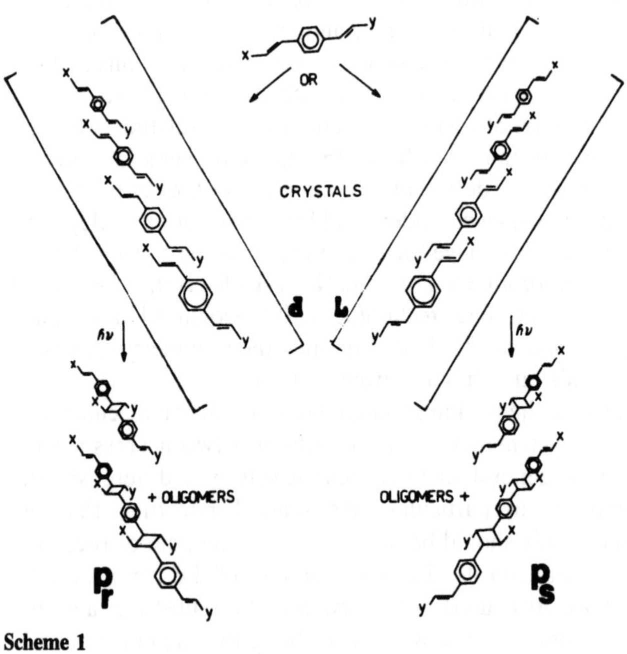

Scheme 1

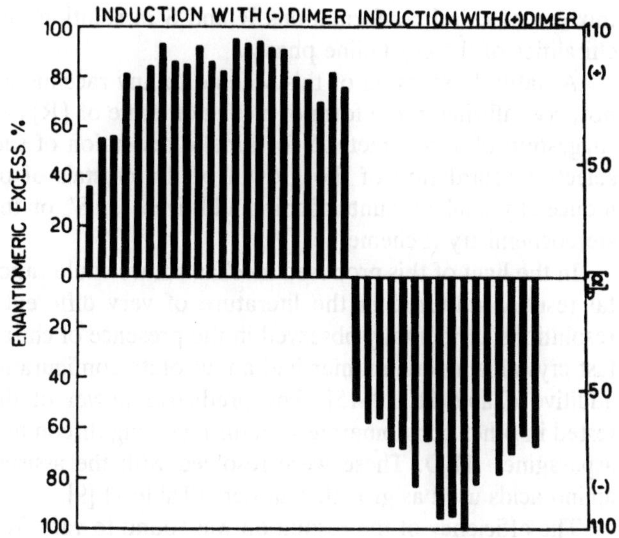

Fig. 1. Enantiomeric excess of the dimers obtained after crystallization of one of the monomers of Scheme 1 in the presence of chiral dimers, under various experimental conditions

that a large excess was formed of the phase enantiomorphic to the one in which the additive itself had been generated (Fig. 1).

The stereochemical similarity between the additive and the crystal structure of one of the enantiomorphic substrates was found to be of paramount importance [8], while parameters like temperature, concentration or nature of the medium had only a quantitative effect on the induction in this system. Further kinetic and mechanistic studies resulted in the formulation of a mechanism according to which the additive is enantioselectively adsorbed in small amounts at the surface of the growing crystal of the same absolute configuration. The adsorption of the chiral additive causes a drastic decrease in the rate of growth of this same crystal, thus shifting the crystallization equilibrium towards the unaffected enantiomorphous phase. This is illustrated in Scheme 2, where the achiral monomer is represented as a fast racemizing

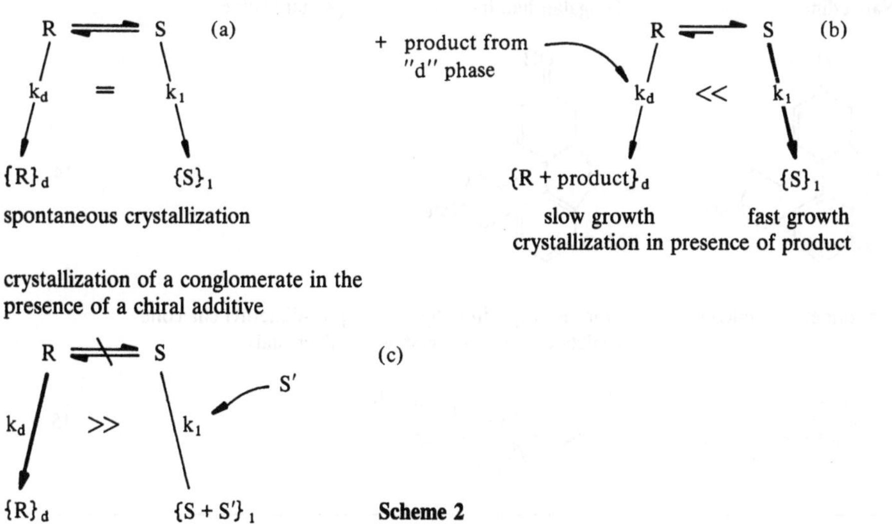

Scheme 2

mixture of chiral conformations R and S in solution, and $\{\ \}_d$ $\{\ \}_l$ represent the chiralities of the crystalline phases.

A natural extension of this scheme to any racemic mixture, fast racemizing or not, crystallizing in the form of a conglomerate of $\{R\}_d$ and $\{S\}_l$ crystals, led to the suggestion of a new method for kinetic resolution of such systems. It is based on selective retardation of the growth of the crystals of one enantiomorph, say S, induced by small amounts of chiral additives, say S′, on the enantiomorph of related stereochemistry (Scheme 2c) [9].

In the light of this proposed mechanism we could rationalize various experimental results, scattered in the literature of very different fields, describing kinetic resolutions accidentally observed in the presence of chiral additives, and where the fast crystallizing enantiomer had an absolute configuration opposite to that of the additive (Table 1) [10–15]. The predictive power of the hypothesis was further tested in other conglomerate systems including threonine, glutamic acid · HCl, and asparagine · H_2O. These were resolved with the assistance of a variety of other amino acids used as growth retarders (Table 1) [9].

The efficiency of the resolution was found to vary from one system to another, and to depend upon the nature and concentration of the additive. In the best cases, [(R, S) glutamic acid + (S) lysine; (R, S)-threonine + (S) glutamic acid; (R, S) asparagine + (S) aspartic acid] resolution with quantitative enantiomeric excess was

Table 1a. Resolutions of conglomerates in the presence of chiral additives, which are described in the literature and are in accordance with our rule

Conglomerate*	Chiral additive	Enantiomer precipitating first in excess	Ref.
GLU	(S)ASP, (S)LEU	(R)GLU	10
GLU	(S)GLU-methyl ester	(R)GLU	11
$Cu(ASP)_2$	(S)GLU, (S)ALA	(R)$Cu(ASP)_2$	12
$NaNH_4$ tartrate	D(+)malic acid	D(−)$NaNH_4$tartrate	13
Narwedine	(−)galanthamine	(+)narwedine	
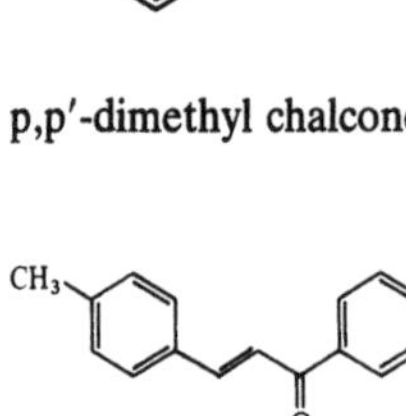	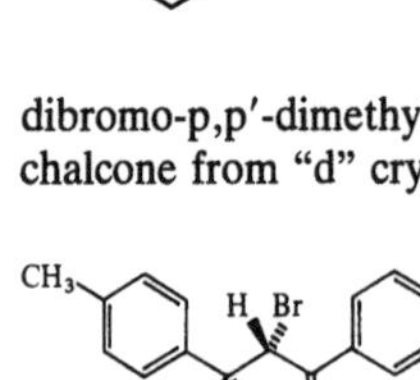		14
p,p′-dimethyl chalcone	dibromo-p,p′-dimethyl chalcone from "d" crystals	p,p′-dimethyl chalcone "l" crystals	
			15

Table 1b. Resolutions of conglomerates accomplished in our laboratory designed on the basis of the "Rule of Reversal"

Conglomerate	Chiral additive	Enantiomer precipitating first in excess	Ref.
phenylene-diacrylates	phenylene-diacrylates dimers from "d" crystals	phenylene-diacrylates "l" crystals	8
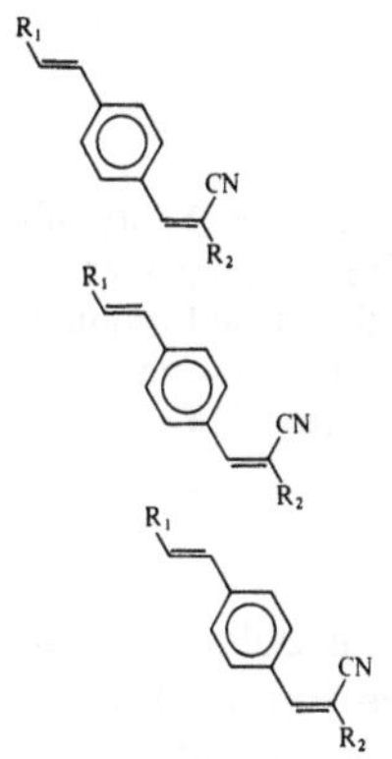	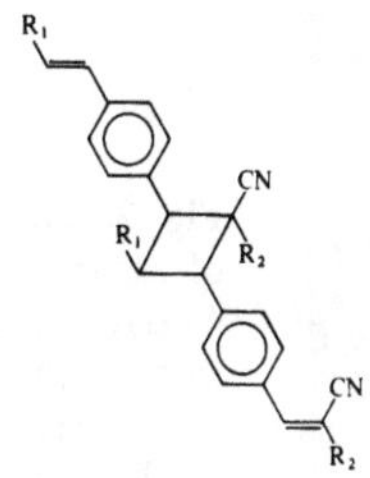		
THR	(S)GLU, (S)GLN, (S)ASN, (R)CYS**, (S)PHE, (S)HIS, (S)LYS, (S)ASP	(R)THR	9
GLU · HCl	(S)LYS, (S)ORN, (S)HIS, (S)SER, (S)THR, (R)CYS, (S)TYR, (S)LEU	(R)GLU	9
ASN · H_2O	(S)GLU, (S)ASP, (S)SER, (S)GLN, (S)LYS, (S)ORN, (S)HIS	(R)ASN	9
pHPGpTS*	(S)PG*, (S)TYR, (S)pMPG*, (S)PHE, (S)Dopa, (S)MeDopa	(R)pHPG	9
HIS · HCl	(S)TRP, (S)PHE	(R)HIS	9
phenylhydracrylic acid	(S)phenyllactic acid	(R)phenylhydracrylic acid	
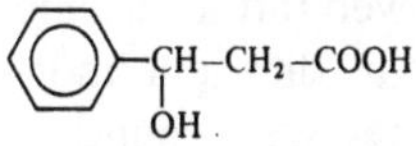	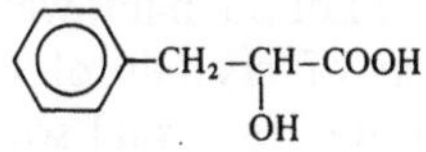		9

* Abbreviations: pHPGpTS – p-hydroxyphenylglycine p-toluene sulphonate; PG – phenylglycine; pMPG – p-methoxyphenylglycine

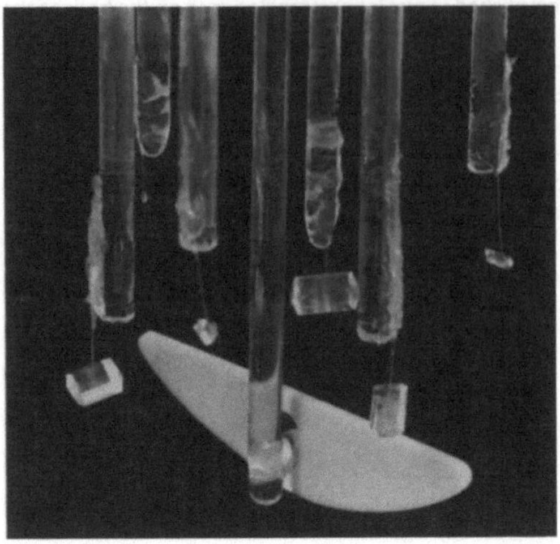

Fig. 2. Three (R) and three (S) crystals of asparagine · H_2O grown together for 45 days under conditions near to equilibrium in the presence of (S) serine. The large crystals are (R) asn · H_2O and the small ones are (S)

achieved. The growth of the affected crystals in these instances was delayed up to several days with respect to that of the unaffected ones. In all the systems investigated we could demonstrate that the additive is occluded throughout the bulk of the affected crystals in amounts ranging typically from 0.05 to 1.5% w/w of substrate; it is found in much smaller amounts, if any, in the crystals of the enantiomorph.

The inhibiting effect of the additive on the growth of the affected enantiomorph has been proven also by direct comparison of the size of $\{R\}_d$ and $\{S\}_l$ single crystals grown in parallel from seeds under conditions close to equilibrium (Fig. 2). Although there are indications that the same kind of inhibition may also influence the crystal nucleation rate, such an effect is very difficult to isolate and quantify.

The most convincing proof of the adsorption-growth inhibition process is associated with changes in the morphology of the affected crystals. The morphology of a crystal is determined by the relative rates of growth of its various faces. Since the growth rates of only those crystal faces where adsorption of inhibitor took place are affected, crystals growing in the presence and absence of additives are expected to display different habits. Indeed, in the resolution experiments it was systematically observed that the appearance of crystals of normal habit was followed (after variable time depending on the system) by that of crystals of distinctly different morphologies: when the two types are separated by visual sorting they were found to consist of the unaffected and affected enantiomers respectively. For asparagine the use of different additives consistently yielded crystals with different morphologies (Fig. 3). For (R, S) threonine (grown in the presence of say (S) glu) and (R, S) glutamic acid · HCl (grown in the presence of (S) lys · HCl) the inhibition of growth is so strong that the affected (S) enantiomer appeared as fine powder coating the well-formed crystals of the (R) enantiomorph (Fig. 4) [16].

This phenomenon offers the possibility of modifying the classical Pasteur experiment and of extending it to systems which undergo spontaneous resolution but do not display hemihedrism [17]. Another kind of visual differentiation of the enantiomorphs was obtained by crystallizing a conglomerate of uncolored crystals in the presence of a resolved colored "tailor-made" inhibitor. When the plate-like crystals

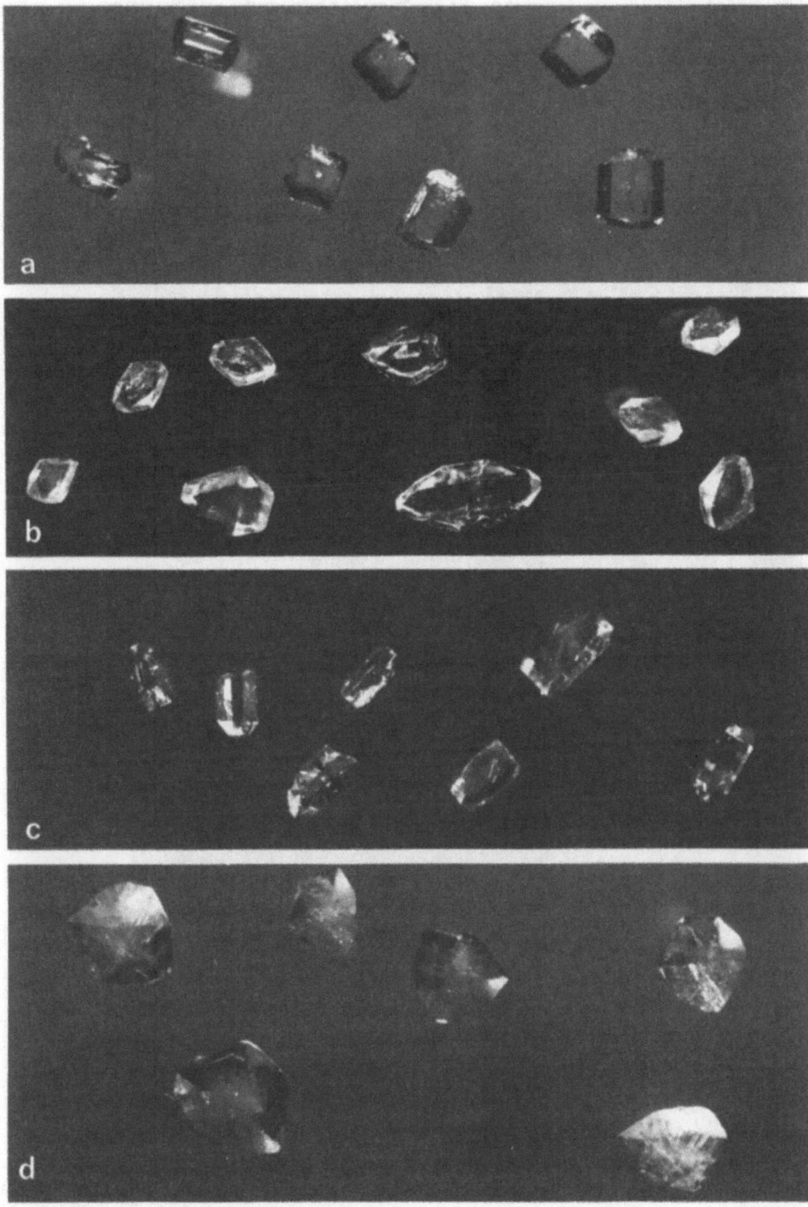

Fig. 3a–d. Crystals of (S) asparagine · H_2O grown in the presence of additives: **a** none, or (R) additives; **b** (S) glutamine; **c** (S) serine; **d** (S) ornithine. In each case the crystals were obtained from (R, S) asn as a mixture of (R) asn crystals displaying a morphology as in **a**, and (S) asn crystals as in **b**, **c**, **d** or others, depending on the additive

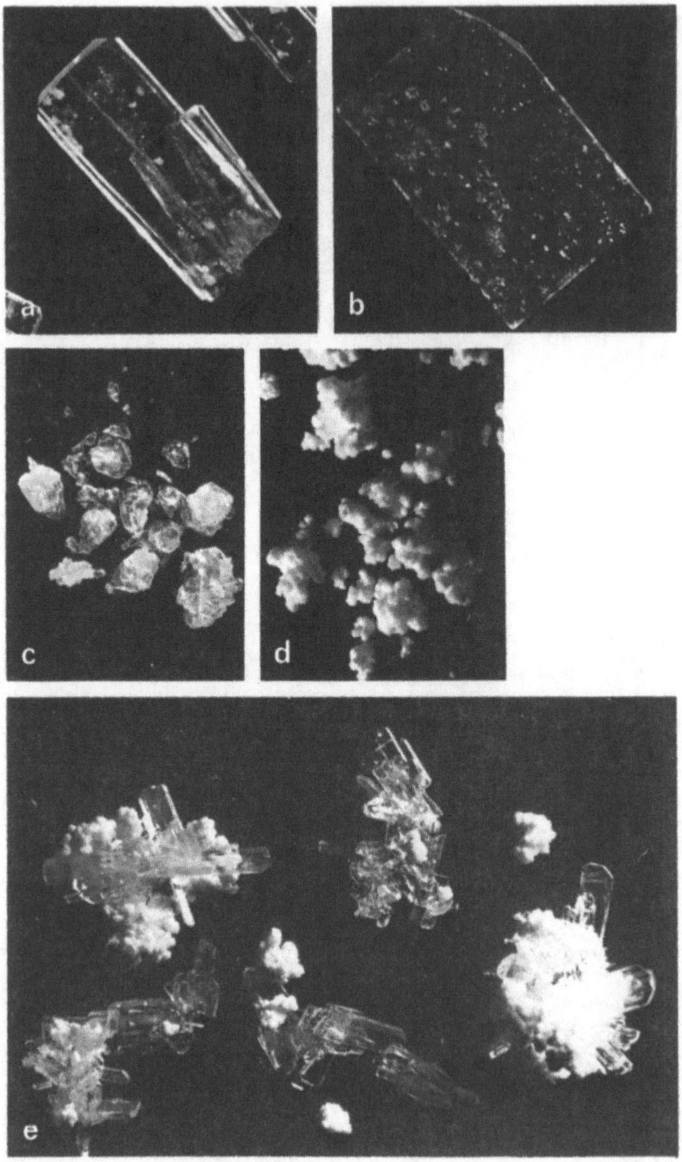

Fig. 4a–e. Crystals of (S) glutamic acid · HCl grown in the presence of increasing amounts of additive lysine: **a** none or (R) lys; **b** + 2 mg/ml (S) lys; **c** + 5 mg/ml (S) lys; **d** + 50 mg/ml (S) lys; **e** crystals of (R) and (S) glu · HCl grown in the presence of (S) lys: the plates are the (R) enantiomer while the powder is the (S) enantiomer

of (R) and (S) glu · HCl are grown in the presence of yellow N-(2,4-dinitrophenyl)-(S)-lysine as additive, uncolored crystals of (R) glu · HCl precipitate first, followed by yellow ones of the (S) enantiomer.

The stereochemical correlations between the enantiomorphic crystals and their chiral inhibitors, namely that the additive affects only the enantiomer of the same absolute configuration, provides us with a new method for the determination of absolute configuration on a relative scale. This is revealed independently by morphological changes and enantioselective occlusion of additives. The method is related to the quasi-racemate technique of Fredga [18], but has the advantage of a wider applicability.

3 Morphological Crystal Engineering

The dramatic morphological changes associated with the growth of organic crystals in the presence of additives reveal a high degree of specificity in the interaction of the foreign material with the different structured surfaces of the crystalline matrix. The morphological changes therefore have a direct bearing on the mechanism of the adsorption-inhibition process on a molecular level.

In general, when growth is inhibited in a direction perpendicular to a given face, the area of this face is expected to increase relative to those of other faces of the same crystal (Scheme 3). Differences in the relative surface areas of the various

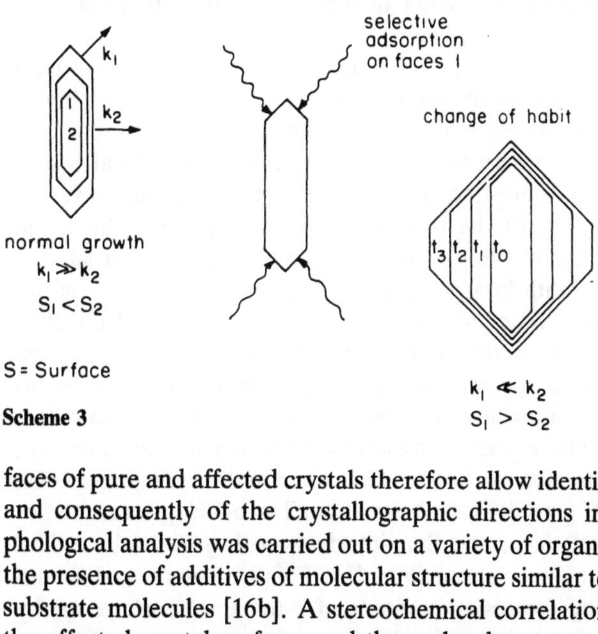

Scheme 3

faces of pure and affected crystals therefore allow identification of the affected faces and consequently of the crystallographic directions involved. This type of morphological analysis was carried out on a variety of organic compounds crystallized in the presence of additives of molecular structure similar to those of the corresponding substrate molecules [16b]. A stereochemical correlation between the structures of the affected crystal surfaces and the molecular structure of the inhibitor could be deduced in each system. We could infer that the additive may be adsorbed only at those faces in which the part of the adsorbate that differs from that of the substrate points away from the crystal interior. Once adsorbed, the additive inhibits the regular deposition of oncoming layers of substrate molecules, slowing down the

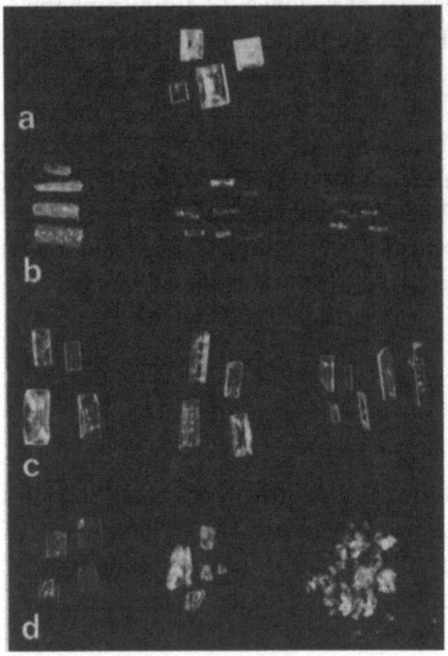

Fig. 5a–d. Crystals of benzamide: **a** pure; or crystallized in the presence of increasing amounts (from left to right) of **b** benzoic acid; **c** *o*-toluamide; **d** *p*-toluamide

growth perpendicular to that face and leading to a concomitant relative increase in its surface area. A similar mechanism was invoked by Smythe [5b, c] in his elegant studies on the changes in habit of the crystals of sucrose grown in the presence of oligosaccharides as additives [19].

Once this mechanism was established, it became possible to exploit it in order to systematically modify the morphology of crystals by tailoring additives which bind at a preselected face and thus inhibit growth in a predictable manner.

This approach is illustrated here for benzamide, which crystallizes from ethanol in the form of platelike crystals elongated in *b* (Fig. 5a). The main packing feature of the structure (Fig. 6) consists of the typical ribbon motif [20a] of hydrogen-bonded cyclic dimers interlinked by NH···O bonds along the *b* axis. Retardation of growth along *b* was achieved with benzoic acid as additive. The substitution of a molecule of benzamide at the end of the ribbon motif by benzoic acid results in replacement in the *b* direction of an NH···O bond by a repulsive O···O interaction, thus retarding growth in this direction and inducing the growth of bars elongated in *a* (Fig. 5b). Inhibition of growth along *a* was accomplished by adding *o*-toluamide to the crystallization solution. The crystals consistently grew as bars elongated in *b* (Fig. 5c). The additive *o*-toluamide can be easily adsorbed in the hydrogen-bonding chain without interfering with growth in the *b*-direction. The *o*-CH_3 group emerges, however, away from the (100) face, and thus interferes with growth along the *a* direction, in which the dimers are stacked (Fig. 6b). Finally, thinner and thinner plates are obtained by adding increasing amounts of *p*-toluamide, whose methyl substituent perturbs the already weak van der Waals interlayer interactions between phenyls in the *c* direction (Fig. 5d, 6b) [21].

Similar morphological changes were induced in crystals of other primary amides and other organic molecules, including carboxylic acids, amino acids, dipeptides and steroids. Our interpretation of the results was complemented by atom-atom poten-

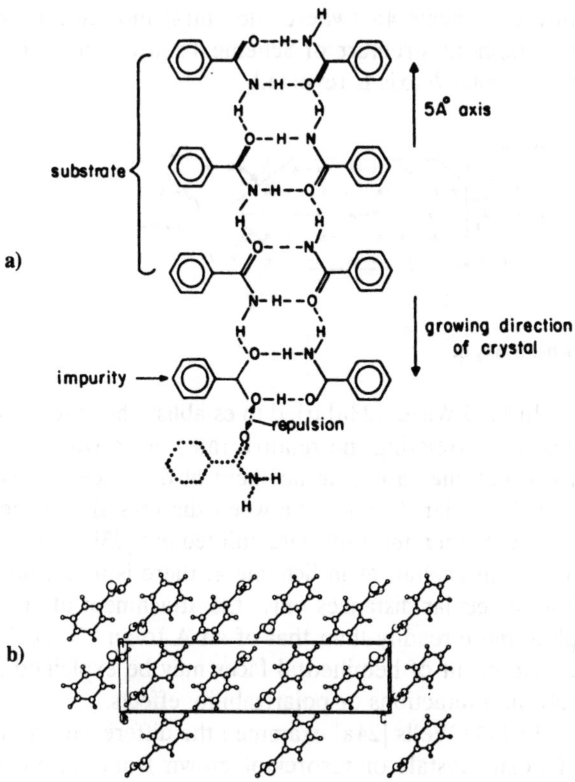

Fig. 6a. Schematic representation of the benzamide structure showing a benzoic acid molecule inserted at the end of the ribbon motif of hydrogen bonded dimers along the *b* axis; **b** packing arrangement of benzamide viewed along the *b* axis

tial calculations which pinpoint the nature of substrate-additive interactions on the basis of the established mechanism.

The theoretical and empirical considerations allow us both to understand and to predict the induced changes in crystal habit. We shall now show how this knowledge was applied to the solution of certain fundamental problems in stereochemistry.

4 Direct Assignment of Absolute Configuration of Chiral Molecules: Application to Polar and Centrosymmetric Crystals

After the recognition of chirality in molecules by Biot, the direct assignment of the absolute configuration of a chiral molecule in a chiral crystal which develops hemihedral faces has challenged the chemical crystallographer.

The essence of the problem is that conventional X-ray diffraction analysis of chiral crystals (*i.e.* assumption of Friedel's law) does not permit deduction of the absolute orientation of the chiral molecule with respect to the crystal axes or to its faces [22]. In other words, it is not possible to distinguish between the polar structure of Scheme 4a (where the chiral molecule is represented by X–A) and the enantiomeric structure of Scheme 4b in which the orientation of X–A with respect to the polar *b* axis is reversed.

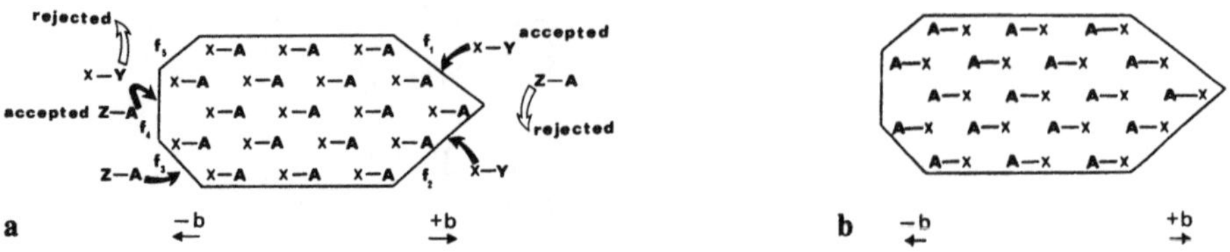

Scheme 4a, b.

In 1949 Waser [23a] tried to establish the absolute configuration of (D) – tartaric acid by correlating the relative rates of growth of the hemihedral (hkl) and ($\bar{h}\bar{k}\bar{l}$) faces with the ease of attachment of the "free" molecule at either face in terms of intermolecular distances between the crystal and the molecule to be attached. In fact, as Turner and Lonsdale pointed out [23b], given an asymmetric molecule X–A in a polar crystal, as in Scheme 4, there is no *a priori* reason why, on the basis of intermolecular distances only, the attachment of an X to an "A face" should take place more readily than that of an A to an "X face". The observed differences in development of hemihedral faces may be explained primarily in terms of surface-solvent interactions or polarizability effects.

In 1949 Wells [24a] examined the differences in the habit and growth directions of polar crystals of resorcinol grown from various solvents. He interpreted the solvent effect in terms of different interactions of the solvent molecules with the resorcinol molecules at various exposed faces of the crystal, which has hydrophilic OH groups at one end and hydrophobic benzene groups at the other. However, in the absence of a basic and general understanding of the mechanism involved in the effect, this explanation remained an intuitive interpretation of the experimental results, and in fact the subject remains controversial to the present day [24b, c, d]. Based on controlled morphological modifications through adsorption of tailor-made additives, we developed a new method for the direct assignment of the absolute configuration of chiral molecules. We shall discuss first the case of polar crystals [25].

Polar crystals: Scheme 4 describes an appropriate polar crystal belonging to space group $P2_1$. The unique axis of the crystal is parallel to the X–A direction. Faces *f*1 and *f*2 delineate the crystal in the polar $+b$ direction and *f*3, *f*4, *f*5 in the $-b$ direction. The faces within each pair *f*1, *f*2, and *f*3, *f*5, are homotopic since they are related by 2-fold symmetry. Since the crystal is polar the structures of the faces at the $+b$ and $-b$ ends are different. By application of the two-step mechanism of adsorption-inhibition, an inhibitor X–Y will bind selectively at faces *f*1 and *f*2, and once bound, will retard growth along $+b$ and possibly other directions but not along $-b$, in the crystal of Scheme 4a. In an analogous way an inhibitor Z–A will hinder the growth of faces *f*3, *f*4 and *f*5 but not of *f*1 and *f*2 (the opposite would happen in the structure of Scheme 4b). Such retardation will be associated either with an increase in the areas of the inhibited faces with respect to the unaffected ones or with the

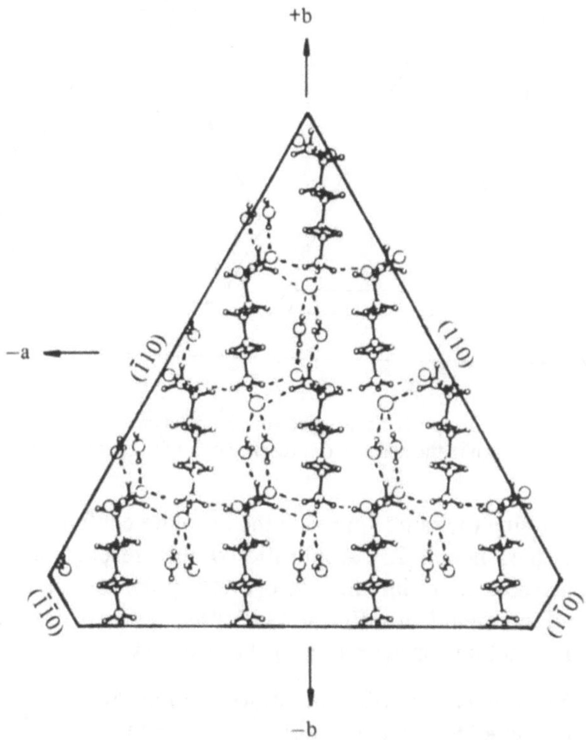

Fig. 7. Packing arrangement of the crystals of (S) lysine · HCl · 2 H_2O viewed along the *c* axis, as delineated by the observed (hk0) crystal faces

appearance of new faces on the affected side of the crystal. The observed morphological differences between crystals grown in the presence and in the absence of the additive thus establish the direction of the substrate molecule X–A with respect to the polar axis. Consequently the absolute configuration of the crystal and that of the chiral molecular constituents can be derived.

We shall briefly illustrate here the case of lysine · HCl and cinnamoylalanine. (S) or (R) lys · HCl crystallizes from water as a dihydrate in a monoclinic structure of space group $P2_1$. The packing arrangement delineated by the crystal faces of the unaffected crystal is shown in Fig. 7. In the crystal the lysine molecules are aligned parallel to the *b* axis with the NH_3^+–CH–COO^- moiety emerging from the + *b* end of the crystal and the ε-NH_3^+ pointing towards − *b*. In agreement with the analysis presented above, additives with a modified carboxyl or α-amino group, like lysine methyl ester, indeed inhibited growth in the + *b* direction, inducing development of the (010) face (Fig. 8). Conversely, additives which bear a modified ε-amino group, like norleucine or norvaline, inhibited growth along the −*b* direction, with a concomitant pronounced increase in the areas of the (1$\bar{1}$0) and ($\bar{1}\bar{1}$0) faces (Fig. 8c).

The observed changes in morphology implied that the occluded inhibitor would be distributed anisotropically along the polar *b* axis of the crystal. This expectation was experimentally confirmed for the crystals of (S)-lysine · HCl grown in the presence of (S)-norleucine. According to the chromatographic analysis of material taken from the + and − *b* ends of the crystal, the additive was occluded preferentially at the − *b* side (Fig. 9).

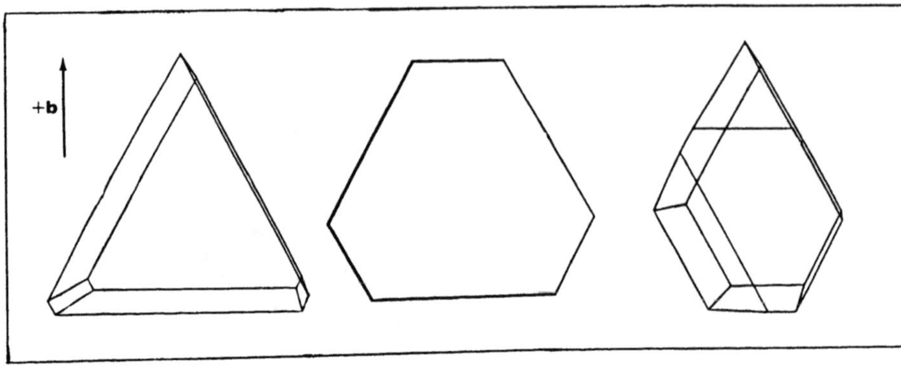

a b c

Fig. 8a–c. Computer-drawn pictures of typical crystals of (S) lysine · HCl · 2 H_2O viewed along the *c* axis, grown in the absence or presence of additives: **a** pure; **b** (S) lys methyl ester; **c** (S) norleucine

Using the same approach the absolute configuration of E-cinnamoyl alanine was assigned as well [25]. Application of the present analysis to the morphological difference noticed by Smythe [5b, c] on the polar crystals of sucrose grown in the presence and in absence of raffinose (and other di- and trisaccharides), enabled us to assign the absolute configuration of these sugars.

Enantiopolar crystals: The above method of determining absolute configuration is not limited to polar crystals but is applicable as well to chiral molecules which induce morphological changes in centrosymmetric crystals composed of racemic mixtures or meso compounds [26, 27]. The method is based on the fact that, unlike chiral crystals, in centrosymmetric ones the orientations of the constituent molecules with respect to the crystal axes are unambiguously assigned by conventional X-ray analysis. The known orientation of the enantiomers in such crystals can be exploited for the direct assignment of the absolute configuration of chiral resolved additive molecules, provided that this structural information can be transferred to such

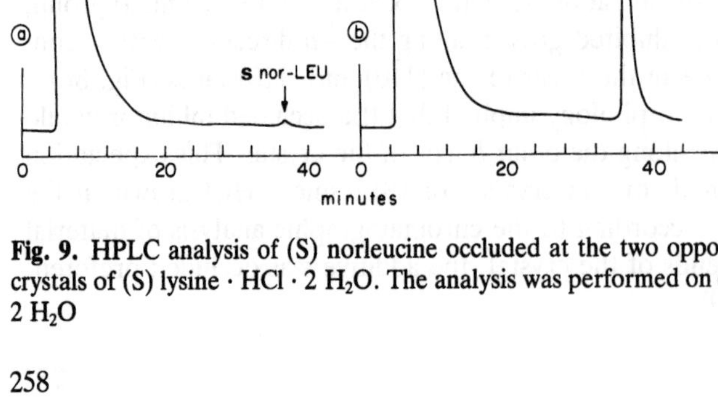

Fig. 9. HPLC analysis of (S) norleucine occluded at the two opposite sides of the polar *b* axis of crystals of (S) lysine · HCl · 2 H_2O. The analysis was performed on equal amounts of (S) lys · HCl · 2 H_2O

molecules. The absolute configuration of the resolved additives is determined through the morphological changes they selectively induce on the enantiotopic faces of appropriately selected crystals.

A prerequisite for application of this method is that within the centrosymmetric racemic crystal a specific functional group attached to an (R) molecule will point towards the face $f\bar{1}$ but not towards $f1$ (Scheme 5). By symmetry, the same func-

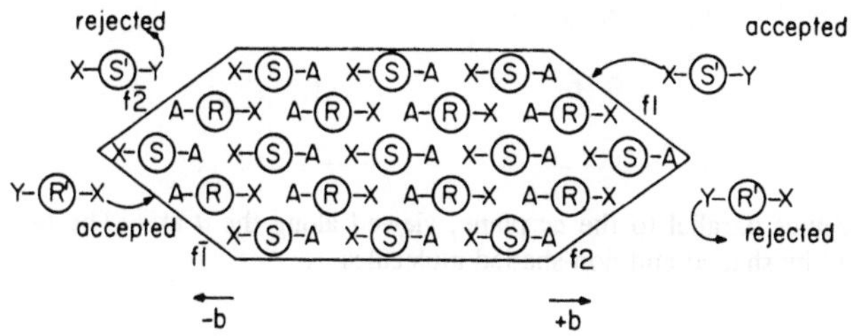

Scheme 5

tional group attached to an (S) molecule will emerge at the enantiotopic face $f1$, but not at $f\bar{1}$. It is useful here to regard centrosymmetric crystals containing chiral molecules as enantiopolar. We define an enantiopolar crystal as one comprising two enantiomeric sets of intermeshed polar crystal structures related to each other by symmetry elements of the second kind, e.g. a centre of inversion or a glide plane. Crystallization of such compounds in the presence of a chiral additive (R′), appropriately designed so that it will fit in the site of an (R) molecule on the growing crystal faces $f\bar{1}$ or $f\bar{2}$, but not on the enantiotopic faces $f1$ or $f2$, will hinder growth along the $-b$ direction but not along $+b$. By virtue of symmetry, the enantiomeric additive (S′) will inhibit growth of faces $f1$ and $f2$, while racemic (R′, S′) will inhibit growth along both directions $+b$ and $-b$.

These principles were applied to the monoclinic crystals of (R, S) serine [26, 27b] and glycine [27]. Here we shall describe the assignment of the absolute configuration of the α-amino acids, through the morphological changes these molecules induce in the crystalline [α] form of glycine. The glycine molecule, although non-chiral, has two prochiral hydrogens, H(R) and H(S) (by analogy with the chirality of the natural amino acids). In its α-modification, space group $P2_1/n$ (Fig. 10), the glycine molecules form in the crystal two enantiopolar sets, which are represented in Fig. 10 by shaded and non-shaded atoms respectively. All the molecules in the non-shaded set have the C–H(R) bond aligned with the b axis and pointing towards $+b$, so that C–H(R) emerges from the (010) face, while C–H(S) lies almost perpendicular to it in the ac molecular layer. Only an amino acid additive of chirality R, whose side chain will be parallel to the C–H(R) bond, can be adsorbed at the (010) face and subsequently hinder the crystal growth perpendicular to it. By virtue of symmetry, an (S) amino acid can be adsorbed only at the $(0\bar{1}0)$ face.

Pure glycine crystallizes from water in the form of bipyramids, with the b axis perpendicular to the base of the pyramid (Fig. 11a). It was observed that all natural (S)-amino acids, apart from proline, induce a dramatic morphological change at the $-b$ side of the crystals, with the appearance of pyramids exhibiting a predominant

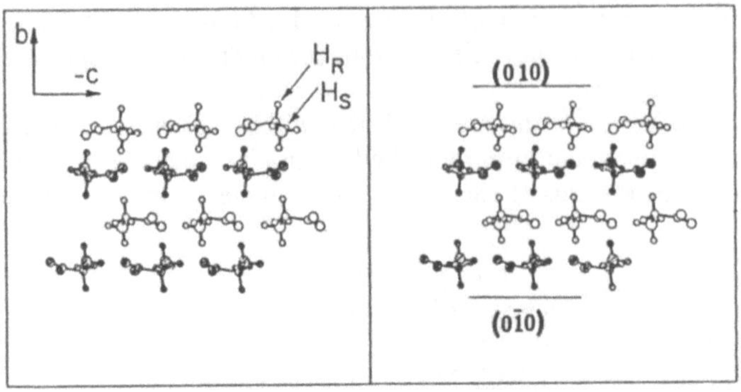

Fig. 10. Layers of glycine molecules parallel to the *ac* plane, viewed along the *a* axis. The two enantiopolar sets are represented by shaded and non-shaded molecules .

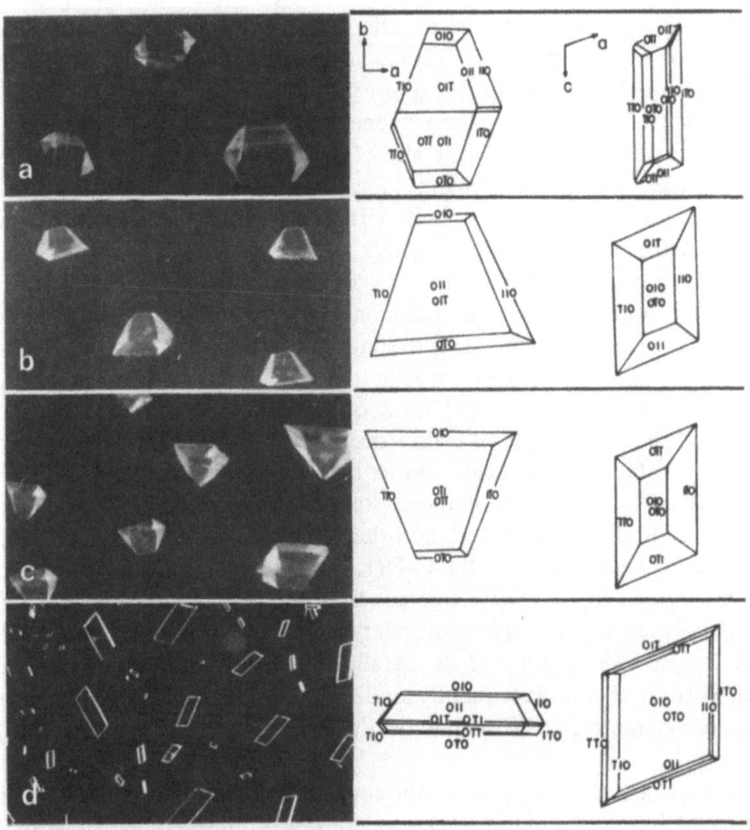

Fig. 11a–d. Comparison between crystals of glycine grown in the absence and presence of additives: **a** pure; **b** (S) additive; **c** (R) additive; **d** racemic additive

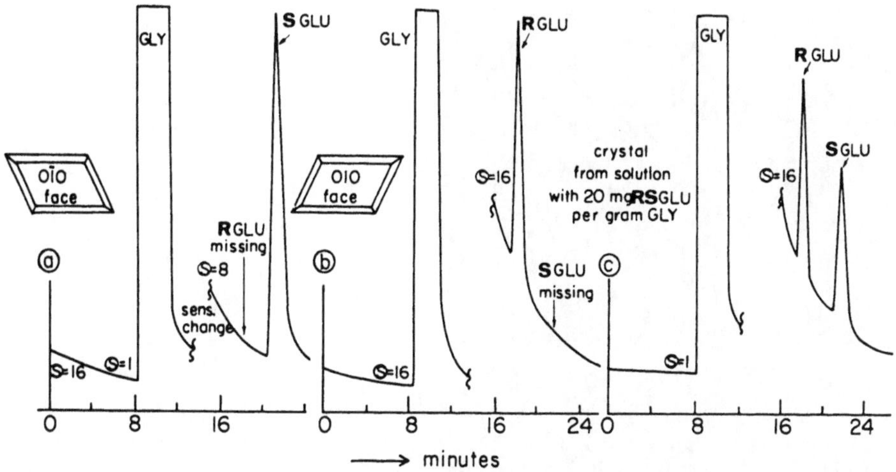

Fig. 12a–c. Enantiomeric distribution of occluded (R, S) glutamic acid in the {010} plate-like crystals of glycine by HPLC: (S) denotes arbitrary sensitivity scale 1, 8, 16. **a** Only additive (S)-glu is present in material shaved from the (01̄0) face of the crystal; **b** only additive (R)-glu is present in material shaved from the (010) face; **c** relative distribution of (R, S) glu in the remaining whole crystal

(01̄0) face (Fig. 11b). All (R) amino acids induce the development of a large (010) face (Fig. 11c), while a racemic additive causes crystallization of {010} plates (Fig. 11d).

Amounts of 0.02–0.2% of the additives were found in the bulk of the glycine crystals. HPLC analysis of these occluded α-amino acids inside single plate-like crystals of glycine demonstrated a total enantioselective segregation along the b axis (Fig. 12); as expected, the (R) amino acids prevail at the $+b$ half of the crystal and the (S) amino acids at the $-b$. This segregation suggests that the overall symmetry of the crystal has been lowered from $P2_1/n$ to $P2_1$.

The changes in crystal habit of glycine were studied by atom-atom potential energy calculations; the energy results were completely consistent with the abovementioned observations [27a, 28].

5 A New Model for Spontaneous Generation of Optical Activity

The results reported for the glycine system acquire further interest as the {010} plate-like crystals of glycine, grown in the presence of other (R, S) α-amino acids, have a tendency to float at the air/water interface with the b-axis perpendicular to the interface [29]. Each such floating crystal will grow either along its $+b$ or $-b$ direction, depending upon whether face (010) or (01̄0) is pointing towards the water solution. Thus, if a single crystal of glycine exposes its (010) face to a solution of glycine containing a mixture of (R, S) amino acids, only the (R) acids will be occluded, and vice versa. HPLC analysis of the enantiomeric content of the occluded amino acids indeed confirmed these expectations. Such resolutions of (R, S) leu, and (R, S) val are shown in Fig. 13.

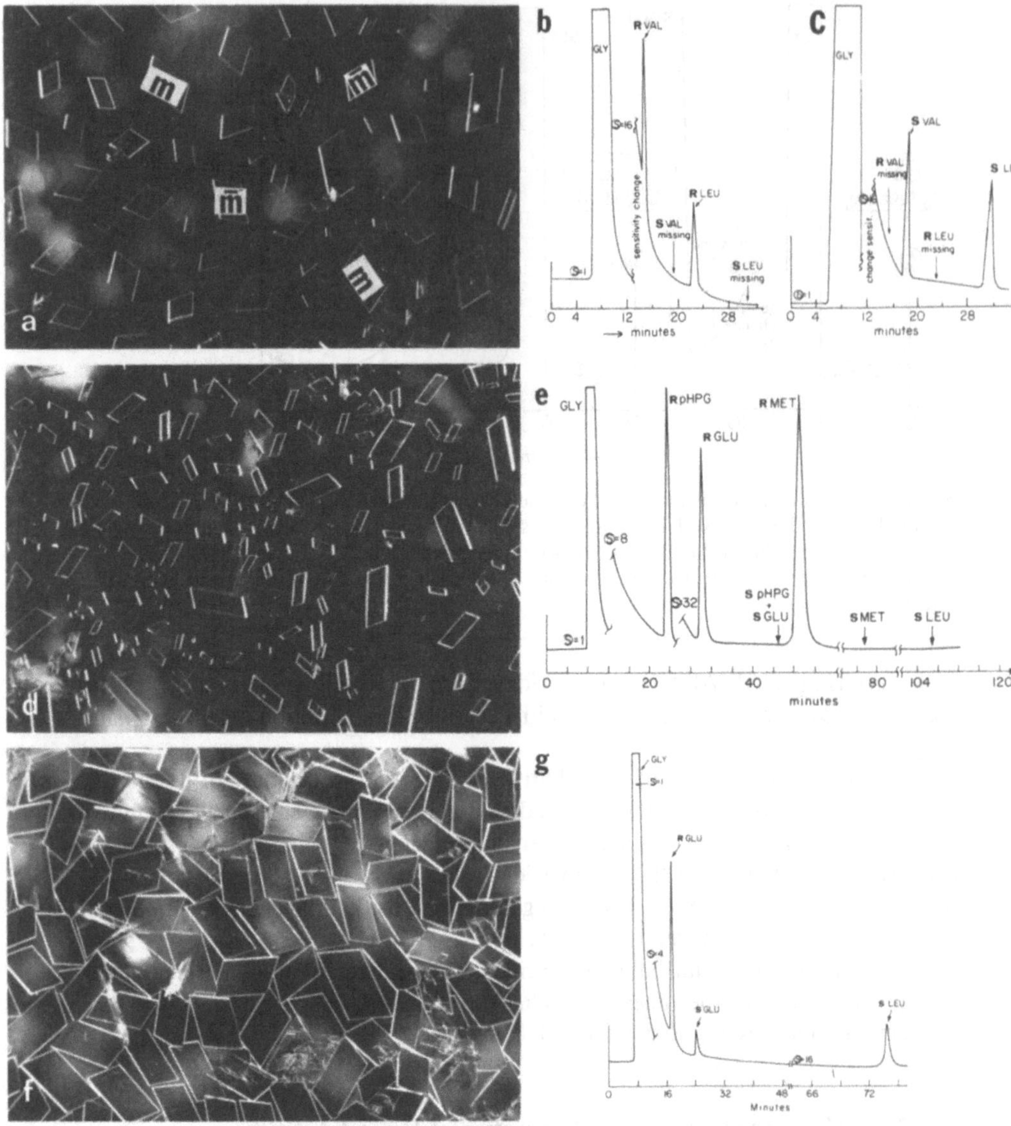

Fig. 13. a Floating plate-like crystals of glycine grown from a water solution (4.4 M) containing racemic 1% wt/wt (R, S) leu and 1% wt/wt (R, S) val. The crystals display both enantiomorphous morphologies, labelled *m* and $\bar{m}$; **b** HPLC analysis of crystals *m* with the face (010) exposed to solution, after washing with water; **c** HPLC analysis of glycine crystals $\bar{m}$ with the $(0\bar{1}0)$ exposed to the solution after washing with water; **d** Floating plate-like crystals grown in the presence of 1% wt/wt (S)-leu and racemic p-hydroxyphenylglycine (0.5% wt/wt), racemic glu (1%) and racemic met (0.5%). All the crystals float with their $(0\bar{1}0)$ faces exposed to air and exhibit the same enantiomorphous morphology; **e** HPLC analysis of the enantiomers of p-hydroxyphenylglycine (pHPG), glu, met and leu occluded inside the crystals of (d). Only the R enantiomers of pHPG, glu and met are found; **f** Crystalline crust of glycine grown at the air/water interface. The water solution (4.4 M) contains 1% wt/wt (S)-leu and 3% wt/wt racemic glu; **g** HPLC analysis of the same crust indicating an enantiomeric excess of 70% of occluded (R) glu

It is clear that in a racemic mixture there is equal probability for a given crystal to float on its (010) or (0$\bar{1}$0) face. We have observed, however, that if glycine is crystallized in the presence of small amounts of resolved amino acids bearing hydrophobic side chains, like (S) leucine, (S) phenylalanine, or (S) α-amino octanoic acid, the floating crystals of pyramidal shape always expose their (0$\bar{1}$0) face to the air. In fact when crystals were grown in the presence of equal amounts of two amino acids with side chains of different character and opposite absolute configuration, for example (S) leucine with (R) glutamic acid [or (R) alanine, or (R) serine], all the resulting plates had their (0$\bar{1}$0) faces directed to the air. This indicates that stereospecifically adsorbed leucine molecules are capable, even in tiny amounts, of orienting the glycine crystals by virtue of the hydrophobic character of their side chains.

The tendency of leucine molecules to orient floating glycine crystals and the enantioselective adsorption at the face exposed to the solution were exploited to induce resolution of racemic amino acids by crystallization of glycine at interfaces in the presence of resolved leucine; for example glycine, grown in the presence of racemic hydrophilic amino acids plus small amounts of (S) leucine, yielded plates all floating with their (0$\bar{1}$0) faces pointing upwards, thus selectively occluding the (R) enantiomers (Fig. 13d, e). Furthermore, even under conditions where polycrystalline crusts of glycine were formed (Fig. 13f, g) resolution occurred with an enantiomeric excess higher than 70% (Fig. 13g) in the examples studied.

The experiments described above provide us with a possible model for the generation and amplification of optical activity in an initially symmetrical world. Let's assume that the first plate-like crystal of glycine is formed from the "prebiotic soup" floating with its (010) face upward. During growth the crystal will occlude exclusively the (S) amino acids present in the "soup", thus leaving the solution enriched in the (R) enantiomers. The generated excess of hydrophobic (R) amino acids will in turn direct the floating of later-formed crystals with their (010) faces again upward, thus ever further enriching the solution with (R) enantiomers.

6 Prospects for the Future

In the present study we have attempted to show that it is possible with the assistance of tailor-made additives to control both growth and dissolution of organic crystals. A remarkable aspect of this process is that it provides us with the ability to navigate a given additive through a given face into a selected subset of the symmetry-related sites of the crystal. Thus, for example, in the case of benzamide, benzoic acid may be occluded at only one of the two centrosymmetrically related sites at the end of the ribbon motif of Fig. 6a; the overall symmetry of the crystal is thus reduced from the centrosymmetric $P2_1/n$ to the chiral $P2_1$. Such "induced chirality" is currently being exploited for the performance of spontaneous asymmetric synthesis. Furthermore, this type of chiral discrimination in centrosymmetric crystals is presumably manifested also in inorganic crystals, including minerals, which might have a bearing on the spontaneous separation of enantiomers in nature.

The present approach might provide, further, an effective method for pinpointing fine intermolecular interactions in crystals, (e.g. O···O repulsion in the benzamide-benzoic acid system, Sect. 3) as revealed by the morphological changes they induce.

The role of the tailor-made additives is probably not limited to the growth stage, but includes prenucleation aggregates which are expected to structurally resemble the mature crystal. An extension in this direction might be of importance in the resolution of enantiomers by crystallization.

Growth and dissolution of crystals are considered as reciprocal processes, which can be interchanged by altering the degree of saturation of the solution. This implies that an inhibitor of growth of a given crystal face must in principle affect the rate of dissolution of that same face. Preliminary studies along these lines are most promising [30].

The present experimental studies were also complemented by atom-atom potential energy calculations. In some cases the morphology of a crystal was computed in terms of intermolecular interactions between substrate molecules in the crystal bulk, and additive at the crystal solvent interface, yielding results which are in perfect agreement with the experimental observations.

So far we have confined ourselves to additives which resemble the host molecule. This was done in order to facilitate the extraction of detailed mechanistic information. We can now systematically reduce this similarity, provided that strong specific interactions are preserved at interfaces between substrate and additive. This approach provides a unique entry to the study of solvent effects on crystal morphology. It might, further, be of particular relevance for the better understanding of crystallizations taking place at structured biological interfaces, such as bones, teeth and shells [31], and in the inhibition of ice formation in fish living at temperatures below the freezing point of water [32].

In summary, we see that the old, but until recently empirical, technique of modifying crystal growth by the addition of impurities has a wide range of consequences both in theoretical and practical chemistry and in materials science [33]. This field of research may be expected to expand rapidly in the future.

Acknowledgement

We thank the U.S.-Israel Binational Foundation, Jerusalem, the donors of the Petroleum Research Fund administered by the American Chemical Society, the Volkswagen Stiftung and the Israel Academy of Science and Humanities for financial support of this work.

One of us (L.A.) is the recipient of the Helena Rubinstein Career Development Chair.

We are indebted to Prof. M. D. Cohen for a critical reading of the manuscript. We are grateful to Edna Gati and Marianne Idelson for carrying a heavy load of the experimental work. We thank Dr. S. Weinstein for the important cooperation in the HPLC analyses.

References

1. Pasteur, L. (1848) Ann. Phys. *24,* 442
2a. Buckley, H. E., in "Crystal Growth", Chapman and Hall, London 1951
2b. Hartman, P., in "Physics and Chemistry of the Organic Solid State", Ed. by Fox, D., Labes, M. M. and Weisberger, A., Interscience, New York, 1963, Vol. *1,* p. 369 and Vol. *2,* p. 873
2c. Hartman, P., "Crystal Growth", North-Holland Publishing Co., Amsterdam and London, 1973, p. 367
2d. Kern, R. (1953) Bull. Soc. Fr. Miner. Crystallogr. *76,* 391
2e. Wells, A. F. (1946) Phil. Mag. *37,* 180; (1946) *37,* 217; (1946) *37,* 605
3. Miles, F. D. (1941) Proc. R. Soc. London Ser. A *132,* 266
4. For relevant publications on role of impurities on crystal growth see:
4a. Cabrera, N., Vermilyea, D. A., in "Growth and Perfection of Crystals", Doremus, R. H., Roberts, R. W. and Turnbull, D., Eds., Wiley, London, 1958, p. 393
4b. Franck, F. C., *ibid.,* p. 411
4c. Sears, G. W., *ibid.,* p. 441
4d. Dugua, J. and Simon, B. (1978) J. Cryst. Growth *44,* 265
5a. Dunning, W. J. and Albon, N., in "Growth and Perfection of Crystals", J. Wiley & Sons, London, 1958, p. 416
5b. Smythe, B. M. (1967) Aus. J. Chem. *20,* 1115
5c. Smythe, B. M. (1971) Sugar Technology Reviews *1,* 191
5d. Mantovani, G., Gilli, G. and Fagioli, F. (1967) Compt. Rend. XIII, Assembly C115, Falsterbo, Sweden, 289
5e. van Hook, A. (1968) Kristal. Acad. Nauk, S.S.S.R. Inst. Kristallogr. *8,* 45
5f. van Hook, A. (1981) Journal of the A.S.S.B.T. *21,* 130
6a. Green, B. S., Lahav, M. and Rabinovich, D. (1979) Acc. Chem. Res. *12,* 191
6b. Addadi, L., Ariel, S., Lahav, M., Leiserowitz, L., Popovitz-Biro, R. and Tang, C. P. (1980) in Chemical Physics of Solids and their Surfaces, Spec. Period. Reports *8,* 202
7a. Addadi, L. and Lahav, M. (1979) Pure Appl. Chem. *51,* 1269
7b. Addadi, L. and Lahav, M. (1978) J. Am. Chem. Soc. *100,* 2838
7c. Addadi, L. and Lahav, M. (1979) J. Am. Chem. Soc. *101,* 152
7d. Addadi, L., van Mil, J. and Lahav, M. (1982) J. Am. Chem. Soc. *104,* 3422
8a. van Mil, J., Gati, E., Addadi, L. and Lahav, M. (1981) J. Am. Chem. Soc. *103,* 1248
8b. van Mil, J., Addadi, L., Gati, E. and Lahav, M. (1982) J. Am. Chem. Soc. *104,* 3431
8c. Addadi, L., van Mil, J., Gati, E. and Lahav, M. (1981) Macromol. Chem. Suppl. *4,* 37
9. Addadi, L., Weinstein, S., Gati, E., Weissbuch, I. and Lahav, M. (1982) J. Am. Chem. Soc. *104,* 4610
10. Purvis, J. L. (1957) U.S. Patent 2,790,001
11. Fike, H. L. (1960) U.S. Patent 2,937,200
12a. Harada, K. (1965) Nature *205,* 590
12b. Harada, K., Tso, W. (1972) Bull. Chem. Soc. Japan *45,* 2859
13. Ostromisslensky, L. (1908) Chem. Ber. *41,* 3035
14. Barton, D. H. R., Kirby, G. W. (1962) J. Chem. Soc. 806
15. Green, B. S. and Heller, L. (1974) Science *185,* 527
16a. Addadi, L., Gati, E. and Lahav, M. (1981) J. Am. Chem. Soc. *103,* 1251
16b. Addadi, L., Berkovitch-Yellin, Z., Domb, N., Gati, E., Lahav, M. and Leiserowitz, L. (1982) Nature *296,* 21
17a. Bunn, C. W. (1961) Chemical Crystallography, Oxford University Press
17b. Paul, I. C. and Curtin, D. Y. (1982) J. Chem. Rev. *81,* 525
18. Fredga, A. (1960) Tetrahedron *8,* 126 and references cited therein
19. We thank Prof. A. van Hook for bringing the reference of this work to our attention
20a. Berkovitch-Yellin, Z. and Leiserowitz, L. (1980) J. Am. Chem. Soc. *102,* 7677
20b. Berkovitch-Yellin, Z. and Leiserowitz, L. (1982) J. Am. Chem. Soc. *104,* 4052
20c. Berkovitch-Yellin, Z., Ariel, S. and Leiserowitz, L. (1983) J. Am. Chem. Soc. *105,* 765
21a. Berkovitch-Yellin, Z., Addadi, L., Idelson, M., Lahav, M. and Leiserowitz, L. (1982) Angew. Chem. Suppl. 1336

21b. Berkovitch-Yellin, Z., van Mil, J., Addadi, L., Idelson, M., Lahav, M. and Leiserowitz, L. (1985) J. Am. Chem. Soc. *107,* 3111
21c. Addadi, L., Berkovitch-Yellin, Z., Weissbuch, I., Lahav, M. and Leiserowitz, L. (1983) Mol. Cryst. Liq. Cryst. *96,* 1
22. For comprehensive discussions on methods for the determination of the absolute configuration of chiral molecules see:
22a. Dunitz, J. D., "X-ray Analysis and the Structure of Molecules", Cornell University Press, Ithaca, N.Y., 1979, p. 129–148
22b. Mason, S. F., "Molecular Optical Activity and Chiral Discrimination", Cambridge University Press, Cambridge, England, 1982
22c. Addadi, L., Berkovitch-Yellin, Z., Weissbuch, I., Lahav, M. and Leiserowitz, L. in: Topics in Stereochemistry, Ed., Allinger, N., Eliel, E. and Wilen, S., John Wiley and Sons Inc., 1986, *16,* in press
23a. Waser, J. (1949) J. Chem. Phys. *17,* 498
23b. Turner, E. E. and Lonsdale, K. (1950) J. Chem. Phys. *18,* 156
24a. Wells, A. F. (1949) Disc. Faraday Soc. *5,* 197
24b. Davey, R. J. (1976) J. Cryst. Growth *34,* 109
24c. Davey, R. J., in: Current Topics in Material Science, *8,* Kaldis, E., Ed., North Holland Publ., 1982, Chapt. 6
24d. Milisavljevic, B. S., Ph. D. Thesis, E.T.H. Diss. N6898, Zurich
25. Berkovitch-Yellin, Z., Addadi, L., Idelson, M., Leiserowitz, L. and Lahav, M. (1982) Nature *296,* 27
26. Addadi, L., Weissbuch, I., Berkovitch-Yellin, Z., Lahav, M., Leiserowitz, L. and Weinstein, S. (1982) J. Am. Chem. Soc. *104,* 2075
27a. Weissbuch, I., Addadi, L., Berkovitch-Yellin, Z., Gati, E., Weinstein, S., Lahav, M. and Leiserowitz, L. (1983) J. Am. Chem. Soc. *105,* 6615
27b. Weissbuch, I., Ph. D. thesis, Feinberg Graduate School, Rehovot, Israel, 1985
28. Berkovitch-Yellin, Z. (1985) J. Am. Chem. Soc. *107,* 3375
29. Weissbuch, I., Addadi, L., Berkovitch-Yellin, Z., Gati, E., Lahav, M. and Leiserowitz, L. (1984) Nature *310,* 16
30. Shimon, L. J. W., Lahav, M. and Leiserowitz, L. (1985) J. Am. Chem. Soc. *107,* 3375
31. Addadi, L. and Weiner, S. (1985) Proc. Natl. Acad. Sci. USA *82,* 4110
32. de Vries, A. L. (1980) in "Animals and environmental fitness", Gilles, R., editor, Pergamon Press, 583
33. Berg, W. F. (1983) J. Photographic Science *31,* 62

Discussion

List of interventions

1 Luisi
  1.1 Lahav
2 Wüthrich
  2.1 Lahav
3 Hess
  3.1 Lahav
4 Cram
5 Wüthrich
  5.1 Lahav
6 Gil-Av

1 Intervention of Luisi

How are your studies related to the more general question of how optical activity was generated on earth?

1.1 Lahav

It is not possible at present to provide a plausible route for the spontaneous generation of optical activity on earth. There are a number of hypotheses, but at present it is not possible to prove or disprove any of them The experiments that we have performed only demonstrate a simple and possible route for spontaneously obtaining chiral amino acids of a single handedness in the absence of any outside chiral agent. Whether nature has selected such a route or not is impossible to answer at present.

2 Intervention of Wüthrich

There is something that is not quite clear to me. Prof. Luisi asked about the creation of stereoselectivity in a system where there is none to start with, whereas you have been talking about systems in which you did have a stereoselective compound as a starting material which you then use to select the chiral compound. Have I understood this correctly?

2.1 Lahav

What I'm saying is that we have a non chiral system which develops into a chiral one in the absence of any chiral influence to start with.

3 Intervention of Hess

You said this is a closed system? In what respect is it closed? You start with a supersaturated solution of your substrate, which is obviously a precursor substrate, and this is a process wherein you have some feedback loop, so that at least for some time the system is not at equilibrium.

3.1 Lahav

In the present experiment we are dealing with a feedback process. One starts with a single or a small number of oriented glycine crystals, grown at the air/water interface, and the water solution contains a racemic mixture of α-amino acids.

When this crystal of glycine occludes one enantiomer it enriches the water solution with the other. However, once the solution has been enriched with one of the enantiomers, crystals of glycine undergoing subsequent crystallization will be oriented with the same face exposed to the water solution. The eventual orientation of all the glycine crystals depends on the first arbitrary orientation of the floating glycine crystal. If this process occurs in another part of the globe in a different closed system, one might equally well spontaneously obtain the other enantiomers.

4 Intervention of Cram

Crystallization is replication, and like other kinds of replication, crystallization is subject to inhibition. Your results provide a splendid exemplification of this statement.

5 Intervention of Wüthrich

Are there preparative applications of the present study?

5.1 Lahav

There are at least three ways in which the present study can be immediately applied. Firstly, we have already successfully applied our "tailor-made" inhibitors in the kinetic resolution of enantiomers by the process of crystallization. Secondly we are at present testing the ability to monitor the dissolution properties of organic crystals in general and of certain drugs in particular. Finally, our method might find application in the field of the material sciences. The magnitude of many physical properties of polar crystals such as piezoelectricity or pyroelectricity depends on the shape of these crystals. Our ability to shape crystals at will might improve the pyhsical properties of many such polar crystals.

6 Intervention of Gil-Av

When a small enantiomeric excess is initially present even in trace amounts, there is no difficulty in conceiving the generation of enantiomeric domains by one mechanism or another.

If however, there is no such initial excess, formation of a local excess of one antipode remains a statistically controlled phenomenon with the chance of deviation from the racemic ratio decreasing with the number of events. In your mechanism, whatever you do, you need a small enrichment in the beginning, otherwise the chances of the reaction going one way or the other is equal.

Binding and Information Transfer in Conformationally Restricted Peptides

Victor J. Hruby and Mac E. Hadley

Departments of Chemistry and General Biology, University of Arizona, Tucson, Arizona 85721, USA

Introduction

Many peptide hormones and neurotransmitters (both hereafter referred to as hormones) are short, linear, and highly flexible molecules with conformational properties that are dependent on environment. Such properties might be imagined to be valuable for the purposes for which Nature uses them, namely, as molecular switches and chemical messengers for intercellular communication and control of cellular function. In particular, conformational flexibility could be functionally important in several ways. For example, the different conformational states might be related to a number of biological properties of these peptides including:

1) the ability to recognize and interact with several different cellular receptors;
2) the ability to "adjust" to evolutionary changes of receptors and of the peptide itself; and
3) the possibility of recognition by different enzymes leading to rapid biodegradation and turnover, etc.

On the other hand, these properties of peptides pose problems for developing an understanding of the physical-chemical basis for biological recognition (binding; address function) to specific receptors and for their ability to initiate intracellular biological changes (transduction; message function). For example, the question arises whether the conformations found in solution for these chemical messengers have any relevance to their biological activities, and stated more precisely, which conformations of the many possible are important to each of the biological functions of the peptide. A comprehensive answer to this question will require an intimate understanding of the three-dimensional properties of peptide receptors and the associated membrane components needed for biological transduction. Furthermore, these answers will require an intimate knowledge of the interactions(s) of the hormones with their receptors and the mechanism(s) by which binding, transduction, reversal of activity, and other physical-chemical perturbations produce a cellular response. At present our knowledge of peptide hormone membrane receptors is limited. Perhaps our most extensive knowledge is of the insulin receptor [1, 2], but even here only limited information is available and virtually nothing is known of the primary structure or tertiary conformation of the protein components of the receptor complex. This limited state of knowledge will probably remain a problem for some time, though modern molecular biology may make it possible to isolate the gene for the receptor, to clone the gene, and to produce large quantities of a functional receptor.

Despite the complexities and difficulties of the problems posed above, we believe it is possible to develop a rational, systematic approach to an understanding of the three-dimensional, stereoelectronic features which are involved in peptide hormone activity at receptors. This requires the use of one or more biological assays, a further understanding of the nature and complexities of membrane receptor systems, and the development of strategies for determining hormone conformations related to receptor activation. In the latter regard, we are not limited to Nature's solution, that is, we can utilize the power of organic chemistry to design and synthesize peptide analogs which Nature cannot make, but which provide specific structural, conformational, and stereoelectronic properties.

In this paper we will discuss an approach based on conformational and structural considerations which we believe provides the potential for a systematic approach to understanding the physical-chemical basis of peptide hormone action.

Some Biological and Physical-Chemical Considerations in Information Transfer for Peptide Hormones and Neurotransmitters

The mechanism by which peptide hormones initiate a biological response in a target cell has been the subject of intensive investigation during the past twenty years, and a number of comprehensive reviews have appeared recently (for example see [3–9]). Briefly summarized, these studies have provided evidence that the receptor system consists of a multicomponent membrane complex which includes (minimally) a receptor, one or more regulatory proteins required for transduction, and an amplification system (e.g., adenylate cyclase) which generates the intracellular second messengers (Fig. 1). For the purposes of this discussion, the crucial importance of these studies is the recognition that any approach which seeks to understand the relationship of the physical-chemical properties of the hormone to biological activity must fully utilize biological assay systems if the proper biological insights are to be obtained. This problem has been discussed in detail elsewhere [10–14]. In general it

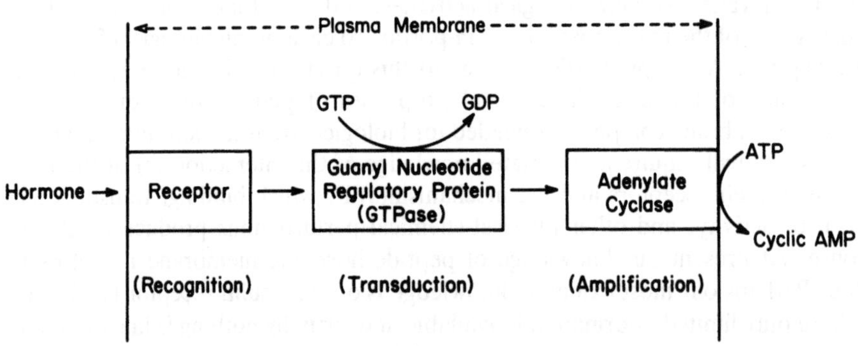

Fig. 1. General scheme for the mechanism of action of many peptide hormones at plasma membrane receptor systems. Such systems include (minimally) a membrane surface receptor at which peptide binding (recognition) occurs, a regulatory protein through which signal transduction occurs, and an amplification system (here represented as adenylate cyclase) which produces the intracellular second messenger (here cyclic AMP) which modulates cellular function

is critical that assay methods and conditions are used which can provide the following:

1) direct evaluation of full dose-response curves;
2) examination of partial agonism or partial antagonism if such exists;
3) full agonist activity over a large concentration range of peptide (preferably >10,000);
4) the possibility for quantitative evaluation of antagonist (inhibitor) activities; and
5) the possibility of quantitative binding studies.

Furthermore, since many (perhaps most) biologically active peptides have receptors on different cell types with different structural requirements, understanding the various structural and conformational properties important to the biological activity of the hormone will require several different assay systems.

In addition, we need to determine those conformational, stereoelectronic and topological properties which are important for biological recognition (binding at specific receptors) and transduction (initiation of the biological response). Classical structure-biological activity investigations approach this problem by determining which amino acid residues of the native hormone are important for biological activity. In this approach, one or more amino acid residue is either deleted or replaced by another residue. Such studies can provide important insights into those aspects of primary structure which are either important or critical for receptor recognition (binding) or for transduction (efficacy). For example, the minimal "message sequence" required for activity can sometimes be determined by sequential deletion of amino acid residues from the N- and C-terminal. In addition, such studies when used in conjunction with bioassays over the full range of the dose-response curve can provide valuable leads for the design of analogs with high potency, high receptor specificity, and with antagonist (inhibitory) activities [10–14]. Generally these studies provide limited understanding of the relationship of three-dimensional structure to biological activity which ultimately form the basis for the biological activity of peptide hormones. However, recently a number of studies have been carried out to examine how conformation-biological activity considerations might lead to peptide analogs with specific biological properties [14–21].

In our own laboratory [14, 20, 22–25] we have utilized conformational, steric and topographical restrictions of peptide hormones to examine this problem. The conformational consequences of these restrictions have been investigated by biophysical methods (NMR, CD, Raman, etc.), and the results of these studies utilized to establish a testable model for the "biologically active" conformations related to binding and information transfer at specific receptors.

Conformational Restriction and the Biological Activities of Peptides: What Can We Learn? – A Working Hypothesis

Any effort which attempts to establish relationships between the conformations and the biological properties of peptide hormones and neurotransmitters directly confronts the complexities of the membrane receptor systems for these molecules. Fortunately during the past two decades or so exceptional progress has been made in

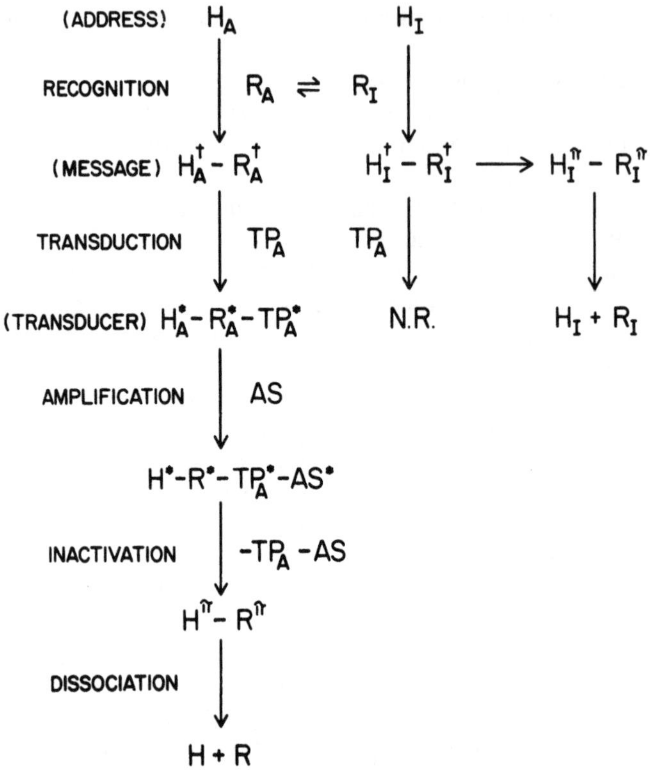

Fig. 2. A representation of a general mechanism by which a peptide hormone or neurotransmitter can participate in information transfer in a membrane receptor system. In this model the hormone agonist (H_A) interacts with the receptor in its agonist state (R_A) to form a complex ($H_A^{\dagger}$–$H_R^{\dagger}$) which can interact with the transducing protein system (TP_A) to give a complex (H_A^*–R_A^*–TP_A^*) that is capable of activating the amplification system (AS) (e.g., adenylate cyclase) which can then modulate intracellular biochemical events via a second messenger such as cyclic AMP. Inactivation of the biologically active system (represented as H^*–R^*–TP_A^*–AS^*) proceeds via a dissociation process in which the hormone-receptor system uncouples from the transducing protein and amplification system. The hormone-receptor complex (H^{π}–R^{π}) eventually dissociates to its original or modified monomeric forms (H + R). The exact molecular equilibria in this system can vary depending on the number of binding and regulatory components, but the mechanism shows the hormone in four different "states" corresponding to the possible utilization of different structural or conformational properties of the ligand for binding (recognition), transduction, amplification, and inactivation-dissociation. See text for further discussion. A hormone inhibitor analog (H_I) can interact with the receptor in its inhibitory state (R_I). In this case, coupling of the hormone-receptor complex ($H_I^{\dagger}$–$R_I^{\dagger}$) to the transducing protein (TP_A) in such a manner as to produce a biologically active complex does not occur, and inhibition of the biological response occurs. Dissociation of the inhibitory state can occur (via the H_I^{π}–R_I^{π} complex) to give the original or modified hormone inhibitor. The complexity of the inhibitory state again will depend on the precise nature of the regulatory components in the system, but the hormone inhibitor can exist in at least three different "states" corresponding to binding (recognition), inhibition of transduction, and dissociation of the inhibitor complex

examining the nature of membrane receptor systems [1–9] and the mechanism(s) of information transfer in these systems (see e.g., [10, 12, 14, 26]).

A simple working model of peptide hormone action derived from these studies is shown in Fig. 2. The basic components of this system include the following: 1) a receptor (R) which can exist either in an agonist state (R_A) or an antagonist (or inhibitory) state (R_I). These states are represented here in equilibrium, but in some cases there is circumstantial evidence that distinct agonist (R_A) and antagonist (R_I) receptor components exist. In many (perhaps all) peptide receptor systems there is a transducing regulatory component referred to here as the transducing protein (TP_A). (In several cases this regulatory component is referred to as the nucleotide regulatory protein N, since it is regulated by a nucleotide such as GTP. In some systems [4, 8, 27, 28] there appears to be a separate and distinct transducing component Ni [not shown here] which apparently interacts with the hormone-receptor system to inhibit transduction.) Though different mechanisms have been proposed for biological amplification in the transduction process, in most cases it appears that the transducing protein (TP_A), when appropriately activated, interacts with the amplification system (AS) necessary for a biological response. In many cases this is the adenylate cyclase system discovered by Sutherland and coworkers over 20 years ago [29]. In other cases, amplification may be provided by opening of a cation channel or some other "second messenger". To "turn off" the system requires an inactivation process which includes dissociation of the hormone-receptor complex and subsequent inactivation of the transducing protein and/or amplification system. Though there may be several levels of complexity in this process [1, 3, 5, 30], it is simply represented here by the inactivation and dissociation steps in Fig. 2.

How do the conformational and dynamic properties of the peptide hormone or neurotransmitter enter into this process? To answer this question we believe several *functionally different* attributes of the peptide must be considered. As shown in Table 1 these include: receptor specificity, potency, efficacy, development of inhibitory properties, and prolongation of biological activities.

To relate these conformational and biological relationships to the "general" mechanism of peptide hormone action we have proposed the following hypotheses:

1. Peptide hormones which have several distinct physiologically and pharmacologically relevant receptors will utilize *different* structural-conformational features of the

Table 1. Possible effects of conformational restriction of peptide hormones and neurotransmitters on biological activities

Receptor specificity	Preference for one of several biologically relevant receptors is shown.
Potency	Increased (or decreased) overall binding and biological activity is manifested.
Efficacy	Maximal (or fractional) stimulation of the full biological response changes.
Inhibition	Binding to receptors occurs (generally competitive with the natural ligand) but transduction does not.
Prolongation	The binding and/or biological response is prolonged relative to the natural ligand.

hormone at each receptor. Classical structure-function studies have provided considerable evidence to support this hypothesis. From the standpoint of conformational restriction, this hypothesis would predict that appropriate conformational restriction can lead to high receptor selectivity. Conversely, the different topochemical requirements can be deduced if receptor-specific, conformationally restricted analogs are available. Of course specificity can apply to both agonists and inhibitors (antagonists).

2. Increased potency will depend on appropriate conformational and dynamic properties of the peptide. We have proposed that hormones utilize different topographical properties for receptor binding (represented in Fig. 2 as $H_A^{\dagger}$) and transduction (represented in Fig. 2 as H_A^*). This can explain how a peptide with very low potency can be fully active (high efficacy), but another peptide with high potency can be a partial agonist (or antagonist) [10, 31].

3. Efficacy will depend on the availability of specific conformational and structural features of the peptide for converting the hormone receptor complex from its initial bound state ($H_A^{\dagger}$–R_A^*) to the transduced state (H_A^*–R_A^*–TP_A^*) necessary for interaction with the amplification system (AS) (Fig. 2).

4. Inhibition can occur by two separate but related mechanisms. a) The structural-conformational requirements for hormone agonist-receptor binding are different from those for hormone antagonist-receptor binding. b) Alternatively, a situation can be imagined where hormone-receptor interaction could occur utilizing the hormone agonist receptor, but in which the hormone possesses conformational or dynamic properties which for thermodynamic or kinetic reasons cannot assume the transduced conformational state. We have previously suggested that both mechanisms may apply to oxytocin antagonists [14, 32].

5. There are now several examples in the literature where peptide hormones have a much longer duration of biological response (prolonged activity) than the native hormone. We have recently hypothesized [23, 25] that the prolonged biological activity of some ligands might be a receptor-related event, and thus a function of structure-conformation properties of the peptide in the transduced (biologically active) state for an agonist, or in the bound hormone-receptor complex in a non-biologically active (inhibitory) state of an antagonist. This hypothesis predicts that appropriate conformational and/or structural modifications which are specific to the bound transduced state of an agonist or the peptide-receptor complex of an antagonist will slow the inactivation process leading to hormone-receptor dissociation. This hypothesis implies that hormone-receptor inactivation and dissociation proceeds via a different pathway than does hormone-receptor binding and transduction. This hypothesis also predicts: 1) That there is no necessary relationship between potency and prolongation of biological agonist or antagonist activity, and 2) that prolongation for a particular analog need not be manifested at all physiologically relevant receptors for the peptide.

We now discuss how we have utilized and tested these hypotheses in the development of peptide hormones and neurotransmitters with high receptor specificity, superagonist potency, prolonged biological activity, and antagonist properties.

Development of Superpotent α-Melanotropin Agonist Analogs

Superpotency in peptide hormone analogs can result by maximizing those conformational and stereoelectronic properties which are necessary for receptor binding and those which are required for transduction. Development of such properties generally requires systematic examination of classical structure-activity relationships. In our analysis of the results from such classical structure-activity studies we have attempted to interpret the results in terms of specific conformational properties which *may* be involved. In favorable circumstances, insights into conformation biological relationships can be derived by examining the consequence of primary structure modification on conformational features using prediction methods such as Chou and Fasman analysis [33]. It should be emphasized that such analyses can at best only provide trends in potential conformation-biological activity relationships. We now briefly discuss here how such ad hoc considerations have led to the development of linear and cyclic α-melanotropin analogs with superagonist activity.

α-Melanotropin (α-MSH) is a linear tridecapeptide, Ac–Ser–Tyr–Ser–Met–Glu–His–Phe–Arg–Trp–Gly–Lys–Pro–Val–NH_2, which has been extensively examined for its structure-activity relationships using the classical approach [13, 34, 35]. In addition to its well-known pigmentation effects [36], α-MSH appears to be involved in a number of other biological processes including fetal development and CNS functioning [37, 38].

Our strategy for the design of conformationally-restricted melanotropin analogs with unique biological activities arose from an early observation that heat-alkali treatment of melanotropic peptides led to a complex mixture of diastereoisomers (racemization) which displayed high potency and prolonged biological activity in amphibian skin melanocyte assays [39–41]. This suggested to us that the racemization may have led to specific *conformational* changes which were particularly favorable for melanocyte receptor binding and transduction. We then examined quantitatively the sites of racemization of α-MSH [42, 43], and these results together with various structural and conformational considerations [44] led us to prepare [Nle^4, D-Phe^7]-α-MSH. As we had predicted, [Nle^4, D-Phe^7]-α-MSH was found to be extraordinarily potent in both the frog skin, lizard skin (Table 2), and tyrosinase [42] assay systems, and to be a superagonist in these systems. In addition, it was found to possess superpotency (> 20 times α-MSH) in the melanoma adenylate cyclase assay system [42]. Even more exciting, we found that [Nle^4, D-Phe^7]-α-MSH had exceptionally prolonged biological activities *in vitro* (hours to days, [42]) and *in vivo* (days to weeks [45]) in amphibian melanocyte assay systems.

We postulated that these extraordinary properties might be associated with the stabilization of a reverse turn conformation in the vicinity of the D-Phe-7 position. Since it was believed that the 4–10 sequence of α-MSH may be primarily responsible for its biological activity at least in pigment cell melanocyte receptors, we decided to investigate a variety of truncated analogs related to [Nle^4, D-Phe^7]-α-MSH. In these studies we removed the N-terminal tripeptide to give 4–13 analogs and the N- *and* C-terminal tripeptides to provide the 4–10 analogs.

Several important features of α-MSH structure-activity relationships were revealed by these studies (Table 2). First, replacement of methionine-4 by the pseudo-isosteric amino acid norleucine is completely compatible with hormone-

Table 2. Relative potencies of α-MSH and its analogs in the frog *(Rana pipiens)*, lizard *(Anolis carolinensis)* skin assays, and the melanoma adenylate cyclase assay[a]

Compound	Frog skin assay	Lizard skin assay	Melanoma AC assay
1. α-MSH	1.0	1.0	1.0
2. [Nle^4, D-Phe^7]-α-MSH	60.0	5.0	~25.0
3. [Nle^4]-α-MSH	2.3	1.5	2.0
4. Ac-[Nle^4]-α-MSH_{4-13}-NH_2	1.0	1.0	2.0
5. Ac-[Nle^4]-α-MSH_{4-10}-NH_2	0.002	0.06	0.09
6. Ac-[Nle^4, D-Phe^7]-α-MSH_{4-10}-NH_2	0.02	10.0	7.7
7. Ac-[Nle^4]-α-MSH_{4-11}-NH_2	0.002	1.0	1.0
8. Ac-[Nle^4, D-Phe^7]-α-MSH_{4-11}-NH_2	0.20	8.0	-
9. [Cys^4, Cys^{10}]-α-MSH	10–100[b]	2.0	3.0
10. Ac-[Cys^4, Cys^{10}]-α-MSH_{4-13}-NH_2	10–100[b]	0.6	1.1
11. Ac-[Cys^4, Cys^{10}]-α-MSH_{4-10}-NH_2	0.06	0.003	0.016
12. Ac-[Cys^4, Cys^{10}]-α-MSH_{4-11}-NH_2	0.16	0.07	0.16

[a] All potencies are relative to α-MSH in dose-response assays

[b] The cyclic peptides *9* and *10* have much higher minimum effective dose potencies in the frog skin system, being about 1000 times that of α-MSH

receptor interaction and transduction in all three assay systems examined (compare compounds *1* and *3*, Table 2), and leads to a slight (2-fold) increase in potency [46]. Second, replacement of Phe-7 with D-Phe-7 leads to increased potency in all three assay systems for both the full hormone and fragment analogs (compare *1, 2; 5, 6;* and *7, 8;* Table 2). The effects are most dramatic in α-MSH_{4-10} (*5* and *6*, Table 2) where about 100-fold increases in potency occur in the lizard skin and melanoma adenylate cyclase assays [47] after D-Phe-7 substitution. These findings support our hypothesis that replacement of L-Phe-7 by D-Phe-7 stabilizes conformational properties of α-MSH which are important to α-MSH binding and/or transduction. Reproducible binding studies which would help distinguish these possibilities have not yet been developed. However, the parallel nature of the dose-response curves suggest that the increased potency may be due to more efficacious binding of the D-Phe-7 analogs to their receptors. Third, different receptor systems for α-MSH, even closely related ones such as the frog and lizard skin melanocyte systems, have quite different structural and perhaps conformational requirements for activation by hormone. In fact, careful examination of the data for compounds *1–8* (Table 2) reveals that the lizard and melanoma adenylate cyclase assay systems show striking similarity. On the other hand, none of the receptors appears to require, or even utilize to any significant extent, the 1–3 residues (Ser–Tyr–Ser) of α-MSH in the hormone-receptor interaction. However, all of them to varying degrees lose significant potency (10–500 fold) when the C-terminal tripeptide (Lys–Pro–Val) is also removed (Table 2). This loss of potency can be more than compensated for by the D-Phe-7 substitution in the lizard skin and melanoma adenylate cyclase systems (*6*, Table 2), but only partially in the frog skin system.

These and related results provided us with the opportunity to test our tentative conformational model for melanotropin activity in frog skin melanocytes using conformational restriction. The following factors were considered: 1) the Phe-7 residue is apparently critical for the hormone to assume the correct biologically active conformation at the frog melanocyte receptor; 2) it is well known that D-amino acid residues in peptides can stabilize reverse turn conformations [48], and thus it appeared reasonable to propose that a reverse turn conformation in the vicinity of Phe-7 might be important to the biologically active conformation [49]; 3) examination of conformational models, and Chou and Fasman analysis, suggested that the tetrapeptide fragment of α-MSH, –His–Phe–Arg–Trp–, had a reasonable probability of existing in a reverse turn conformation; 4) examination of structure-function studies suggested the possible close proximity of the Lys-11 and the Gly-5 (or perhaps His-6) residues in the biologically active conformation; 5) conformations utilizing D-Phe-7 as part of a α-turn, or a C_7 turn, or a γ-turn all appeared to be reasonable.

Examination of the possible conformational models indicated that the Met-4 and Gly-10 residues were in reasonable proximity. We therefore decided to prepare a cyclic analog to test our conformational hypothesis. For this purpose we prepared [$\overline{Cys^4, Cys^{10}}$]-$\alpha$-MSH (*9,* Table 2) in which the Met-4 ($-CH_2-CH_2-S-CH_3$) and Gly-10 (–H) side chains are replaced by the "pseudoisosteric" $Cys^4 \sim\sim Cys^{10}$ bridging side chain ($-CH_2-S-S-CH_2-$) group. [$\overline{Cys^4, Cys^{10}}$]-$\alpha$-MSH was found to have superpotency in the frog skin assay system, being about 10–100 times more potent than the native hormone in this system over the linear portion of the dose-response curve. In addition, the peptide sometimes exhibits activity at concentrations as low as 10^{-14} M (about 1000 times the minimum effective dose potency of α-MSH), but linear dose-response curves cannot be obtained. The reasons for this are unclear, but may reflect variability with season and from animal to animal, and perhaps variability in compartmentalization, receptor sensitivity, and/or non-specific adsorption properties of the cyclic peptides.

In the lizard skin and melanoma adenylate cyclase assay systems, the cyclic analog *9* (Table 2) displayed high potency. Again, as in the linear analogs, removal of the N-terminal tripeptide Ser–Tyr–Ser from the cyclic analog *9* had virtually no effect on the potency of the cyclic analog in the frog skin receptor system, but in this case did lead to slight losses in potency in the lizard skin and melanoma membrane assay systems (*10,* Table 2). On the other hand, additional removal of the C-terminal sequence Lys–Pro–Val– from the cyclic analog to give Ac-[$\overline{Cys^4, Cys^{10}}$]-$\alpha$-$MSH_{4-10}$-$NH_2$ (*11,* Table 2) provided an analog with greatly decreased biological activities in all assay systems [50]. These activities are significantly restored by addition of the Lys-11 residue to give Ac-[$\overline{Cys^4, Cys^{10}}$]-$\alpha$-$MSH_{4-11}$-$NH_2$ (*12,* Table 2), and fully restored by further addition of the Pro-12 residue (unpublished results).

These studies support our model of the putative conformation of α-MSH at the melanotropin receptor. However, it is clear that the 4–10 residues by themselves are insufficient for superagonist activity, but also require related interactions of the Lys-11 and Pro-12 residues with the 23-membered ring moiety (unpublished results). Work is currently in progress to refine our conformational model and to further develop receptor specific melanotropin analogs.

Receptor Antagonists: Development of Oxytocin and Glucagon Antagonist Analogs

The development of a receptor antagonist for a peptide hormone requires that the analog bind effectively to the receptor but that some structural, conformational, or dynamic property of the peptide necessary for transduction be absent or inaccessible. Therefore, it is necessary to distinguish those stereostructural features important for binding and those critical for transduction. For this reason, we have emphasized the importance of careful assay analysis in the development of a conformational model of antagonist activity in 1-penicillamine-containing analogs of oxytocin [14, 20, 49–53] and glucagon (see below). The former studies have been recently reviewed [13, 32], and we present here a brief outline which illustrates some of principles we discussed earlier in this paper.

Though oxytocin possesses a 20-membered ring, H–Cys–Tyr–Ile–Gln–Asn–Cys–Pro–Leu–Gly–NH_2, it has considerable conformational flexibility especially in aqueous solution [54, 55]. On the other hand, we have shown by NMR, CD, and other studies that Pen-1-containing oxytocin antagonist analogs have quite specific conformational features and have utilized these features more recently to develop oxytocin antagonist analogs with prolonged *in vivo* biological activities [25]. Since these studies provide some insights into the value of conformational restrictions, a brief summary of our findings to date is given.

Though oxytocin has considerable conformational flexibility [54, 55], Walter and coworkers utilized a proposed conformation for oxytocin in DMSO and careful analysis of structure-function relationships to propose a model for conformation-agonist activity relationships at the uterine receptor [56], which was later slightly modified [15]. This model has been tested in several ways [15, 57–61], and while some aspects are no longer tenable others have provided reasonable starting points for further conformation-activity studies. The two major suggestions of Walter et al. were: 1) that the Ile-3 and Gln-4 and Pro-7 and Leu-8 residues were at the corners of β-turns and were the major sites of binding to the uterine receptor ([56]; the latter reverse turn was dropped in the later model [15]; 2) that Tyr-2 and Asn-5 side chain group were in close proximity with the phenol hydroxyl group of Tyr-2, and that the Asn-5 carboxamide group is in a hydrogen bonding or other noncovalent interaction in the "biologically-active" state. Subsequent studies, especially by Roy et al. [57–59], have shown that the hydrogen bonding scheme of the model was not important to biological activity. On the other hand, nearly all the evidence supports the idea that the side chain groups of positions 3, 4, 7 and 8 are important for receptor recognition and binding [14–15, 60]. The importance of the conformational relationships between the Tyr-2 and Asn-5 side chain were tested in collaboration with Walter and Smith [61] and the results support the Walter model. However, we have recently shown that [cLeu^2]oxytocin is a weak but fully active agonist, suggesting the need for some modification of the model [62].

On the other hand, the evidence we have obtained with [1-penicillamine]-oxytocin antagonist analogs strongly supports a central role of the Tyr-2 and Asn-5 side chain for the transduction process (see [14] for a review). In particular we have shown, primarily by NMR and CD methods, that [Pen^1]oxytocin (replacement of the Cys-1 residue (–CH_2–S– side chain) by a Pen-1 residue (–$C(CH_3)_2$–S– side chain)

Table 3. Conformational restrictions and structural features apparently related to antagonist activity in oxytocin analogs

Conformational or structural change	Conformational and structural effects	Examples of oxytocin antagonists	Ref.
1. β,β-dimethyl (or larger alkyl) in half-Cys-1 position	a) S–S dihedral angle ≥ 110°	[Pen^1]OT	86
	b) one (or two) C_7 turns	[Pen^1, Leu^2]OT	53
	c) tripeptide side chain restricted	[Pen^1, Thr^4]OT	87
	d) Tyr-2 side chain restricted	[Pen^1]OT	52
	e) Asn-5 side chain restricted	[Pen^1, Leu^2]OT	53
2. Pen-1 + Leu-2	a) Leu-2 side chain pseudo-axial	[Pen^1, Leu^2]OT	53
3. D-Amino Acid – position 2	a) Stereostructural relation of side chain changed	[D-Tyr^2]OT	61
4. D-p-Me(Et)–Phe-2 or D-p-MeO(EtO)–Tyr-2	a) Side chain more bulky	[β-MPA^1, Phe(Et)^2]OT [Pen^1, Phe(Et)^2]OT	88
5. Pen-1 + Thr-4	a) Side chain in position 4 conformationally restricted	[Pen^1, Thr^4]OT	87
	b) Asn^5 side chain restricted	[Pen^1, Phe^2, Thr^4]OT	87
6. $\Delta^{3,4}$-Pro^7	a) Side chain restricted, more π character	[Pen^1, $\Delta^{3,4}$-Pro^7]OT	
7. Orn^8	a) Neutral to basic amino acid	[β,β-Et$_2$-β-MPA^1, Orn^8]OT	64

[63], an antagonist analog under some conditions, has a greatly restricted 20-membered disulfide-containing ring with two adjacent C_7 turns involving Asn-5 to Ile-3 and Ile-3 to Cys-1 hydrogen bonds, and a 110–115° disulfide dihedral angle of right-handed chirality ([14] and [32] for reviews). An interesting aspect of these studies was our observation that in the Pen-1 oxytocin antagonists, the conformations of side chain groups were also severely limited especially at the Tyr-2 and Asn-5 side chains. For example, in [Pen-1]oxytocin and several related analogs, the side chain conformation of Tyr-2 that would place the aromatic ring over the 20-membered disulfide ring pointing toward the Asn-5 side chain is completely excluded [51, 52]. In other cases, when the position 2 side chain group had greater flexibility, Asn-5 side chain conformations were "frozen out" [53]. These and other findings which are summarized in Table 3 for oxytocin antagonists led us to suggest that conformational restriction, especially at appropriate side chain moieties, provides an excellent approach to obtaining peptide hormone antagonists. These findings also appear to be consistent with the suggestion of Walter and coworkers that the relationship of the Tyr-2 and Asn-5 side chains in oxytocin is important to the "biologically active conformation", though our recent studies [62] demonstrate that interaction of the Tyr-2 phenol hydroxy group with the Asn-5 carboxamide group is not necessary for transduction. We are now systematically examining how conformational restrictions in cyclic peptide analogs can be utilized to specifically restrict side chain conformations, and the effect of such restrictions on binding and transduction. Most intriguing are observations that certain amino acid substitutions in positions 2 and 8 that lead

to large *decreases* in potency in agonist analogs lead to large *increases* in potency in antagonist analogs [25, 64, 65].

The development of a glucagon antagonist which lowers blood glucose levels in diabetic animals [66] illustrates the importance of careful binding studies in conjunction with full dose-response analysis of activity. In addition these studies provide another illustration of the manner in which steric changes which cause conformational perturbations apparently can lead to antagonist analogs.

Glucagon is a 29-peptide with the structure H–His–Ser–Gln–Gly–Thr–Phe–Thr–Ser–Asp–Tyr–Ser–Lys–Tyr–Leu–Asp–Ser–Arg–Arg–Ala–Gln–Asp–Phe–Val–Gln–Trp–Leu–Met–Asn–Thr–OH. Numerous structure-function studies (e.g. [67–69]) have shown that the C-terminal portion is primarily important for receptor binding while the N-terminal portion is primarily important for transduction. Despite earlier reports to the contrary, it has recently been shown that $glucagon_{1-6}$ [70] and $glucagon_{1-21}$ [71] have little or no agonist or antagonist activity in the liver plasma membrane assay system. On the other hand, highly purified [1-des-histidine]-glucagon is a partial agonist [72] and [des-His^1–Ser^2–Gln^3–Gly^4]glucagon ($glucagon_{5-29}$) has no agonist activity, but does bind to liver plasma membrane glucagon receptors [71]. Based on these and other structure-function studies we and others prepared a number of analogs with various steric substituents at the N^α His-1 or N^ε Lys-12 amino groups (Table 4). Many of these analogs are partial agonists, and circular dichroism studies [73] suggested that the secondary structure was disrupted by these substitutions. Based on these and other partial agonists, many of which appear to bind better than they activate adenylate cyclase [67–69], we decided to synthesize glucagon analogs with larger substituents at the N^α or N^ε amino groups. Our strategy here was to further disrupt the steric relationships between the N-terminal residue(s) and residue(s) around Lys-12 that we thought might be related to the transduction process. As shown in Table 5, these analogs showed antagonist activities in the *in vitro* liver plasma membrane adenylate cyclase assay, and the

Table 4. Biological activities of partial agonist analogs of glucagon

Compound	Hepatic adenylate cyclase[a]	Hepatic receptor binding[b]	Ref.
Glucagon	100(100)	100	
N^α, N^ε-Bis-4-Mercapto-butyramidino-glucagon	90(7.5)	10.	89
N^α, N^ε-Bisacetamido-glucagon	80(7.5)	10.	89
[des-His^1]glucagon	60(2.0)	12.	72
[N^α-CAR-His^1]glucagon	27(5.5)	–	73
[des-His^1][$HArg^{12}$]glucagon	22(0.1)	–	73
[N^α-CAR-His^1, N^ε-CAR-Lys^{12}]-glucagon	17(0.4)	–	73

[a] An activity of 100 represents full stimulation of the adenylate cyclase system by the compound under standard assay conditions; numbers less than 100 represent partial stimulation. Number in parenthesis represent potency of compound relative to glucagon (100) at half maximal stimulation

[b] Number represents relative potency of compound to compete with [^{125}I]-glucagon for receptors on liver plasma membranes

Table 5. Biological activities of glucagon antagonists

Compound	Hepatic adenylate cyclase pA_2[a]	Ref.
[N^α-CAR-His[1], N^ε-TNP-Lys[12]]glucagon	6.4	73
[des-His[1]][N^ε-PTC-Lys[12]]glucagon	7.2	90
[des-His[1]][N^α-TNP-Ser[2], HArg[12]]glucagon	7.4	73
[N^α-TNP-His, HArg[12]]glucagon	8.2	73

[a] pA_2 is the negative logarithm to the base 10 of the average molar concentration of an antagonist that will reduce the response of 2 X units of glucagon activity to X units of glucagon activity

[N^α–TNP–His[1], HArg[12]]glucagon analog has recently been shown to lower blood glucose levels in diabetic rats [66].

These studies suggest that modulation of steric relationships between side chain moieties believed to be important for transduction can lead to antagonist analogs. Since these changes must certainly have their conformational effects, appropriate conformational restrictions which lead to similar topographical relationships of side chain group should also lead to antagonist analogs.

Conformational Restrictions for Receptor Specific Analogs. Development of δ Receptor Selective Enkephalins

One of the major potential results of conformation restriction in peptide hormones and neurotransmitters [14, 16–18, 20] is the development of analogs with high specificity for only one of several physiologically relevant receptors. The peptide neurotransmitters [Met[5]]enkephalin (H–Tyr–Gly–Gly–Phe–Met–OH) and [Leu[5]]enkephalin (H–Tyr–Gly–Gly–Phe–Leu–OH) are excellent examples of peptides for which appropriate conformational restriction should lead to useful and interesting analogs. First, these peptides have been found to interact with several classes of opioid receptors including the mu (μ), delta (δ), kappa (κ), and epsilon (ε) receptors [74–77], but very few receptor-specific peptide analogs or even nonpeptide opiate ligands have been prepared. Thus, the physiological roles for these different receptors are not well understood. Second, very extensive structure-activity studies have been made, with several thousand analogs already examined. Most of these analogs have been linear peptides, and the receptor specificity, if any, has been for the mu receptor. Numerous conformational studies have been made using these linear analogs and the native hormones, and these studies have led to conflicting models of solution conformation and of proposed receptor-bound conformations (see [77–79] for recent reviews). We therefore decided to investigate whether cyclic, conformationally-restricted analogs with high receptor specificity might be developed to examine these problems. Interesting related studies by Schiller and coworkers have also appeared [18, 79, 81].

Structure-function studies showed that highly potent enkephalin analogs could be prepared by substituting the Gly-2 residue with a variety of D-amino acids, and that the Met-5 (or Leu-5) residue could be replaced by a variety of L- or D-amino

Table 6. Cyclic enkephalins. Assays for delta (δ) vs. mu (μ) receptor specificity

Analog	IC_{50} (GPI) nM	IC_{50} (MVD) nM	IC_{50} (GPI) / IC_{50} (MVD)	[^{3}H]NAL nM	[^{3}H]DADLE nM	$IC^{\mu}_{50}/IC^{\delta}_{50}$
1. [D-Cys^2, L-Cys^5]EA	1.51 ± 0.03			9.4	30.2	0.31
2. [D-Cys^2, D-Cys^5]EA	0.75 ± 0.01			5.2	15.4	0.34
3. [D-Pen^2, L-Cys^5]EA	118 ± 19	3.6 ± 0.67	32.8	73.4 ± 15	3.35 ± 0.15	21.9
4. [D-Pen^2, D-Cys^5]EA	117 ± 21	16.8 ± 3.1	6.9	162 ± 35	7.20 ± 1.8	22.6
5. [D-Pen^2, L-Cys^5]E	213 ± 63	0.32 ± 0.3	666	178 ± 16	11.7 ± 1.2	15.2
6. [D-Pen^2, D-Cys^5]E	1350 ± 340	6.37 ± 1.2	212	157 ± 74	26.0 ± 0.5	6.0
7. [D-Pen^2, Cys^5, Thr^6]E	228 ± 52	0.60 ± 0.6	380	146 ± 25	5.2 ± 0.7	28.1
8. [D-Cys^2, L-Pen^5]E	39.9 ± 15	0.75 ± 0.05	53.2	52.7 ± 2.3	5.4 ± 0.1	9.8
9. [D-Cys^2, D-Pen^5]E	66.7 ± 1.3	0.13 ± 0.06	513	22.2 ± 2.8	3.5 ± 0.8	6.3
10. [D-Pen^2, L-Pen^5]E	2720 ± 50	2.50 ± 0.03	1090	4100 ± 800	10.0 ± 2.2	410.0
11. [D-Pen^2, D-Pen^5]E	6930 ± 124	2.19 ± 0.30	3160	3230 ± 670	16.2 ± 0.9	200.0

acid residues with retention of high potency. On the other hand, both Tyr-1 and Gly-3 were highly sensitive to any structural change. We therefore decided to conformationally restrict by cyclization at the Met-5 and Gly-2 positions. In view of our success with cyclic α-MSH analogs, we first thought to prepare [D-$\overline{Cys^2, Cys^5}$]enkephalin. A thorough search of the literature revealed, however, that this analog had already been made by Sarantakis [80], and about this time Schiller et al. also prepared these compounds and examined their biological properties [81]. We therefore decided to restrict the conformation of the peptide even further by use of the gem dimethyl effect previously used so successfully in oxytocin. Thus we prepared a number of D-Pen-2, D(L)-Cys-5 cyclic enkephalin and enkephalinamide analogs, and the corresponding D-Cys-2 and D(L)-Pen-5 analogs (14-membered ring systems). As shown in Table 6, these compounds were found to have very high delta receptor selectivity as measured by the guinea pig ileum (μ receptor) vs. mouse vas deferens (δ receptor) assays, *and* by receptor binding activities in the rat brain assay system. In fact, the [D-$\overline{Pen^2, L\text{-}Cys^5}$]-enkephalin is among the most potent delta selective enkephalin analogs known [82]. On inspecting conformational models of these peptides, we were surprised to find that the Cys-2 (or Cys-5) β-hydrogens were exposed and not sterically hindered. Thus, though we had tried several times unsuccessfully to prepare [$\overline{Pen^1, Pen^6}$]-oxytocin, we decided to try and synthesize the 14-membered, bis-penicillamine analogs of enkephalin.

These syntheses were readily accomplished and, as shown in Table 6, these cyclic bis-penicillamine analogs are extraordinarily δ receptor specific [83], being more selective than any previously reported δ selective analog. Currently we are examining the physiological function of delta receptors and have found, for example [84], that δ receptors in the brain can mediate analgesia but do not affect gut transit, while in the spinal cord δ receptors appear to affect both functions. Conformational analysis of the rigid 14-membered ring analogs should provide critical insights into the biologically active conformation at δ receptors, and aid in the development of even more selective δ receptor agonists and antagonists. In addition, since these analogs interact so poorly with μ (and κ) receptors, the differences in conformational requirements at δ, μ, and κ receptors may become apparent.

Prolonged Biological Activity. A Possible Feature of Peptide Hormone-Receptor Interaction?

Efforts to develop peptide hormone and neurotransmitter agonist and antagonist analogs with prolonged biological activity have generally centered on developing analogs which are stable to degradative enzymes (use of D-amino acids, N-methylamino acids, etc.), increasing the lipophilicity of the peptide, or substituting peptide bonds by "amide bond replacements". All of these approaches are basically aimed at pre-receptor events so as to increase the half-life of the peptide or its distribution into membrane environments.

One other strategy, however, apparently has received little attention, namely, the possibility of developing prolongation as part of the peptide-receptor interaction [25]. Since the dissociation constant $K = k_{off}/k_{on}$, analogs with high potency and strong binding interactions with the receptor should lead to long k_{off} times and prolonged activity. However, what we wish to suggest is that prolongation might also result from conformational and structural features of the peptide important only to the receptor-hormone complex in the transduced (biologically active) state of an agonist, or to the bound complex in a non-biologically active (inhibitory) state of an antagonist. In terms of our simplified mechanism of hormone action (Fig. 2), we are suggesting that in addition to different features of peptide structure and conformation being important for binding and efficacy (transduction), yet other features may be important for reversal of the binding-transduction or binding-antagonist processes (in Fig. 2 we have referred to this as the π state). If this hypothesis has any merit, it should be possible to obtain peptide analogs with prolonged activity irrespective of agonist potency, and irrespective of efficacy (even weak agonist or partial agonists may have prolonged activity). Furthermore, prolongation may be receptor specific and need not apply for a particular analog to all physiological receptors for the peptide. In addition there need not be any relationship between enzymatic stability (resistance to inactivation) and prolonged activity. We would like to report here some evidence which tentatively supports this hypothesis.

Recently we have obtained a series of oxytocin analogs with prolonged activity. [Pen^1, Leu^2]oxytocin is a moderately weak oxytocin antagonist *in vitro* when administered simultaneously with oxytocin in the uterine system. However, when administered at the height of uterine contraction induced by oxytocin, it showed highly prolonged antagonist activity [65]. In fact, in this case, antagonist action occurs for over an hour in the presence of oxytocin. A possible explanation for this effect is that interaction of this analog with its receptor in the post-transduced state produces an analog-receptor antagonist complex which dissociates slowly, leading to prolonged antagonist activity. However when agonist is present with antagonist a seemingly competitive inhibition occurs during the time of the assay (about 10 min) which presumably is primarily a function of the relative binding interaction of the two peptides at the receptor. It should also be noted that while replacement of Tyr-2 by Leu-2 in oxytocin leads to a 1200 fold *decrease* in oxytocic *agonist* potency, a similar replacement of Tyr-2 by Leu-2 in the antagonist [Pen^1]oxytocin analog leads to a 3 fold *increase* in oxytocic *antagonist* potency. Clearly the agonist and antagonist utilize different features of oxytocin structure for their interaction with uterine receptors. Recently we have developed a series of Pen-1 oxytocin analogs with extraordinarily prolonged biological activities. These analogs contain the conformationally restrictive β-branched amino acid Thr in position 4, a lipophilic aromatic amino acid in position 2, the conformationally restricted $\Delta^{3,4}$-proline residue in position 7, and ornithine in position 8. In the agonist series, Orn-8 decreases the potency of oxytocin at the uterine receptor by over 12-fold, but appears to increase antagonist potency at least 10-fold [64]. In any case, we have found (unpublished) that [Pen^1, Phe^2, Thr^4, Δ-Pro^7, Orn^8]-oxytocin, [Pen^1, $Tyr(OMe)^2$, Thr^4, Orn^8]oxytocin and [Pen^1, $Phe(Me)^2$, Thr^4, Orn^8]oxytocin all have exceptionally prolonged (hours) *in vivo* antagonist activity.

While the above results are promising, more convincing results have emerged from our studies with α-MSH analogs. As previously noted, [Nle^4, D-Phe^7]-α-MSH has extraordinarily prolonged biological activities *in vitro* (days, 42) and *in vivo* (weeks, 45) in the frog skin melanocyte system, and we have recently shown that the fragment analog, Ac-[Nle^4, D-Phe^7]-α-MSH_{4-11}-NH_2 has similar properties [85].

Table 7. Relative potency, prolongation, and serum stability of α-Melanotropin analogs in the *in vitro* frog and lizard skin receptor systems[a]

Compound	Potency in frog	Prolongation in frog	Potency in lizard	Prolongation in lizard	Serum stability
1. α-MSH	1.0	No	1.0	No	No
2. [Nle^4, D-Phe^7]-α-MSH	60.0	Yes	5.0	Yes	Yes
3. Ac-[Nle^4, D-Phe^7]-α-MSH$_{4-11}$-NH$_2$	0.16	Yes	8.0	Yes	Yes
4. Ac-[Nle^4, D-Phe^7]-α-MSH$_{4-10}$-NH$_2$	0.02	Minor	10.0	Yes	Yes
5. Ac-[D-Phe^7]-α-MSH$_{5-11}$-NH$_2$	0.01	No	1.0	No	Yes
6. Ac-α-MSH$_{5-11}$-NH$_2$	0.0002	Yes	0.004	No	Yes
7. Ac-[Tyr^4]-α-MSH$_{4-10}$-NH$_2$	0.0002	Yes	0.0006	No	Yes
8. [$\overline{\text{Cys}^4, \text{Cys}^{10}}$]-$\alpha$-MSH	10–100[c]	No	~2.0	No	Yes
9. Ac-[$\overline{\text{Cys}^4, \text{Cys}^{10}}$]-$\alpha$-MSH$_{4-13}$-NH$_2$	10–100[c]	No	0.6	No	Yes

[a] All potencies are relative to α-MSH which is defined as equal to 1.0
[b] N.D. = not determined
[c] Minimal effective dose potency ~ 1,000. See Table 2

These peptides also have extraordinary stability to serum enzymes and to purified proteolytic enzymes such as trypsin and α-chymotrypsin. However, we have recently developed a series of linear and conformationally restricted cyclic α-MSH analogs in which there is no relationship between potency and prolongation, and which are prolonged at one physiological receptor for MSH but not another. The activities of some of the most relevant analogs are summarized in Table 7.

As shown in Table 7 there is no relationship between potency and prolonged activity. [Nle^4, D-Phe^7]-α-MSH is a superagonist with prolonged activity; [$\overline{\text{Cys}^4, \text{Cys}^{10}}$]-$\alpha$-MSH is also a superagonist, but it does not have prolonged activity. On the other hand, both Ac-α-MSH$_{5-11}$-NH$_2$ *(6)* and Ac-[Tyr^4-α-MSH]$_{4-10}$-NH$_2$ are exceptionally weak agonists and yet they have prolonged activities. Furthermore, prolonged activity at one receptor type does *not* necessarily lead to prolongation at another receptor type. For example, while Ac-[Nle^4, D-Phe^7]-α-MSH$_{4-10}$-NH$_2$ has very prolonged activity at the lizard skin melanocyte receptor, it has only minor prolongation in the frog skin melanocyte system. On the other hand Ac-α-MSH$_{5-11}$-NH$_2$ is very prolonged in the frog skin bioassay, but has no prolonged effects in the lizard assay. Clearly these two physiologically similar receptor systems have quite different structural and presumably conformational requirements for prolonged biological activity. Finally, prolonged activity apparently has no relationship to enzymatic stability. We have examined the stability of these analogs to serum enzymes (Table 7) and to trypsin and α-chymotrypsin (not shown). As shown in Table 7, compounds which are completely stable to serum enzymes such as Ac-[Nle^4, D-Phe^7]-α-MSH$_{4-10}$-NH$_2$ and Ac-[$\overline{\text{Cys}^4, \text{Cys}^{10}}$]-$\alpha$-MSH$_{4-13}$-NH$_2$ are not prolonged in their activity.

Concluding Remarks

In this paper we have attempted to illustrate how careful examination of the probable conformation and stereoelectronic properties of peptide hormone and neurotransmitter analogs in conjunction with extensive bioassay data over the entire dose-response range can lead to a rational basis for the design and synthesis of these important molecules which possess specific binding and transduction properties. Utilization of conformational restriction by side chain-side chain covalent cyclization, transannular stabilization of conformation by geminal dialkyl substitution in cyclic systems, etc., provide excellent opportunities for investigating the topographical requirements for peptide hormone and neurotransmitter interactions. Obviously it would be desirable to have large quantities of the receptor available in a biologically relevant form for biophysical studies of the hormone-receptor system, and for examining the changes in conformation and dynamics which accompany hormone binding, information transfer, and reversal of these processes. In the meantime the approach outlined in this paper which we and other laboratories have pursued during the past years [11–20] offers a useful intermediate step in this development. In addition to the need for further knowledge of the physical-chemical properties of the receptor system, further progress in understanding the chemical-physical basis of information transfer will depend on careful examination of chemical-physical and biological properties of well-chosen conformationally restricted peptide analogs with unique and highly disparate biological activities. Perhaps the critical need in this area is a much closer interaction and collaboration between the physical and biological scientists interested in this problem. Under these circumstances we have considerable optimism about further development in our understanding of information transfer by biological messengers such as peptide hormones and neurotransmitters.

Acknowledgements

The support of the U.S. Public Health Service, the National Science Foundation, and Merck, Sharp & Dohme is gratefully acknowledged. The authors are particularly grateful to their many students and colleagues whose creative ideas and extraordinary laboratory skills made this report possible.

References

1. Current Views on Insulin Receptors, Andreani, D., DePirro, R., Lauro, R., Olefsky, J., Roth, J., eds., Academic Press, London, 1981
2. Czech, M. P., Massague, J. (1982) Fed. Proc. *41,* 2719–2723 and references therein
3. Ross, E. M., Gilman, A. G. (1980) Ann. Rev. Biochem. *49,* 533–563
4. Cooper, D. M. F. (1983) Current Topics Memb. Transp. *18,* 67–84
5. Rodbell, M. (1980) Nature (London) *284,* 17–22
6. Limbird, L. E. (1981) Biochem. J. *195,* 1–13
7. Tutwiler, G. F., Bowden, C. R. Kiorpes, T. C., Tuman, R. W. (1983) Ann. Rep. Med. Chem. *18,* 193–202
8. Stadel, J. M., deLean, A., Lefkowitz, R. J. (1982) Adv. Enzymol. *53,* 1–43
9. Lever, O. W., Jr., Chang, K.-J., McDermed, J. D. (1983) Ann. Rep. Med. Chem. *18,* 51–60

10. Rudinger, J., Drug Design, Vol. 2, Ariens, J., ed., Academic, New York, pp. 319–419, 1971 and references therein
11. Rudinger, J., Pliska, V., Krejci, I. (1972) Recent Prog. Horm. Res. *28,* 131–166
12. Schwyzer, R., Peptides 1972, Hanson, Jakubke, eds., North-Holland, Amsterdam, pp. 424–436, 1973
13. Schwyzer, R. (1977) Ann. N.Y. Acad. Sci. *297,* 3–26
14. Hruby, V. J., Topics in Molecular Pharmacology, Burgen, A. S. V., Roberts, G. C. K., eds., Elsevier/North-Holland, Amsterdam, pp. 99–126, 1981
15. Walter, R. (1977) Fed. Proc. *36,* 1872–1878
15a. Blundell, T., Wood, S. (1982) Ann. Rev. Biochem. *51,* 123–154
16. Marshall, G. R., Gorin, F. A., Moore, M. L. (1978) Ann. Rep. Med. Chem. *13,* 227–238 and references therein
17. Veber, D., Peptides: Structure and Biological Function, Gross, E., Meienhofer, J., eds., Pierce Chem. Co., Rockford, IL, pp. 409–419, 1979
18. Schiller, P., Peptides: Structure and Function, Hruby, V. J., Rich, D. H., eds., Pierce Chem. Co., Rockford, IL, pp. 269–278, 1983 and references therein
19. Chipens, G. I., Krikis, A. J., Polevaya, L. K., Biophysical and Biochemical Information Transfer in Recognition, Vassileva-Popova, J. G., Jensen, V., eds., Plenum Press, New York, pp. 23–48, 1979
20. Hruby, V. J., Perspectives in Peptide Chemistry, Eberle, A., Geiger, R., Wieland, T., eds., S. Karger, Basel, pp. 207–220, 1981
21. Hruby, V. J. (1982) Life Sciences *31,* 189–199 and references therein
22. Hruby, V. J., Mosberg, H. I., Sawyer, T. S., Knittel, J. J., Rockway, T. W., Ormberg, J., Darman, P., Chan, W. Y., Hadley, M. E. (1983) Biopolymers *22,* 517–530
23. Hruby, V. J. (1984) Conformationally Directed Drug Design, Vida, J. A., Gordon, M., eds., ACS Monograph Series, No. 251, Washington D.C., pp. 9–27
24. Hruby, V. J., Mosberg, H. I. (1982) Peptides *3,* 329–336
25. Hruby, V. J., Knittel, J. J., Mosberg, H. I., Rockway, T. W., Wilkes, B. C., Hadley, M. E., Peptides 1982, Bláha, K., Malon, P., eds., Walter de Gruyter, Berlin, pp. 19–30, 1983
26. Ariens, E. J., Beld, A. J., Rodrigues de Miranda, J. F., Simonis, A. M., The Receptors, Vol. 1, O'Brien, R. D., ed., Plenum Press, New York, pp. 33–91, 1979
27. Smigel, M. D., Northrup, J. K., Gilman, A. G. (1982) Rec. Prog. Hormone Res. *38,* 601–624
28. Hilldebrandt, J. D., Sekura, R. D., Cordina, J., Iyengar, R., Manclark, C. R., Birnbaumer, L. (1983) Nature *302,* 706–709
29. Sutherland, E. W., Rall, T. W., Menon, T. (1962) J. Biol. Chem. *237,* 1220–1227
30. Hollenberg, M. D., Cuatrecasas, P., The Receptors, Vol. 1, loc. cit., pp. 193–214, 1979
31. Hruby, V. J. (1982) Molec. Cellul. Biochem. *44,* 49–64
32. Hruby, V. J., Mosberg, H. I., Hormone Antagonists, Agarwal, M. K., ed., Walter de Gruyter & Co., Berlin, pp. 433–473, 1982
33. Chou, P. Y., Fasman, G. D. (1978) Annu. Rev. Biochem. *47,* 251–276
34. Medzihradszky, K., Recent Developments in the Chemistry of Natural Carbon Compounds, Bogner, R., Bruckner, R., Szanty, C., eds., Publ. House Hungarian Acad. Sci., Budapest, pp. 207–250, 1976
35. Hruby, V. J., Wilkes, B. C., Cody, W. L., Sawyer, T. K., Hadley, M. E. (1984) Peptide Protein Rev. *3,* 1–64
36. Hadley, M. E., Bagnara, J. T. (1975) Amer. Zool., Suppl. 1, *15,* 81–96
37. Evered, D., Lawrenson, G., eds., Peptides of the Pars Intermedia, Pitman Medical, London, 1981
38. O'Donohue, T. L., Dorsa, D. M. (1982) Peptides *3,* 353–395
39. Smith, P. E., Graeser, J. B. (1924) Anat. Rec. *27,* 187
40. Pickering, B. T., Li, C. H. (1964) Arch. Biochem. Biophys. *104,* 119–127
41. Bool, A. M., Grey II, G. H., Hadley, M. E., Heward, C. B., Hruby, V. J., Sawyer, T. K., Yang, Y. C. S. (1981) J. Endocrinol. *88,* 57–65
42. Sawyer, T. K., Sanfilippo, P. J., Hruby, V. J., Engel, M. H., Heward, C. B., Burnett, J. B., Hadley, M. E. (1980) Proc. Natl. Acad. Sci. USA *77,* 5754–5758
43. Engel, M. H., Sawyer, T. K., Hadley, M. E., Hruby, V. J. (1981) Anal. Biochem. *116,* 303–311
44. Sawyer, T. K., Hruby, V. J., Hadley, M. E., Engel, M. H. (1983) Amer. Zool. *23,* 529–540

45. Hadley, M. E., Anderson, B., Heward, C. B., Sawyer, T. K., Hruby, V. J. (1981) Science *213*, 1025–1027
46. Hruby, V. J., Sawyer, T. K., Yang, Y. C. S., Bregman, M. D., Hadley, M. E., Heward, C. B. (1980) J. Med. Chem. *23*, 1432–1437
47. Sawyer, T. K., Hruby, V. J., Wilkes, B. C., Draelos, M. T., Hadley, M. E., Bergsneider, M. (1982) J. Med. Chem. *25*, 1022–1027
48. Chandrasekaran, R., Lakshminarayanan, A. V., Pandya, U. V., Ramachandran, G. N. (1978) Biochim. Biophys. Acta *303*, 14–27
49. Sawyer, T. K., Hruby, V. J., Darman, P. S., Hadley, M. E. (1982) Proc. Natl. Acad. Sci. USA *79*, 1751–1755
50. Knittel, J. J., Sawyer, T. K., Hruby, V. J., Hadley, M. E. (1983) J. Med. Chem. *28*, 125–129
51. Meraldi, J.-P., Yamamoto, D., Hruby, V. J., Brewster, A. I. R., Peptides: Chemistry, Structure and Biology, Walter, R., Meienhofer, J., eds., Ann Arbor Sci. Publ., Ann Arbor, MI, pp. 729–736, 1975
52. Meraldi, J.-P., Hruby, V. J., Brewster, A. I. R. (1977) Proc. Natl. Acad. Sci. USA *74*, 1373–1377
53. Mosberg, H. I., Hruby, V. J., Meraldi, J.-P. (1981) Biochemistry *20*, 2822–2828
54. Kotelchuck, D., Scheraga, H. A., Walter, R. (1972) Proc. Natl. Acad. Sci. USA *69*, 3626–3630
55. Brewster, A. I. R., Hruby, V. J. (1973) Proc. Natl. Acad. Sci. USA *70*, 3806–3809
56. Walter, R., Schwartz, I. L., Darnell, J. H., Urry, D. W. (1971) Proc. Natl. Acad. Sci. USA *68*, 1355–1359
57. Roy, J., Roy, U., Gazis, D. C., Schwartz, I. L. (1979) Proc. Natl. Acad. Sci. USA *76*, 3309–3313
58. Roy, J., Johnson, M., Gazis, D., Schwartz, I. L. (1980) Int. J. Peptide Protein Res. *16*, 55–60
59. Gazis, D., Roy, J., Roy, U., Glass, J. D., Schwartz, I. L., Peptides 1982, loc. cit., pp. 809–814, 1983
60. Hruby, V. J., Mosberg, H. I., Viswanatha, V. (1982) J. Am. Chem. Soc. *104*, 837–841
61. Hruby, V. J., Upson, D. A., Yamamoto, D. M., Smith, C. W., Walter, R. (1979) J. Am. Chem. Soc. *101*, 2717–2721
62. Hruby, V. J., Rockway, T. W., Viswanatha, V., Chan, W. Y. (1983) Int. J. Peptide Protein Res. *21*, 24–34
63. Schulz, H., du Vigneaud, V. (1966) J. Med. Chem. *9*, 647–650
64. Bankowski, K., Manning, M., Sato, J., Haldar, J., Sawyer, W. H. (1980) Int. J. Peptide Protein Res. *16*, 382–391
65. Hruby, V. J., Deb, K. K., Yamamoto, D. M., Hadley, M. E., Chan, W. Y. (1979) J. Med. Chem. *22*, 7–12
66. Johnson, D. G., Goebel, C. U., Hruby, V. J., Bregman, M. D., Trivedi, D. (1982) Science *215*, 1115–1116
67. Hruby, V. J. (1982) Molec. Cellul. Biochem. *44*, 49–64 and references therein
68. Epand, R. M., Rosselin, G., Hoa, D. H. B., Cote, T. E., Laburthe, M. (1981) J. Biol. Chem. *256*, 1128–1132 and references therein
69. England, R. D., Jenkins, W. T., Flanders, K. C., Gurd, R. S. (1983) Biochemistry *22*, 1722–1728 and references therein
70. Pelton, J. T., Trivedi, D., Hruby, V. J. (1983) Life Sciences *33*, 1307–1314
71. Frandsen, E. K., Grønvald, F. C., Heding, L. G., Johansen, N. L., Lundt, B. F., Moody, A. J., Markussen, J., Vølund, A. (1981) Hoppe Seyler's Zeit. Physiol. Chem. *362*, 665–677
72. Lin, M. C., Wright, D. E., Hruby, V. J., Rodbell, M. (1975) Biochemistry *14*, 1559–1563
73. Bregman, M. D., Trivedi, D., Hruby, V. J. (1980) J. Biol. Chem. *255*, 11725–11733
74. Lord, J. A. H., Waterfield, A. A., Hughes, J., Kosterlitz, H. W. (1977) Nature *267*, 495–499
75. Wolzin, B. L., Pasternak, G. W. (1981) Proc. Natl. Acad. Sci. USA *78*, 6181–6185
76. Chang, K.-J., Cuatrecasas, P. (1979) J. Biol. Chem. *254*, 2610–2618
77. Olson, G. A., Olson, R. D., Kastin, A. J., Coy, D. H. (1982) Peptides *3*, 1039–1072
78. Miller, R. J., Cuatrecasas, P. (1978) Vitamins and Hormones *36*, 297–382
79. Schiller, P. W., DiMaio, J. (1982) Nature *297*, 74–76
80. Sarantakis, D., U. S. Patent 4,148,789, 1979
81. Schiller, P. W., Eggimann, B., DiMaio, J., Lemieux, C., Nguyen, T. M.-D. (1981) Biochem. Biophys. Res. Commun. *101*, 337–343

82. Mosberg, H. I., Hurst, R., Hruby, V. J., Galligan, J. J., Burks, T. F., Gee, K., Yamamura, H. I. (1983) Life Sciences *32,* 2565–2569
83. Mosberg, H. I., Hurst, R., Hruby, V. J., Gee, K., Yamamura, H. I., Galligan, J. J., Burks, T. F. (1983) Proc. Natl. Acad. Sci. USA *80,* 5871–5874
84. Porecca, F., Mosberg, H. I., Hurst, R., Hruby, V. J., Burks, T. F. (1983) Life Sciences *33,* Suppl. I, 457–460
85. Wilkes, B. C., Sawyer, T. K., Hruby, V. J., Hadley, M. E. (1983) Int. J. Peptide Protein Res. *22,* 313–324
86. Hruby, V. J., Deb, K. K., Fox, J., Bjarnason, J., Tu, A. T. (1978) J. Biol. Chem. *253,* 6060–6067
87. Hruby, V. J., Mosberg, H. I., Peptides: Synthesis, Structure, Function, Rich, D. H., Gross, E., eds., Pierce Chemical Co., Rockford, IL, pp. 375–378, 1981
88. Lebl, M., Barth, T., Servitová, L., Slaninová, J., Jŏst, K., Peptides 1982, Bláha, K., Malon, P., eds., W. de Gruyter, Berlin, pp. 457–460, 1983
89. Wright, D. E., Rodbell, M. (1980) Eur. J. Biochem. *111,* 11–16
90. Bregman, M. D., Hruby, V. J. (1979) FEBS Letters *101,* 191–194

Communications

Chiral Recognition in Partitioning Systems

E. Gil-Av

The Weizmann Institute of Science, Rehovot, Israel

The separation of antipodes by chromatography has developed into an important tool for resolution in recent years [1 – 3]. The newly introduced procedures utilize various chromatographic modes, such as GC, LC, HPLC, TLC [4] and DCCC [5], and imply by necessity at least one chiral phase (stationary or mobile).

Some results of this work, particularly significant for the understanding of chiral recognition, will be considered in this lecture. The discussion will focus mainly on the scope of the phenomena observed, the magnitude of the stereoselective effects, possible mechanism of the chiral differentiation processes and the types of intermolecular interactions involved. A review as brief as this one cannot be comprehensive, and it will in fact be limited mainly[1] to the work carried out in the laboratory of the author. However, it should be pointed out that our research over a period of nearly 20 years has dealt with a relatively wide range of resolution phenomena based on a variety of intermolecular interactions and having different mechanisms.

The parameters basic to this discussion are the resolution coefficients (α), defined as the ratio of the retention volumes, measured from the air or solvent peak in GC, and of the capacity factors in HPLC. From these data, readily obtained from the chromatograms (see e.g. Fig. 1), also the difference in free energy of solution or adsorption of the enantiomers can be calculated by the relationship $\Delta\Delta G = -RT\ln\alpha$.

Prior to the mid-sixties some instances of the resolution of racemates by chromatography on chiral solids were known, e.g., the optical enrichment of Troeger's base on D-lactose achieved by Prelog and Wieland [6]. Some notice was taken of these early observations, but it was generally thought that such processes would be inefficient and limited in scope. As to the separation of enantiomers on chiral liquids, serving as stationary phases in GC, the intuitive feeling of chemists was that the differences in solubility of antipodes could not possibly be sufficient to permit resolution. However, when a determined effort was initiated to investigate this approach, it was soon found that these preconceptions were erroneous.

[1] In citing references, an attempt was made to include the names of as many scientists as possible who have contributed to the recent development of the topic, but not to cover the literature in detail

The Scope of Chiral Recognition by Partitioning

Resolution of optical isomers by chromatography is ascribed to the rapid and reversible formation in the column of diastereomeric complexes between the chiral component of the phase (selector) and the chiral solutes (selectands). If these selector-selectand associates differ sufficiently in their free energy of formation, the resulting differences in partitioning coefficients will lead to the separation of the enantiomers.

It has been shown that such stereoselective complexes can be based on different types of intermolecular forces, e.g., hydrogen bonding, charge transfer (CT) complexation, ionic and dipole-dipole attractions, or combinations of two or more of such interactions.

The reported scope of the phenomenon is continuously expanding, as new examples of stereoselectively interacting selector/selectand pairs of known or yet unknown types are being found. It should be mentioned that modern, highly efficient columns allow chiral differentiation to be detected with certainty, even where the effects are very small and the resolution factors are close to unity.

Some of the most important selectors are derived from amino acids, and their stereoselective effects are due mainly to their capacity for hydrogen bonding. They include diamides, $R_1CONHCH(R_2)CONHR_3$ [7, 8], N-acyl-dipeptide esters, carbonylbis(N-L-amino acid esters), and monoamides, e.g., N-lauroyl-S-α-(1-naphthyl)ethylamine [9]. Recently, hydrogen bonding phases have been developed with multiple dipeptide, tripeptide or diamide moieties linked covalently to either a central ring system [10] or a polymeric backbone [11].

The compounds initially resolved on these solvents consisted of amines, amino alcohols and α-, β- and γ-amino acids, containing a primary amine linked, in general, directly to the asymmetric carbon. For occurrence of chiral recognition the amine function has to be converted into an amide and the carboxylic group into an ester (or also into an amide).

A second family of substances resolvable on hydrogen bonding phases does not intrinsically possess a primary amine, but has a function which can be derivatized to a nitrogen containing moiety with both an acceptor and a donor site. Such compounds are, e.g., α-substituted carboxylic acids, $RCH(X)CO_2H$, where X = alkyl, aryl, Cl, Br, OH or O-acyl, and the carboxylic group is derivatized to an amide.

The scope of chiral recognition on these hydrogen bonding phases (mainly on diamides) has recently been shown to extend also to alcohols in the form of their carbamates, and to ketones derivatized to their oximes [12]. Furthermore, selectands which do not contain nitrogen, such as some free alcohols [10], α-hydroxy esters [11], perfluorinated acyl derivatives of sugars [13] and aromatic diols, have also been resolved on diamides.

Selectors capable of charge transfer (CT) complexation are another type of solvents which show chiral differentiation for a wide range of compounds. For instance, R(−)-2-(2,4,5,7-tetranitro-9-fluorylideneaminooxy) propionic acid (TAPA, I) resolves (by HPLC) polyaromatic hydrocarbons such as helicenes [14, 15], partially hydrogenated helicenes, 1,12-disubstituted benzo[c]-phenanthrene, napthophanes [16], naphthyl ethers [17], and diol epoxides of polyaromatic hydrocarbons.

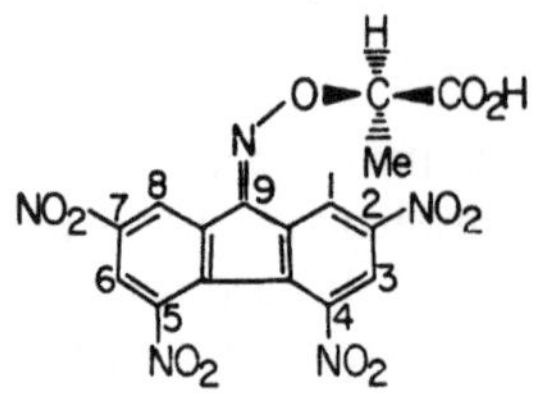

R(—)-TAPA

(I)

By incorporating into the stationary phase a chiral charge transfer reagent also capable of hydrogen bonding, such as 2,2,2-trifluoro-1-[(10-methyl)-9-anthryl] ethanol, Pirkle and coworkers [18] have resolved substances which either contain a CT complexing moiety in their molecule, e.g. aromatic sulphoxides and lactones, or possess a function through which such a moiety can be introduced, e.g. alcohols, mercaptans, amines, amino alcohols, amino acids, hydroxy acids and esters.

Another very important class of chiral partitioning systems involves coordination of the selector and selectand to a metal. Notable examples are the Cu(II) complexes of α-amino acids, such as L-proline [2, 19] and N,N-di-n-propyl-L-alanine [20] and of similar zinc and nickel derivatives, e.g. Ni(L-proline n-octylamide)$_2$ [21]. With N,N-di-n-propyl-L-alanine, used as additive to the mobile phase in HPLC, a particularly wide range of compounds could be resolved, including free α-amino acids, their dansyl and ester derivatives, dipeptides, α-amino caprolactame, as well as α-hydroxy acids [22].

Two cyclic olefins were resolved by coordination with dicarbonyl Rh(I)-3-trifluoroacetyl-1R-camphorate [23, 24]. Starting with this observation, Schurig and coworkers [25] have systematically explored stereoselective partitioning on optically active β-diketonates of transition metals, and have extended the use of these selectors to the chiral differentiation of underivatized ketones and alcohols and to heterocyclic compounds such as oxiranes, aziridines, azetidines, thiiranes, tetrahydrofuranes, etc.

These examples should suffice to indicate that there is a large number of optically active molecules known to interact with each other with at least some degree of chiral differentiation, and that many more such pairs remain to be discovered[2].

Magnitude of the Stereoselective Effect

As mentioned in the introduction, it was assumed at one time that stereoselective effects, if they occurred at all, would be very small. The recent studies on chromatographic resolution showed, quite on the contrary, that there are certain selector/selectand pairs which have relatively large resolution coefficients for enantiomers.

Thus, on N-lauroyl-L-valine t-butylamide, the two enantiomeric N-TFA O-methyl esters of leucine emerge (Fig. 1) seven minutes apart, and for the iso-

[2] One family of optically active mesophases, the smectogenous carbobis(N-L-α-amino acid esters) (see above), have been found to show chiral differentiation [26]

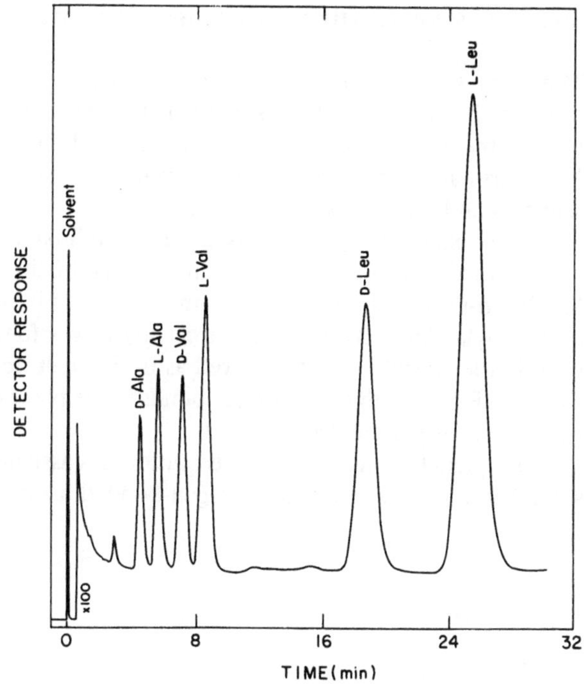

Fig. 1. Chromatogram of the N-trifluoroacetyl methyl ester of alanine, valine and leucine (enriched in the L-isomer). Chromatographic conditions: column, 1.9 m × 4 mm I.D., containing 60–80 mesh Chromosorb W AW coated with 10% N-lauroyl-L-valine t-butylamide; temperature, 110 °C [8]

propyl ester the α value at 0 °C has been estimated by extrapolation to be equal to 3.5–4.0, corresponding to $\Delta\Delta G = -0.8$ Kcal.

It is of interest to point out that this is the order of magnitude for the difference in interaction with liposomes, containing cerebroside sulfate, of all-L and all-D leu-enkephalins (2.5–3.0), as reported by R. Schwyzer and coworkers [27].

Still higher resolution factors were found in systems in which a diastereomeric metal complex is responsible for chiral recognition. Thus, on Cu(L-proline)$_2$ dissolved in the mobile phase (in HPLC), the enantiomers of valine were resolved with $\alpha = 4.8$ at room temperature ($\Delta\Delta G = -0.95$ Kcal) [19]; Davankov [28] has reported for proline on N-n-heptyl-L-hydroxyproline $\alpha = 10.9$ ($\Delta\Delta G = -1.4$ Kcal).

The above α values for partitioning systems overlap to a large extent with those found in host-guest interactions involving chiral crown ethers prepared from the rigid 2,2′-dihydroxy-1,1′-binaphthyl by Cram and coworkers [29]. Whether such values as $\alpha = 30$ – the highest resolution factor measured by Cram [30] – could be equalled in systems, as discussed in this lecture, remains an open question. As to the contribution of rigid covalent structures to chiral recognition, it has to be pointed out that, although in partitioning phenomena cyclic moieties may enhance stereoselectivity (see: the above data of Davankov for proline derivatives) the presence of such rings is not a sine qua non for achieving marked effects.

The many data which have accumulated through chromatographic studies of resolution permit: a) to single out systems which seem suitable for more detailed investigations of the intimate molecular mechanism of chiral recognition by methods such as X-ray crystallography; and b) to correlate the influence of structural factors of selectors and selectands on stereoselectivity.

a1) An example of a selector/selectand pair linked by hydrogen bonds, which was chosen for the investigation of the mechanism of chiral recognition, is N-lauroyl-S-(or -R-)-α-(1-naphthyl)-ethylamine interacting with lower molecular weight N-acyl-α-arylethylamines. By X-ray spectroscopy it was found that the selector arranges itself in the crystalline state, as well as in the melt, in a 5-Å translational array, as shown in Fig. 2, with the hydrogen bonds in the plane of the paper and the naphthyl rings oriented perpendicularly to it.

To explain the mechanism of resolution it is assumed that the selectand (guest or solute) intercalates into the hydrogen-bonded array of the host (selector) matrix.

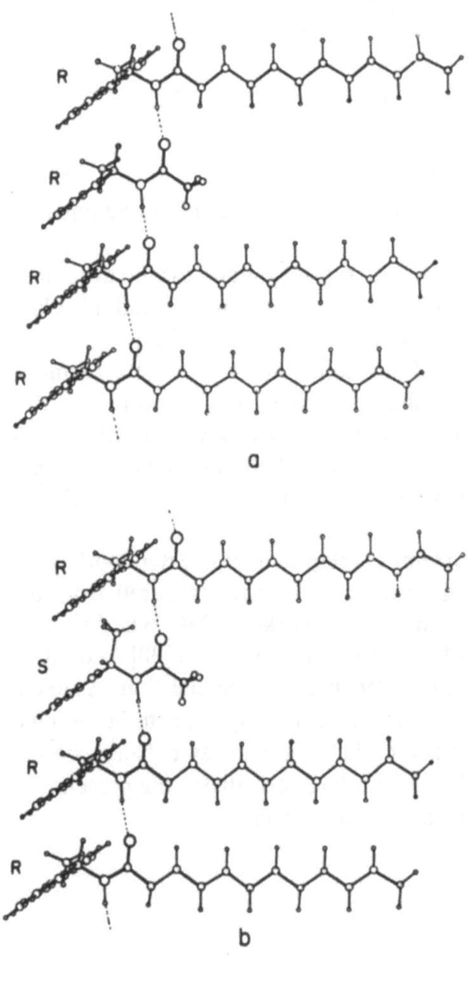

Fig. 2a, b. Intercalation of a chiral guest molecule, N-trifluoroacetyl-α-phenylethylamine, into a H-bonded stack of (R)-N-lauroyl-α-(1-naphthyl)ethylamine: **a** guest of R-configuration, **b** guest of S-configuration [9]

According to the configuration of the guest, two diastereomeric arrangements having different stabilities result. When host and guest have the same configuration (Fig. 2a), the original motif of hydrogen bonding is preserved, the aromatic groups make plane-to-plane contacts and the hydrogen atom linked to the asymmetric carbon is wedged[3] between two aromatic rings as in the original stack. The methyl group, on the other hand, points away from the stack.

For unequal configuration of host and guest (Fig. 2b), intercalation, while maintaining the host structure, requires that the hydrogen and the methyl groups linked to the asymmetric carbon of the solute be interchanged. The methyl group, thus wedged between the two naphthyl groups, would lead to severe overcrowding, i.e. there is a worse fit between the host and the guest in this case than in the case of the RR combination. These predictions of the model are in agreement with the experimentally found difference in selector-selectand interactions. Other arguments and evidence supporting the intercalation model are given in Ref. 9.

a2) The high chiral differentiation achieved through the indermediacy of coordination to metals has led to intense research on the mechanism of resolution involved, notably by Davankov and coworkers [30]. The ideas put forward by the Russian group are based on studies of the relative stability in solution of diastereomeric complexes of type Cu(L-Lig.)$_2$ and Cu(L-Lig.)(D-Lig.), X-ray structure determination of suitable model complexes, assignment of the coordination number, and observations on the effect of solvents. The behavior of bifunctional α-amino acids on a Cu-proline resin with an aqueous eluant is explained by ligand exchange [31], taking into account results obtained in the study of N-benzyl copper complexes serving as low molecular weight models. As illustrated below, in the resin complex

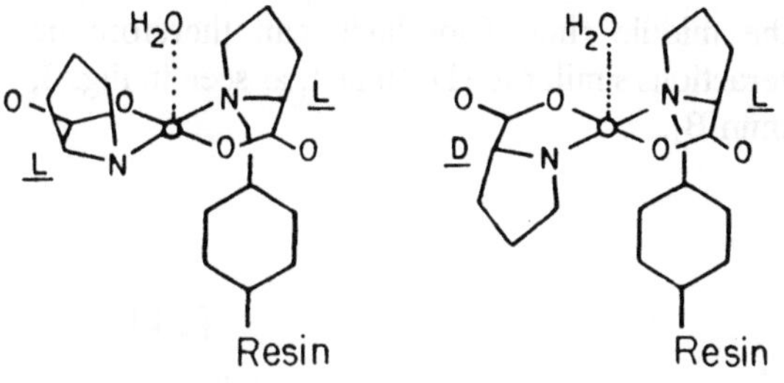

only one of the apical positions can be occupied, since the other is sterically hindered by the phenyl group linked to the polystyrene backbone. When L-proline (the selectand) enters the coordination sphere (complex on the left), it can be seen that interference occurs with the water molecule bonded apically to the metal. This is not the case (structure on the right) for D-proline, for which in addition stability is increased by hydrophobic interactions with the phenyl group. These considerations readily explain the chromatographic experiments in which the D-isomers emerge after the L-compounds.

In this ligand exchange mechanism the occupancy of the apical position is thus seen to play a determining role. It also explains [31] the reversed order of emergence

[3] If the atoms had been drawn with their Van der Waals radii, it would have become immediately apparent by inspection of Fig. 3 that the hydrogen linked to the asymmetric carbon fully occupies the space between the rings

found on the above resin for trifunctional amino acids (L-eluting after the D-isomer). For the latter compounds, the third function (e.g., OH in allo-Hyp) can displace the water molecule from the inner coordination sphere, so that the LL diastereomer becomes more stable than the DL form.

Ligand exchange is, however, not the only mechanism conceivable, when chiral differentiation proceeds through formation of diastereomeric metal complexes. On the basis of certain observations in systems in which chiral complexes such as Cu(N,N-dialkyl-α-amino acids)$_2$ are added to the mobile phase, it has been suggested [32] that the selectand coordinates to the apical position forming compounds of type Cu(N,N-dialkyl-α-amino acid)$_2$ (selectand). This hypothesis is particularly attractive for explaining the resolution of α-hydroxy acids. These acids coordinate to copper with stability constants many orders of magnitude lower than α-amino acids, so that ligand exchange in the plane of coordination can hardly occur.

Furthermore, Yoneda and coworkers [33] have produced X-ray structure evidence that the resolution of Δ and Λ metal coordination complexes by solutions of sodium tartrate can be related to the relative stability of outersphere associates of selector and selectands.

b) The following comments illustrate some structure-selectivity relationships as found for selector/selectand pairs interacting by charge transfer complexation. They also exemplify the potential of chromatography for the detection and study of chiral recognition by small biological molecules.

The TAPA (I) molecule is characterized by a tricyclic aromatic system having electron acceptor properties and a chiral handle attached to the central ring. A biological molecule, such as riboflavin (Vitamin B$_2$, II), has certain analogous features, i.e., an isoalloxazine ring known to act as an electron acceptor and an asymmetric substituent (D-ribityl) in the middle ring. Riboflavin can therefore be expected to show stereoselective interactions similar to (I). In fact, as seen in Fig. 3, helicenes are well resolved on Vitamin B$_2$.

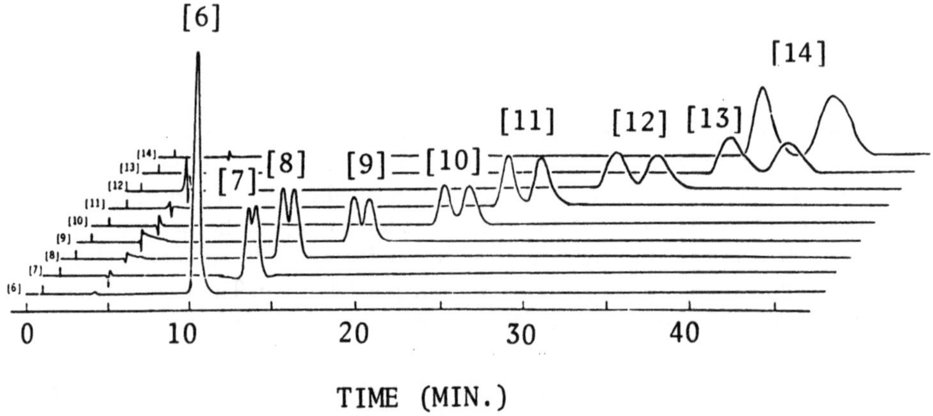

Fig. 3. Resolution of carbohelicenes by high pressure liquid chromatography on silica gel coated with riboflavin and CH_2Cl_2/n-hexane as eluant. The figures in brackets give the number of benzene rings in the helicene molecule. For resolution [6]-helicene requires different conditions from those required by the higher homologs

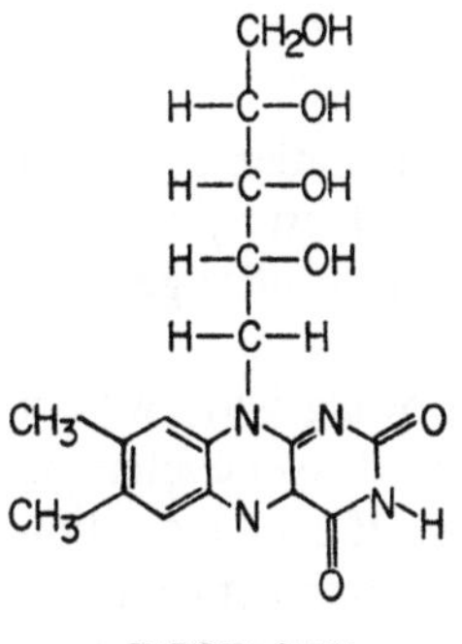

(II)

Further stretching the above by analogy, nucleosides and nucleotides were also studied [34]. It was found that purine but not the monocyclic pyrimidine derivatives showed chiral differentiation for helicenes however, the stereoselectivity was lower than on (II).

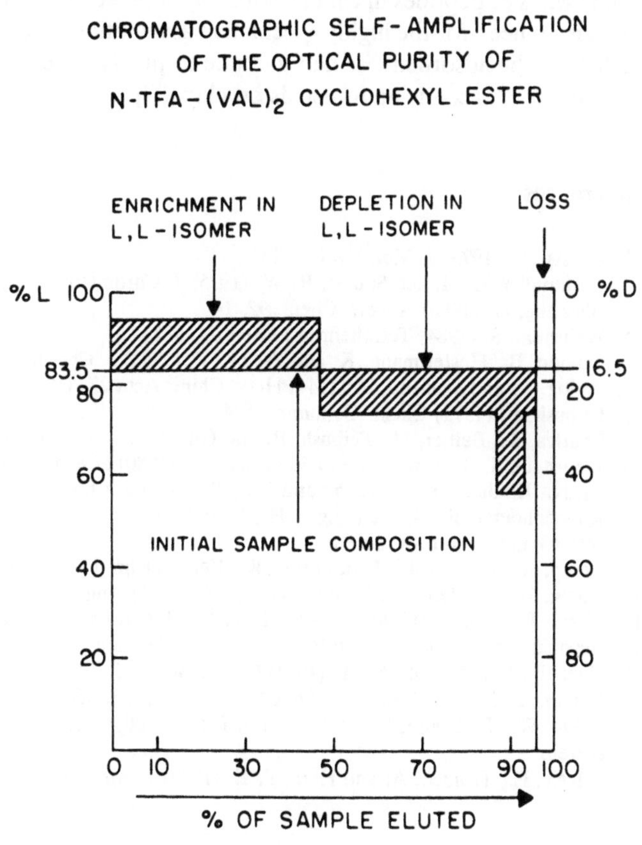

Fig. 4. Composition of fractions eluted with n-hexane/ethyl acetate (95/5) from a column 80 × 1.5 I.D. packed with silica gel; sample load 550 g

Final Comments

It has been shown in our laboratory [35] that α-amino acid derivatives, such as N-lauroyl-L-valine t-butylamide and N-TFA-val-valine-cyclohexyl ester, of 70% optical purity, could be enriched to 90–100% by chromatography on silica gel with an organic eluant in the absence of a chiral reagent (see Fig. 4).

These stereoselective effects can be explained in the following way: with a sample containing an excess of one of the enantiomers, an adsorbed layer containing a dissymmetric mixture of molecules will form on the silica gel. If self-association (e.g., to LL and LD "dimers") takes place with appropriate diastereomeric effects, one of the enantiomers will be retained more strongly than the other on the above layer, with consequent differences in their overall rates of migration.

Enrichment of the optical activity of enantiomeric mixtures containing one antipode in excess – without resorting to an external chiral reagent – is well known to be feasible by crystallization, but not in that of adsorption[4].

The results reported here might be relevant to the problem of the amplification of optical activity under primordial conditions, if the occurrence of such enrichment could be demonstrated in geochemically feasible systems, e.g. aqueous solutions of amino acids or peptides in contact with clay. For studies based on the present model, there is no need for the highly speculative assumption of chiral inorganic sites on the surface of the adsorbent, as had to made to justify experiments on amplification by adsorption of racemic amino acids on clay [37].

References

1. Gil-Av, E. (1975) J. Mol. Evol. *6,* 131
2. Lochmüller, C. H. and Souter, R. W. (1975) J. Chromatogr. *113,* 283
3. Blaschke, G. (1980) Angew. Chem. *92,* 14
4. Weinstein, S. (1984) Tetrahedron Letters *25,* 985
5. Domon, B., Hostettmann, K. and Prelog, V. (1982) J. Chromatogr. *250,* 149
6. Prelog, V. and Wieland, P. (1944) Helv. Chim. Acta *27,* 1127
7. Feibush, B. (1971) Chem. Commun., 544
8. Charles, R., Beitler, U., Feibush, B. and Gil-Av, E. (1975) J. Chromatogr. *112,* 121
9. Weinstein, S., Leiserowitz, L. and Gil-Av, E. (1980) J. Amer. Chem. Soc. *102,* 2768
10. Oi, N., Kitahara, H., Inda, Y. and Doi, T. (1981) J. Chromatogr. *213,* 141
11. Koppenhoefer, B., Allmendinger, H., Nicholson, G. J. and Bayer, E. (1983) J. Chromatogr. *260,* 63 and references therein
12. König, W. A., Benecke, I. and Ernst, K. (1982) J. Chromatogr. *253,* 267 and references therein
13. König, W. A., Benecke, I. and Bretting, H. (1981) Angew. Chem. Int. Ed. Engl. *20,* 693
14. Mikeš, F., Boshart, G. and Gil-Av, E. (1976) J. Chromatogr. *122,* 205
15. Newman, H., Helder, R. and Wynberg, H. (1976) Rec. Trav. Chim. Pays. Bas. *95,* 211
16. Haenel, M. and Staab, H. A. (1973) Chem. Ber. *106,* 2203
17. Klemm, L. H. and Reed, D. (1960) J. Chromatogr. *3,* 364
18. Pirkle, W. H., House, D. W. and Finn, J. M. (1980) J. Chromatogr. *192,* 143 and references therein
19. Gil-Av, E., Tishbee, A. and Hare, P. E. (1980) J. Amer. Chem. Soc. *102,* 5115

[4] While this manuscript was in preparation, a paper appeared, reporting a similar observation for optically enriched nicotine [36]

20. Weinstein, S. (1982) Angew. Chem. Suppl. *425,* 21
21. Tapuhi, Y., Miller, N. and Karger, B. L. (1981) J. Chromatogr. *205,* 325
22. Benecke, I. (1984) J. Chromatogr. *291,* 155
23. Schurig, V. (1977) Angew. Chem. *89,* 113
24. Schurig, V. and Gil-Av, E. (1976/77) Isr. J. Chem. *15,* 96
25. Schurig, V. and Bürkle, W. (1982) J. Amer. Chem. Soc. *104,* 7573
26a. Corbin, J. A. and Rogers, L. B. (1970) Anal. Chem. *42,* 974
26b. Lochmüller, C. H. and Souter, R. W. (1973) J. Chromatogr. *87*, 243 and subsequent publications
26c. Suzuki, S., Hobo, T., Watabe, K. and Ishikawa, H. (1981) Bunseki Kagaku *30,* 479
27. Schwyzer, R., This Symposium
28. Davankov, V. A., Bochkov, A. S., Kurganov, A. A., Roumeliotis, P. and Unger, K. K. (1980) Chromatographia *13,* 677
29. Kyba, E. B., Koga, K., Sousa, L. R., Siegel, M. G. and Cram, D. J. (1973) J. Amer. Chem. Soc. *95,* 2692 and subsequent papers
30. Cram, D. J. and Leiserowitz, L. (1985) Isr. J. Chem. *25,* 334
31. Davankov, V. A. (1980) Advances in Chromatography, Eds. Giddings, J. C., Grushka, E., Cazes, J. and Brown, P. R., Vol. 18, p. 139, M. Dekker, Inc. N.Y.
32. Weinstein, S., personal communication
33. Kishi, Y., Kuramoto, M. and Yoneda, H. (1976) Chem. Lett., 339
34. Kim, Y. H., Tishbee, A. and Gil-Av, E. (1980) J. Amer. Chem. Soc. *102,* 5915
35. Charles, R. and Gil-Av, E. (1984) J. Chromatogr. *298,* 516
36. Kundy, K. C. and Crooks, P. A. (1983) J. Chromatogr. *281,* 17
37. Friebele, E., Shimoyama, A. and Ponnamperuma, C. (1981) Origin of Life, Proc. 3rd ISSOL Meeting, Jerusalem, June 1980, Ed. Wolman, Y., p. 337, D. Reidel Publishing Co, Dordrecht, Holland

Discussion

List of interventions

1 Intervention of Wulff

Can your method for resolution of enriched racemic mixture be applied to those forming both racemates and conglomerates? There should be a difference since the interaction between (R)- and (S)-configurations should be decisive.

1.1 Gil-Av

These are initial experiments, which will be published shortly. So far we have had experience only with amino acid derivatives. The mechanism is one of self-association or self-recognition by partitioning, but not through conglomerate formation.

1.2 Wulff

In conglomerates and in racemates the interaction between D- and L-isomers is dissimilar and you should have a strong difference of interaction in order to have a separation.

1.3 Gil-Av

I agree. The stereoselectivity observed in our experiments could indeed be related to the difference in stability of pertinent crystalline species.

Designing Activity and Receptor Selectivity in Cyclic Peptide Hormone Analogs

Horst Kessler

Institut für Organische Chemie der J. W. Goethe-Universität, Niederurseler Hang, D-6000 Frankfurt 50, Germany

Peptide hormones often use only a small section of about five amino acid residues in the middle of a chain to transduce information (Fig. 1).

To achieve maximum recognition this so called "active region" should be surrounded three-dimensionally by its receptor. Hence it is logical to assume that this active part forms a reversed turn [1]. On the other hand the positions of the amino acids in the sequence are not typically involved in such turns according to the Chou-Fasman-rules [2]. One might therefore speculate that the rest of the hormone has mainly the function of keeping the active part in the correct conformation for molecular recognition.

We believe that small cyclic peptides can provide good models for achieving the active conformation because they inherently contain loops. Playing with the chirality

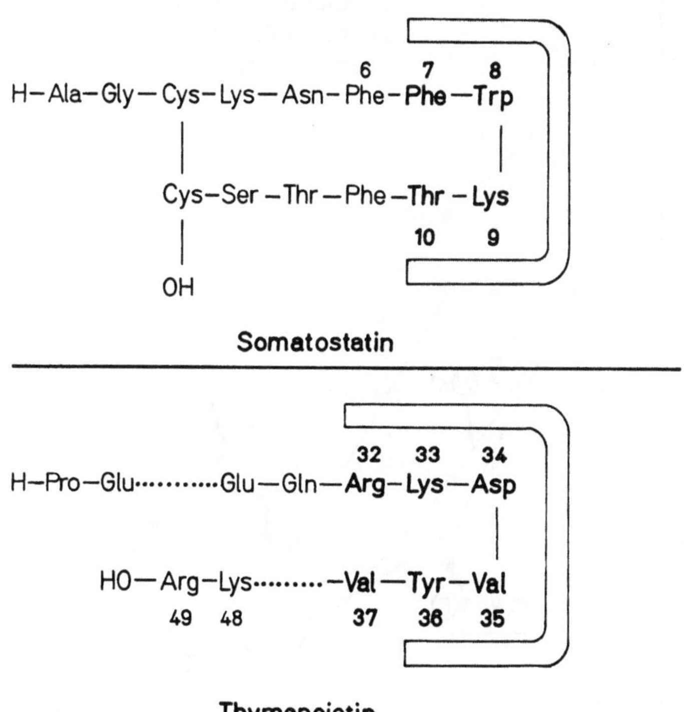

Fig. 1. Active part of the peptide hormones somatostatin and thymopoietin

Table 1. Cytoprotection activity against intoxication of rat liver cells by phalloidine [7]

No.	Structure	CD_{50} [μmol]
1	somatostatin (natural)	109
2	cyclo[Pro–Phe–D-Trp–Lys–Thr–Phe][a]	25
3	cyclo[Pro–Phe–Trp–Lys–Thr]	
4	cyclo[Pro–Phe–D-Trp–Lys–Thr]	
5	cyclo[Pro–Phe–Trp–Lys–D-Thr]	inactive
6	cyclo[Pro–Phe–D-Trp–Lys–D-Thr]	
7	cyclo[Pro–Phe–D-Trp–Lys–Thr–Gly]	
8	cyclo[Gly–Phe–D-Trp–Lys–Thr–Pro]	low activity
9	cyclo[D-Pro–Phe–D-Trp–Lys–Thr–Gly]	
10	cyclo[Gly–Phe–D-Trp–Lys–Thr–D-Pro]	89
11	cyclo[Phe–Phe–Trp–Lys–Thr–D-Phe][b]	14
12	cyclo[Phe–Phe–D-Trp–Lys–Thr–Phe][a, b]	5.7
13	cyclo[D-Phe–Phe–D-Trp–Lys–Thr–Phe][a, b]	9.7
14	cyclo[D-Phe–Phe–Trp–Lys–Thr–D-Phe][b]	5.0

[a] These peptides were first synthesized by the Merck-Sharpe-Dohme group [13, 14]. No. 2 is very active in growth hormone release inhibition

[b] These peptides only show small activity for growth hormone release inhibition [15]

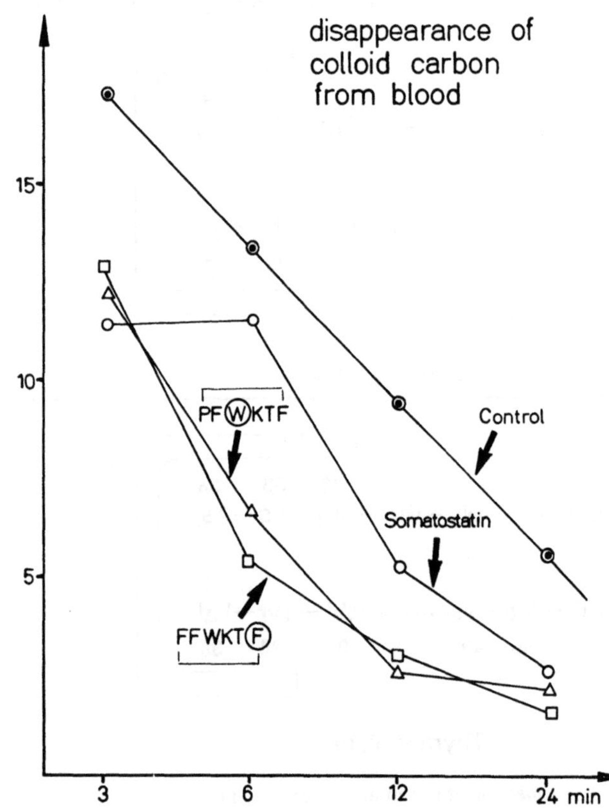

Fig. 2. Stimulation of clearance from colloid carbon in mice by somatostatin analogs

Table 2. Biological activity of linear and cyclic pentapeptides derived from thymopoietin (see text)

peptide	stimulation index PFC-Test		PHA-Test
control	221	0	1.00
H–Arg–Lys–Asp–Val–Tyr–OH (TP5)	298	0	1.05
cyclo[–Arg–Lys–Asp–D-Val–Tyr–]	401	+(+)	1.89
H–Arg–Lys–Glu–Val–Tyr–OMe	?	(+)	?
cyclo[Arg–Lys–Glu–D-Val–Tyr–]	687	+++	1.92
H-D-Lys–Arg–Glu–Val–Tyr–OMe	?	++	?
cyclo[-D-Lys–Arg–Glu–Val–Tyr–]	317	0	1.15

of the bridging amino acid(s) one can manipulate the conformation of the active part and thus elucidate the relationship between conformation and biological activity. Cyclic peptides turned out to have several advantages over linear ones or even the intact hormone itself. These relate to

- higher activity
- higher receptor selectivity
- prolongation of action and life time *in vivo*
- reduction of conformational freedom [3].

The last point enables us to study their conformation in different surroundings by NMR methods [1].

We have synthesized a number of cyclic penta- and hexapeptides containing the sequence Phe–D,L-Trp–Lys–Thr (FWKT) bridged by Pro, Gly, Phe in D- or L-configuration. Their conformations were analyzed by homo- and heteronuclear two-dimensional NMR techniques [4–6] and related to their biological activities. Cytoprotection of mouse liver cells against intoxication by phalloidin was used as the bioassay [7]. It turned out that there are compounds exhibiting more than 20-fold increase of cytoprotection activity than natural somatostatin but the ability to inhibit growth hormone release is almost completely gone.
Cyclic somatostatin derivatives also show stronger stimulation of macrophage activity than the somatostatin itself [8] (Fig. 2).

Other examples of higher activity in cyclic structures than in the linear peptides are demonstrated in cyclic analogs of the active sequence of thymopoietin (TP5: H–Arg–Lys–Asp–Val–Tyr–OH). Cyclo-TP5 stimulates maturation of T-lymphocytes almost to the same degree as the intact thymopoietin molecules [9]. In our plaque-forming cell test system (PFC) [10] this activity is on the lower limit to detect the activity of TP5. Cyclization and changing the chirality of Val to D-Val increases the activity dramatically (Table 2).
An even stronger effect is found when Asp is replaced by Glu in the cyclic TP5 analog. The resulting cyclic pentapeptide is the most active compound so far known in the PFC test and the phytohemagglutinin assay (PHA) [10]. Conformational studies of this peptide by ^{1}H, ^{13}C, and ^{15}N NMR spectra demonstrate [11] that the cyclus prefers a conformation which is totally different from the most stable conformation of the linear TP5 in solution [12].

In conclusion, cyclic peptide derivatives are powerful tools in providing a better understanding of hormone-receptor interaction and recognition. The information from those experiments might be useful for drug developments.

References

1. Kessler, H. (1982) Angew. Chem. Int. Ed. *21,* 512
2. Chou, P. Y., Fasman, G. D. (1974) Biochem. *13,* 222
3. See extensive citation in V. Hruby's paper
4. Kessler, H., Bernd, M., Damm, I. (1982) Tetrahedron Lett. *23,* 4685
5. Kessler, H., Eiermann, V. (1982) Tetrahedron Lett. *23,* 4689
6. Kessler, H., Bernd, M., Kogler, H., Zarbock, J., Sorensen, O. W., Bodenhausen, G., Ernst, R. R. (1983) J. Am. Chem. Soc. *105,* 6944
7. This work was done in collaboration with Prof. Dr. G. S. Rao, Klinische Biochemie, Universität Bonn, and Prof. Dr. K. H. Usadel, II. Medizinische Klinik, Universität Heidelberg – Mannheim; see also Ziegler, K., Frimmer, M., Kessler, H., Damm, I., Eiermann, V., Koll, S., Zarbock, J. (1985) Biochim. Biophys. Acta *845,* 86
8. This work was done in collaboration with Prof. Dr. S. Szabo, Harvard Medical School, Boston
9. Goldstein, G., Scheid, M. P., Boyse, E. A., Schlesinger, D. H., Van Wauwe, J. (1979) Science *204,* 1309
10. This work was done in collaboration with Drs. R. Obermeier and H. Müllner, Hoechst AG, Frankfurt
11. Kessler, H., Kutscher, B. (1985) Tetrahedron Lett. *26,* 177; Kessler, H., Kutscher, B., Kerssebaum, R., Klein, A., Lautz, J., Obermeier, R., Müllner, H., van Gunsteren, W., Boelens, R., Kaptein, R. (1985) Peptides, Proceed. 9th Am. Pept. Symp., Eds. Kopple, K., Deber, C. (1985), in press
12. Krishna, N. R., Huang, D. H., Chen, D. M., Goldstein, G. (1980) Biochemistry *19,* 5557
13. Veber, D. F., Freidinger, R. M., Schwenk Perlow, D., Paleveda, Jr., W. J., Holly, F. W., Strachan, R. G., Nutt, R. F., Arison, B. H., Homnick, C., Randall, W. C., Glitzer, M. S., Saperstein, R., Hirschmann, R. (1981) Nature *292,* 55
14. Veber, D. F., Peptides, Synthesis-Structure-Function, Eds. Rich, D. H., Gross, E., Pierce Chemical Company, Rockford, Ill., 1981, p. 685
15. Tests were performed by Drs. Sandow and M. Bickel at Hoechst AG, Frankfurt

Concluding Remarks by a Physicist

Hans Frauenfelder

Department of Physics, University of Illinois at Urbana-Champaign, 1110 West Green Street, Urbana, IL 61801 (USA)

As a physicist, I have been impressed and overwhelmed by the intricacies, complexities, and beauties of the material to which I was exposed. In this short "physical summary", I will try to organize some of the lessons that I have learned and indicate some of the directions in which I believe further work should move.

I Organic and Biological Chemistry

Structures built in organic and in biological chemistry have some important similarities, but also some crucial differences, as sketched in Fig. 1. Organic molecules are often "closed" and form two- or three-dimensional systems that are strongly (covalently) linked in a number of places. An active center may be held somewhere within such a system. Biological molecules, in constrast, have a linear primary sequence and folding occurs spontaneously into a loosely coupled and not multiply connected three-dimensional structure. An active center may be formed from the primary sequence or from a small organic molecule that is "taken into the fold". The two different types of structures lead to profound differences in two areas, information and dynamics.

Information. To describe an organic molecule and to provide an unambiguous blueprint for its construction one must indicate where branches and covalent connections occur. An essentially three-dimensional description is required and organization, storage, and transfer of this information may be complicated. A biological molecule, in constrast, is essentially one-dimensional and the information for its construction can be stored on a one-dimensional tape, a magnetic tape on a compu-

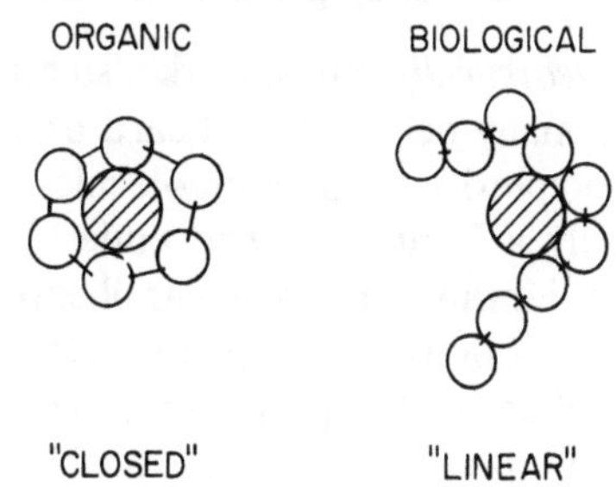

Fig. 1. Organic and biological molecules

ter or a nucleic acid in nature. Storage and transfer of the information is consequently, at least in principle, simple and universal.

Dynamics. The multiple covalent connections in a "standard" organic molecule permit small vibrations, but make most larger motions difficult or impossible. In a sense, an organic molecule is a passive device, even when acting as a catalyst. In contrast, a biomolecule can execute large motions and work like a miniature machine.

II The Physicist's Approach

If a physicist were asked to design a biomolecule for a particular function, he would try to follow the time-honored chain

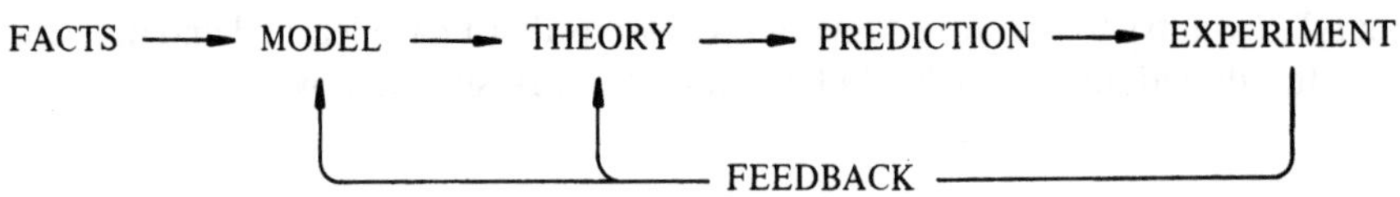

In principle, the design of a biomolecule that performs a certain job would be reduced to finding the correct primary sequence from the full theory. Genetic engineering then could construct the molecule. In practice, however, we are far from such a direct solution. There is no "full" theory of protein structure and function and the overall understanding is probably where nuclear or condensed matter physics stood in about 1930. An enormous amount of experimental and theoretical work remains to be done before we can hope to have a theory with predictive power. The development of such a theory will require close interaction between biologists, biochemists, chemists, and physicists.

Experimentally, we must learn more about the dynamic properties of biomolecules and about the connection between dynamic structure and function. Since no single tool, not even one as powerful as X-ray diffraction or NMR, can look at all aspects of a biomolecular reaction, every available technique will have to be considered. Studies of some simple and elementary biological phenomena will have to be performed with many tools over extremely wide ranges in all parameters, such as temperature, pressure, external fields, and solvents. A completely described elementary process can then take the role of the Balmer series in atomic physics and serve as testing ground for theories.

Theoretically, much work also remains to be done. Excellent progress is being made in many fields, as we heard from Jencks, Karplus, and Scheraga. Considering the complexity of the biomolecules, even the sophisticated computations of today are only a beginning. The underlying reaction theory must be improved, and nonequilibrium and nonlinear phenomena must be considered. As an example, consider the motion of a small molecule such as CO or O_2 through a protein matrix. Such a motion can be approximated as diffusion. In essence it is a random walk, but not of the ordinary kind. It is probably a random random walk in which the CO molecule moves randomly, not on a regular lattice, but on one that is also random and even changes randomly in time. Alternatively, the motion may be described as percolation, in which preferred pathways exist. In order to decide which description is best, the corresponding theories must be worked out *and* very detailed experiments must be performed.

While both organic and bioorganic chemistry are already far advanced and perform spectacular feats, the corresponding physical theory with predictive power is far behind and in a preliminary state, thus offering many exciting challenges.

Some Aspects on Chemical versus Biochemical Recognition and Catalysis

B. Hess

Max-Planck-Institut für Ernährungsphysiologie, Rheinlanddamm 201, D-4600 Dortmund

Whereas modern methodology in chemistry and biochemistry is well developed to assist the curiosity of scientists in unravelling the mysteries of chemical and biochemical recognition and catalysis, their concepts and intuitions in both areas of research are not necessarily converging. The chemist, interested in homogeneous or heterogeneous catalysis, is engaged in the artful design of beautiful structures with powerful functions. He might not be as interested in the equally penetrating efforts of biochemists, who try to understand recognition and catalysis in the most complex and flexible structures of an enzymic machine.

From the study of enzymes the biochemists learn that enzymic recognition and catalysis is a concerted and controlled process of perfect timing. Macroscopic rate constants reflect a large number of elementary steps which are ordered in time: Each state is sufficiently stable over a given transient time to reach a new state until a whole transformation of a given substrate is completed. Here it is important that a time separation between any new transient is achieved and that no state significantly falls back into its previous state. These features might be mechanistically controlled by steric hindrance and cyclic reaction sequences within the protein itself. We do not yet know a complete reaction pathway, mainly because of the restricted knowledge of active site geometry, polarity, and mobility for any given enzyme, which in addition is a highly dynamic structure fluctuating with a large number of modes.

This all can be understood as result of evolution which allowed nature enough time to select the optimum design for each chemical problem an enzyme has to solve. A glance at the tables of known time constants of chemical interactions reveals that the time constants which have been found in chemistry cover a logarithmic scale of time domains in each class of reactions down to the limits of physics. If the transformations involve mass transfer, this limit would be of the order of 10^{11} to $10^{13}\ s^{-1}$. On the other hand, rate constants for hydrolysis are in the range of $10^{-3}\ s^{-1}$. However, the maximum second order rate constants for enzymes are in the order of $10^8\ M^{-1}\ s^{-1}$ with overall turnover numbers of 10^5–$10^6\ s^{-1}$. I should add that in membrane-bound proteins diffusion might be two-dimensional or even one-dimensional, yielding higher second order rate constants of 10^9 or $10^{10}\ M^{-1}\ s^{-1}$.

In summary, it can be said that the rather broad spectrum of rate constants observed in chemistry is strongly reduced in enzymic reactions by the highly ordered local arrangement of interacting groups of host and guest, achieving an excellent balance between high specificity, diffusion control and speed of turnover. I should add that living systems are not restricted to neutral pH and "normal" temperature but have also perfectly adapted their electrostatic properties to live at pH 2 or 10, in 4 M salt, in hydrophobic media, at 100 °C, or at high pressure.

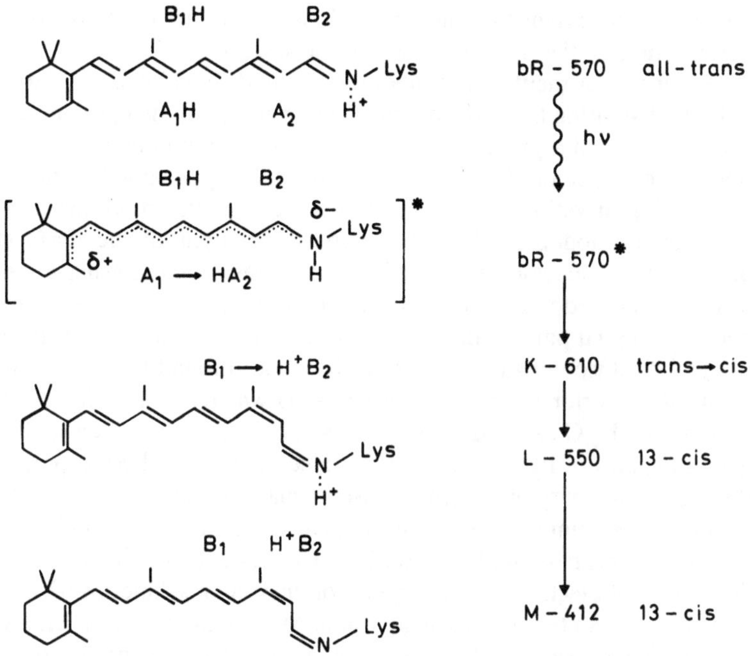

Fig. 1

As an example of perfect timing, I would like to mention the photochemical reaction cycle in bacteriorhodopsin. This retinal protein of MW 26805 dalton functions as a light driven proton pump in salt bacteria. The reaction cycle involves a photochemical isomerization of retinal and the coupled proton transfer through the bacteriorhodopsin molecule from one side of the membrane toward the other. Figure 1 illustrates the isomerization process following quantum absorption by the protonated all-trans-retinylidene Schiff's Base. In the thermal reaction following the excited state, isomerization to 13-cis-retinal occurs, yielding the separation of charges in a concerted reaction sequence with a number of donor-acceptor functions of the opsin moiety around the retinal molecule. The components can be followed spectroscopically as indicated on the right side of the scheme. Finally the M-412 state of a 13-cis-retinal relaxes to the all-trans-state in a process coupled to reprotonation of the Schiff's Base as well as of acceptor functions in the opsin moiety. The time scale of the overall process covers the range of femtoseconds for the first events yielding the excited state, up to nano-, micro- and milliseconds for the rate constants of the following steps.

The figure illustrates the mechanism of intramolecular charge separation. The transition and possible charge recombination might be controlled by either electrostatic charge stabilization or trans-cis-isomerization of the C-13 bond. Obviously, a combination of both principles is realized.

The mechanisms of proton transfer through the protein is again perfectly timed as indirectly reflected in the photocycle kinetics. Here we have to distinguish between coupled proton binding groups which equilibrate with the bulk during the light and/ or the dark state of bacteriorhodopsin and on the other hand proton relay groups in the hydrophobic region of bacteriorhodopsin which are not sensitive to electrostatic perturbation by the aqueous phase of the system.

Recent experiments (in collaboration with M. Engelhard, F. Siebert et al., 1983) indicate that inside the hydrophobic binding groups of the opsin moiety four carboxylate functions of aspartate residues are involved in protonation/deprotonation reactions. Use was made here of infrared spectroscopy applied to thin layers of bacteriorhodopsin within the purple membrane. The incorporation of 13-C-aspartate into bacteriorhodopsin allowed the specific analysis of the C=O stretching frequency of the specifically labeled aspartate residues following a laser flash. This technique showed that protonation and deprotonation reactions of aspartyl residues play an essential part in the proton transfer following the photochemical activation of the retinal isomerization reaction. It was also found that these carboxylates are within the interior of the membrane as shown by the position of the absorption maxima and $H_2^{18}O$ exchange experiments. The experiments indicate that the involvement of chains of hydrogen bonds is indeed unlikely. These experiments illustrate the high complexity of enzyme transport machines although we feel that the light driven proton pump is one of the simplest transport systems in biology.

The membrane-bound bacteriorhodopsin system also provides an example of the illustration of electrochemical aspects of enzymic machineries which were not discussed in the context of this symposium. Membrane-bound enzyme systems are of special interest because they are oriented yielding a vectorial property which was stressed by Peter Mitchell already in the early sixties. The structural mobility of intramembrane molecules is restricted due to the lipid protein interaction. There might be controlled translational motion or rotational motion or in cases such as the purple membrane no gross motion at all.

The low dielectric of the inner layers of membrane proteins being oriented between two water phases generates two surface boundaries and appropriate potentials on both sides of the membrane in terms of surface potential as well as of an overall cross membrane potential. The surface potential is important in controlling pK's of functional groups of the membrane bound proteins on either side. The electrical cross membrane potential generates an electrical field. With the known thickness of the membrane-bound protein system and the electrical potential of the order of 200 mvolt an electric field strength of 500 KV/cm is computed. From this it follows that biomolecules and natural membranes display their functions under a permanent electric field which conditions dipoles and orientation as well as electrostatic reactivity to an appreciable extent. Here, a new world of electrochemical interactions, ion fluxes, ring currents and control is slowly opening up in a large area of biochemical catalysis.

Of course, enzyme machines are not perfect, although the complex process of enzymic recognition is highly selective. On the basis of experiments the magnitude of errors is of the order of 10^{-6}. In the case of phosphokinases the phosphoryl transfer from a donating to an accepting group yields as an error the hydrolysis of phosphoryl groups to inorganic phosphate. Many examples of this imperfection have been described. Prof. Eigen's analysis of the errors implied in the process of replication led to important conclusions in the development of a molecular theory of evolution, and we learn that error corrections in terms of repair mechanisms are available in biological systems.

Finally, I'd like to stress the dynamic aspects which are of utmost importance for an understanding of biochemical processes. Biochemical processes are essentially open in nature. This means that all enzymes are constantly, stochastically or periodically activated by substrates which are produced by the cellular environment or by precursor enzymes and transformed so that products are picked up by other enzymes within a reaction sequence. Biochemical processes are controlled by their cyclic design, by allosteric feedback and by electrochemical coupling. The discovery of the principle of cyclic processes induced the notion that the dynamic coupling of networks of integrated biochemical processes must be extremely complex and nonlinear. Indeed, today we observe a large variety of coupled dynamic states and time pattern formations in biochemical processes from simple periodic reactions to the most complex chaotic states of biochemical turnover.

List of Authors

Dr. L. Addadi
Department of Structural Chemistry, The Weizmann Institute of Science, Rehovot 76100, Israel

Prof. W. S. Bennett, Jr.
Max-Planck-Institut für Biochemie, D-8033 Martinsried bei München

Dr. Z. Berkovitch-Yellin
Department of Structural Chemistry, The Weizmann Institute of Science, Rehovot 76100, Israel

Prof. G. Blaschke
Institut für Pharmazeutische Chemie der Universität, Hittorstr. 58–65, D-4400 Münster

Prof. R. Breslow
Department of Chemistry, Columbia University, New York, N.Y. 10027, USA

Prof. D. L. Cram
Department of Chemistry and Biochemistry, University of California at Los Angeles, Los Angeles, California 90024, USA

Dr. T. J. Franklin
Imperial Chemical Industries PLC, Pharmaceuticals Division, Alderley Park, Macclesfield, Cheshire, United Kingdom

Prof. H. Frauenfelder
Department of Physiology, University of Illinois, Urbana, Illinois 61801, USA

Prof. E. Gil-Av
Department of Organic Chemistry, The Weizmann Institute of Science, Rehovot, Israel

Prof. M. E. Hadley
Department of Chemistry and General Biology, College of Liberal Arts, University of Arizona, Tucson, Arizona 85721, USA

Prof. B. Hess
Max-Planck-Institut für Ernährungsphysiologie, Rheinlanddamm 201, D-4600 Dortmund

Prof. V. J. Hruby
Department of Chemistry and General Biology, College of Liberal Arts, University of Arizona, Tucson, Arizona 85721, USA

Prof. R. Huber
Max-Planck-Institut für Biochemie, D-8033 Martinsried bei München

Dr. K. Jacobson
Department of Organic Chemistry, The Weizmann Institute of Science, Rehovot, Israel

Prof. W. P. Jencks
Department of Biochemistry, Brandeis University, Waltham, MA 02254, USA

Prof. M. Karplus
Department of Chemistry, Harvard University, Cambridge, Massachusetts 02138, USA

Prof. R. M. Kellogg
Department of Organic Chemistry, University of Groningen, Zernikelaan, Groningen, The Netherlands

Prof. H. Kessler
Institut für Organische Chemie der Universität, Niederurseler Hang, D-6000 Frankfurt 50

Prof. B. J. van Keulen
Department of Organic Chemistry, University of Groningen, Zernikelaan, Groningen, The Netherlands

Prof. M. Lahav
Department of Structural Chemistry, The Weizmann Institute of Science, Rehovot 76100, Israel

Prof. J.-M. Lehn
Institut Le Bel, Université Louis Pasteur, 1, rue Blaise Pascal, F-67008 Strasbourg Cedex

Prof. L. Leiserowitz
Department of Structural Chemistry, The Weizmann Institute of Science, Rehovot 76100, Israel

Dr. R. A. Lerner
Scripps Clinic and Research Foundation, La Jolla, CA, USA

Prof. P. L. Luisi
Institut für Polymere, ETH-Zentrum, CH-8092 Zürich

Dr. M. Minárik
Institut für Organische Chemie, Universität Düsseldorf, D-4000 Düsseldorf

Dr. J. van Mil
Department of Structural Chemistry, The Weizmann Institute of Science, Rehovot 76100, Israel

Dr. D. Oberkobusch
Institut für Organische Chemie, Universität Düsseldorf, D-4000 Düsseldorf

Prof. A. Patchornik
Department of Organic Chemistry, The Weizmann Institute of Science, Rehovot, Israel

Prof. H. A. Scheraga
Baker Laboratory of Chemistry, Cornell University, Ithaca, New York 14853, USA

Prof. R. Schwyzer
Institut für Molekularbiologie und Biophysik, ETH Zürich, CH-8093 Zürich

Prof. M. P. Strub
Department of Organic Chemistry, The Weizmann Institute of Science, Rehovot, Israel

Prof. A. R. Ubbelohde
University of London, Imperial College of Science and Technology, Department of Chemical Engineering and Chemical Technology, London SW7 2AZ, United Kingdom

Prof. I. Weissbuch
Department of Structural Chemistry, The Weizmann Institute of Science, Rehovot 76100, Israel

Prof. R. J. P. Williams
Wadham College, Inorganic Chemistry Laboratory, University of Oxford, Oxford OX1 3QR, United Kingdom

Prof. G. Wulff
Institut für Organische Chemie, Universität Düsseldorf, D-4000 Düsseldorf

Prof. K. Wüthrich
Institut für Molekularbiologie und Biophysik, ETH Zürich, CH-8093 Zürich

Chairmen

Prof. E. Katchalski-Katzir
The Weizmann Institute of Science, Rehovot, Israel

Prof. V. Prelog
Laboratorium für Organische Chemie, ETH Zürich, Universitätsstr. 16,
CH-8006 Zürich

MIX
Papier aus verantwortungsvollen Quellen
Paper from responsible sources
FSC® C105338

If you have any concerns about our products,
you can contact us on
ProductSafety@springernature.com

In case Publisher is established outside the EU,
the EU authorized representative is:
Springer Nature Customer Service Center GmbH
Europaplatz 3, 69115 Heidelberg, Germany

Printed by Libri Plureos GmbH
in Hamburg, Germany